# Understanding Social Statistics

# Understanding Social Statistics

## A Student's Guide to Navigating the Maze

Lance W. Roberts
Jason Edgerton
Tracey Peter
Lori Wilkinson

**OXFORD**
UNIVERSITY PRESS

# OXFORD
### UNIVERSITY PRESS

Oxford University Press is a department of the University of Oxford.
It furthers the University's objective of excellence in research, scholarship,
and education by publishing worldwide. Oxford is a registered trade mark of
Oxford University Press in the UK and in certain other countries.

Published in Canada by
Oxford University Press
8 Sampson Mews, Suite 204,
Don Mills, Ontario  M3C 0H5 Canada

www.oupcanada.com

Copyright © Oxford University Press Canada 2015

**Library and Archives Canada Cataloguing in Publication**

Roberts, Lance W., 1950-, author
Understanding social statistics : a student's guide to navigating
the maze / Lance W. Roberts, Jason Edgerton, Tracey Peter, and
Lori Wilkinson.

Includes index.
ISBN 978-0-19-544429-2 (pbk.)

1. Social sciences—Statistical methods—Textbooks. 2. Statistics.
I. Edgerton, Jason, 1970-, author II. Peter, Tracey, 1973-, author
III. Wilkinson, Lori, 1972-, author IV. Title.

HA29.R63 2015          300.1'5195          C2014-907916-8

Cover image: German/Getty Images

Printed and bound in Canada

3 4 5 — 20 19 18

# Dedication

To Charlie — enjoy your journey from one little piece of holy ground to the next.  LWR

To Trina, Tate, and Jordan for always reminding me what really matters.  JDE

To Norah Richards — an expert maze navigator in her own right.  TP

To my mother, Patricia Wilkinson, and my grandmother, Marjorie Wilkinson, two women who have made me truly appreciate the value of my education.  LAW

# Brief Contents

# Contents

# Preface

Welcome! We have successfully taught social statistics to thousands of students just like you. Our approach is organized around the idea of a maze, which is a good place to begin.

## Statistical Maze

As you probably know, a maze is a type of walking puzzle, filled with pathways of many types. Some of the paths are dead ends, while others advance you toward the exit. But you can never see where you are or where you are going because your view is blocked by the tall sides of every path. This restriction is what makes mazes so complex and confusing.

For too many students, studying social statistics is like walking around in a maze. Much of the time students are confused and disoriented. They lack direction because they don't know their location or where they're going. In this situation it is little wonder that the experience of encountering social statistics is frustrating.

Mazes do not have to be confusing or frustrating. All you need is a little help navigating them. If you have ever walked a maze, you've probably seen people jumping up to get a glimpse over the top of the wall. This information helps orient them to where they are and how to proceed. Alternatively, some maze operators offer participants a map upon entry so that they don't have to walk around aimlessly or offer to accompany participants, regularly telling them where they are and where they are going next.

The reason that maze participants can either help themselves or benefit from the help of a map or guide is that mazes have an identifiable structure: they are organized around a pattern. When you know the pattern, navigating the maze is straightforward. Through this book and its associated website, we will provide you with both a map of the statistical maze and directions on how to proceed. With us as your guide, you will have little problem knowing where you are or navigating your way through the apparent complexity. At every turn, you can rest assured that there will be a very clear "You Are Here" sign.

## What's the Point?

Among humans, an activity must have a point or a purpose to be meaningful. When you enter an ordinary maze, your purpose is to find the most efficient route to the exit (preferably with as little help as possible).[1] Entering the statistical maze has a different purpose. It is not enough to navigate your way through. Along the way, you are required to pick up statistical tools and put them in a tool kit. Different regions of the statistical maze contain different tools; when you exit the maze, your tool kit is full. The point of navigating your way through the statistical maze is to *collect a full set of statistical tools and know how to use them.*

---

[1] Clearly, if you entered the maze with a chainsaw and cut your way through the walls of the paths, you would be out in record time. But how meaningful would that be?

# What's the Reward?

Rewards come in two forms, intrinsic and extrinsic. Intrinsic rewards are those that are self-satisfying. Things that are simply enjoyable in themselves, such as observing a colourful sunset or enjoying a warm bath, are intrinsically rewarding. By contrast, other things are rewarding because of what they lead to. These rewards are extrinsic. A driveway cleared of snow so you can leave for your destination is extrinsic, as is the loss of five pounds after you have kept to your diet for a month. Whether any particular pursuit is intrinsically or extrinsically rewarding is a matter of how a person defines the experience. Some people find washing dishes intrinsically rewarding (i.e. they enjoy the process); for others the reward is the finished product.

The same holds true of your journey through the statistical maze and your collection of useful tools. In terms of intrinsic rewards, a remarkable number of students experience a surge in self-confidence and associated self-esteem when they exit the maze. And no wonder. Genuine self-esteem comes from mastering something worthwhile and difficult, which is what your successful journey through the statistical maze entails. Extrinsically, the rewards of gathering a full statistical tool kit are at least threefold. First, when you read reports of research in newspapers, in textbooks, on the Web, or elsewhere, you no longer have to simply believe the results. You will be able to competently examine the evidence and make an independent decision. Second, students who understand social statistics, rather than merely memorize selective components, achieve significantly better course grades. Knowing the pattern of the statistical maze is the key to such understanding. Finally, statistical competence is a widely marketable skill set for gaining employment. Public, non-government, and private sector organizations regularly advertise for persons with a full statistical tool kit.

Some students find the journey intrinsically rewarding, others extrinsically so, and some even both. Whatever your case, we welcome you to the maze.

# How We Will Proceed

This book takes you through the various regions of the statistical maze and lets you gather the statistical tools found in various locations. After working your way through the maze, you will emerge with a full set of statistical tools and the competence for applying them. Before taking the first step, you should understand how we, as your guides, plan to proceed.

Every chapter begins by providing you with coordinates that clearly identify *where you are* in the statistical maze. As you will see, the coordinates come from different combinations of answers to three questions that characterize the pattern of the maze. Knowing where you are in the statistical maze identifies what region you are in. You will always know where you are.

You will also find that different regions contain different statistical tools for you to acquire. The tools are there for the taking, but you need to understand what you are looking at. Therefore, in every chapter, new ideas for you to observe or understand are set in bold and thoroughly explained when they are introduced.

So, at every location in the maze you will know where you are and what statistical concepts and tools you are observing. Now comes the more challenging part. It is not enough that you recognize the statistical tools; you need to be able to understand and use them correctly. To that end, most chapters follow a five-step model. These five steps have guided thousands of students through a successful negotiation of the statistical maze. If you follow them carefully, you too will emerge with a tool kit full of statistical competencies.

Here is a brief overview of the five steps and their rationale.

## STEP 1   Understanding the Tools

All statistical tools are introduced to you before you ever try to use them. The objective of this step is to *understand* the statistical technique under consideration in *everyday language*. This first step is absolutely necessary; without it, statistical techniques are either not understood or used ritualistically. Understanding the basic ideas of a statistical technique is an essential first step to engagement.

## STEP 2   Learning the Calculations

Quantitative statistical techniques are based on calculations. This second step shows you how to perform every calculation required to compute the statistic under consideration. The calculation steps assume little more than basic arithmetic skills, and where there is any question about these, we provide "Math Tips" boxes that remind you of how the operations are performed. This step is important because it takes the mystery out of statistical calculations. You will know exactly how the statistics are computed.

## STEP 3   Using Computer Software

Learning how to calculate specific statistics is important to your understanding of the techniques, but it is not a practical way to proceed regarding real-world data. With the larger sample sizes used in realistic research situations, it is simply too time-consuming and error-laden to perform all the necessary computations by hand. Statistical software is employed instead. This book provides specific descriptions of how to have IBM® SPSS® Statistical software ("SPSS") perform every statistical technique introduced in the book.

## STEP 4   Practice

The first three steps introduce new statistical tools in ordinary language, show the exact steps used to compute the statistics, and provide instructions on how to have statistical software produce the results. The fourth step solidifies the earlier steps through practice. There is no getting around this step. If you want to master statistical tools you need to practise using them. In this respect, mastering social statistics is no different than mastering anything else worthwhile. Time and effort must be invested.

You have probably heard the saying "practice makes perfect." It is more accurate to say that "practice makes permanent." Repetition results in habituation. As everyone knows, both perfect and imperfect habits can be made permanent. To make sure you are on the right track, the practice step includes practice questions to solidify your understanding of both the hand calculations (step 2) and the computer applications (step 3). Then, after you have worked on each of these types of questions, you can check your conclusions in the answer key at the back of the book.[2] If your answer is different than the one provided, you can go to the "solutions" section included on the website. There you will find not just the answers but all the steps in generating the correct answer to both types of practice questions.

## STEP 5  Interpreting the Results

After step 4, your practice will have solidified your competence in computing a wide range of social statistics. For the most part, this means you will be able to generate correct numerical results. This is a significant achievement, but it is not the end of the story. Full statistical competence requires that you be able to state, in ordinary language, what the statistical results mean. Interpretation is essential because it relates the numerical results to the question that guided the research. The final step for every new statistical procedure provides a model of how to interpret the statistical findings in everyday language.

In short, this text will guide you through the regions of the maze in a systematic fashion. In each region, you will take five steps that let you understand, learn, and interpret the statistical tools in that location. As you move through the various regions of the maze, your tool kit of statistical competencies will grow and, by the end, you will emerge with full command of basic social statistics. We trust you will agree that this journey is both challenging and worthwhile.

# The Big Picture

Your entry into the statistical maze is very near. Before proceeding, let's give you a bird's-eye view of your journey. As stated, most chapters use the five-step model to introduce and solidify your competence in specific statistical techniques. At strategic points, however, there are chapters that discuss only concepts and principles, not specific statistical techniques. These chapters prepare you for an upcoming section of the statistical maze by orienting you to the landscape.

In overview, here is the route that your journey will follow, organized in parts.

### Part 1: General Orientation

This part contains three chapters that provide you with the ideas necessary for understanding where social statistics fits into the broader research enterprise. Chapter 1 introduces a perspective on social research and shows the location of social statistics. Chapter 2 shares the logic that underwrites the organization and selection of statistical procedures. This chapter is very important because its contents are used in substantive chapters to determine your location in the statistical maze. Chapter 3 examines calculation and computers. It discusses how the

---

[2] Remember that steps 2 and 3 provide detailed instructions and illustrations of both hand calculations and computer solutions. Therefore, if you follow the steps on the practice questions, checking the "answers" section should be a wonderfully reinforcing experience.

five-step model operates and introduces you to SPSS statistical software. After this general orientation, you are ready to begin moving through the maze.

### Part 2: Univariate Analysis

This part includes five chapters in which you gather statistical tools and techniques related to analyzing one variable at a time. Chapter 4 provides a general orientation to this landscape. Chapter 5 discusses measures of central tendency, while Chapter 6 covers measures of dispersion. Chapter 7 covers techniques for producing charts and graphs, and Chapter 8 brings the material in earlier chapters together in a discussion of normal distributions.

### Part 3: Bivariate Analysis

This part of your journey includes six chapters, all devoted to understanding relationships between two variables. Chapter 9 provides a general orientation, while Chapter 10 shows how relationships are demonstrated in tables and Chapter 11 shows how they are portrayed in scatterplots. Chapter 12 orients you to a family of statistical tools used to summarize relationship information. Chapter 13 discusses the specific tools in this family used for categorical connections, while Chapter 14 does the same for continuous relationships.

### Part 4: Multivariate Analysis

This three-chapter set begins with Chapter 15, which orients you to what it means to consider three or more variables simultaneously. Chapter 16 introduces the statistical tools for conducting multivariate analysis on tables, while Chapter 17 discusses multivariate techniques for continuous variables.

### Part 5: Sampling and Inference

This last section of your journey includes four chapters and covers ideas and techniques related to generalizing statistical results. Chapters 18 and 19 introduce the place and importance of inferential statistics as well as the central ideas underlying the approach. Chapter 20 discusses estimation techniques, while Chapter 21 covers hypothesis testing procedures.

Following this five-part plan takes you systematically through all the basic regions of the social statistics maze. Step by step, your kit of statistical competencies will grow and you will emerge from the maze full of confidence about what you have learned and how to apply it.

We are pleased to be your tour guides on the journey. If you feel like commenting on aspects of your trip, please feel free to send a note to Lance_Roberts@umanitoba.ca. We are happy to hear of your experience.

# A Final Word

Ignorance restricts the world of experience; education makes our realities larger. People who know more about any topic have a broader range of experience. If you know additional languages, know how to play a musical instrument, understand what makes the sun rise in the east, or anything else, your world is bigger and more fulfilling than if you lack these competencies. This is true for every worthwhile competency—including learning about social statistics.

We launch your journey into the statistical maze with the following cartoon. When you emerge from the maze, reread the cartoon and see if it makes you smile.

Enjoy the trip!

LWR
JE
TP
LAW

# Acknowledgements

Creative acts are rarely, if ever, individual endeavours. They are social enterprises. Just as interactions between the book's coauthors improved the product, our efforts have been assisted in many ways at various stages by a wonderful team of publishing professionals who worked backstage. Listing them here is our way of moving them front stage to take a well-deserved bow.

Phyllis Wilson, Managing Editor
Nancy Reilly, Acquisitions Editor
Jodi Lewchuk, Senior Developmental Editor
Leslie Saffrey, Copy Editor
Colleen Ste. Marie, Proofreader
Lisa Ball, Production Coordinator

We thank and applaud you all.

We would also like to acknowledge the book's reviewers, whose feedback was helpful as we worked toward a final product.

Jean Andrey, University of Waterloo
Kenneth MacKenzie, McGill University
Michelle Maroto, University of Alberta
Thomas Varghese, University of Alberta

# PART I
## General Orientation

# 1

# The Location and Limits of Quantitative Analysis

## Overview

Welcome! You are about to embark on a tour through the statistical maze that leads toward mastery of basic social statistics. We trust you will enjoy the tour. The maze contains many paths, and navigating each one requires several steps. Your tour takes you through a variety of locations. We are pleased to be your tour guides. We have successfully guided many students down this path and are confident that, if you follow the steps, you will arrive at the destination. Sometimes you will find yourself at locations where the view is not immediately clear; at other times you will be tired from the effort it takes to reach a particular location. These challenges are part of the journey, and although we can't alleviate all frustrations, we can support both your vision and your efforts.

Before we begin the tour, it is helpful to know where we stand. This chapter provides your bearings. The contents are in two sections. The first provides a model of the social construction of reality and identifies different approaches to making sense of experience. This section locates how the social statistics covered in this book fit into the larger landscape of human understanding. The second section of the chapter talks about variables and causality, which are fundamental concepts to using social statistics.

## 1. Stories and Statistics

### A Little Lesson

Here are two lines of text; examine them carefully.

Line 1:

○ ○ ◐ ❧ ⛢ ♈ → ☉ ◇ ✧ ☘ ⛌ ↑ ⊕ ⟲ ♌ ○ ○ ★ ❦ ↖ ⊙ ℱ ⚘ ⚛ ✧ ⊕ ⟐ ➤ ♌ ∎ ♌ ♯ ❧ ❦ ☆ ⊙ ○ ○ ⟐ ⛢ ⇐ ⟲ ℘ ❦

What do the squiggles on this line mean to you? For most readers, these squiggles are meaningless. This result does not occur because your eyesight is poor. It occurs because, by itself, experience is meaningless.

Line 2:

在相反的情况下, 你可以不必学习阅读汉语就能明白此页中的标记.

We are betting that a large number of you are unable to read this sentence either, but many of those who cannot read it may recognize it as another language. Some of you may recognize it as Chinese.

Now read the following sentence:

*Were it otherwise, you would understand the marks on this page without learning to read English.*

At the concrete level, the squiggles (the words that you see in English) in this sentence are practically no different from those in the two lines of text you just examined. But you experience them differently; they are meaningful, English words.[1]

This little demonstration illustrates a fundamental rule: Without a cultural context, experience is meaningless. This rule applies to all experience. Examine a slide through a powerful microscope. Listen to a piece of Indian music. Chew on a handful of wax worms. In each case what you would see, hear, or taste that may not make sense to you; the experience is bizarre. But for trained histologists, lovers of Hindustani music, or aficionados of grubs, these same experiences are wonderful. It all depends on what cultural context you bring to the situation.

## Applying the Lesson

This lesson guides you to an understanding of social statistics. For many students, social statistics are as foreign, nonsensical, and intimidating as deciphering Line 1 or munching on wax worms. And for good reason. Most students taking social statistics never receive a context that allows them to make sense of what is being taught. Hence, much of what they experience is gobbledygook. The result is they are exposed to the material, but never learn it.

## Goal

This section aims to provide a context in which to make sense of social statistics. At this point, we are not talking about understanding specific social statistics; that comes later. The goal in this section is to situate social statistics as one form for making sense of experience. To accomplish this goal we look around and get our bearings by

- introducing a model containing two levels of experience and their connection;
- situating research methods within the model;
- explaining how statistics fits in the model; and
- identifying how other approaches fit in the model.

By the end of this section you should have a "big picture" in which to locate the specific statistical tools discussed throughout the book.

---

[1]By the way, Line 1 is gobbledygook; Line 2 is not. Line 2 is a Chinese (Mandarin) translation of the English sentence that began this paragraph. For students connected to Chinese culture, Line 2 is just as meaningful as the English counterpart.

## Experience versus Understanding: Two Levels of Experience

It is clear to anyone living in the modern world that the differences among people are very large. The world is full of people who experience, think, feel, and act in a variety of ways. A quick scan of the news reveals the range extending from saints through sinners. The range of human variety is so great that it is difficult to even imagine the lives of some others. For instance, can you imagine what it must be like to be someone who commits familicide?[2] Given the tremendous diversity of human experience, the sociological conclusion that people live in different realities is reasonable. Understanding how differing social realities are constructed begins with appreciating two basic levels of experience, their components, and connection.

The basic level of experience is the concrete, which is the world of physical sensations. You obtain concrete experience by seeing, touching, tasting, smelling, and hearing. The concrete level contains empirical experience. The concrete level of experience is composed of parts called percepts. Percepts are the fundamental elements of sensory experience. When percepts are aggregated they form patterns. For example, a single dot on a page is a percept, while a collection of dots constitutes a pattern. Likewise, when you hear the loud initial "beep" of a garbage truck reversing, you experience a percept; when you hear "beep, beep, beep," you distinguish a pattern.

Two characteristics of concrete experience are worth noting. First, this is the level of experience you share with all other living creatures. Your pet canary experiences life at the concrete level, as does an elephant roaming the forest. Second, the concrete level of experience is meaningless by itself. If life were experienced exclusively at the concrete level, it would be full of sensations but devoid of meaning. That is why, for example, when your parents hear your unfamiliar, loud music they encourage you to "turn down that noise." While, to you, the music may be inspirational, to them it is nonsense. The same phenomenon occurs when you witness how some people are so transfixed by a painting that they return for repeated viewings, while others pass by unaffected. The person who disregards the painting is not blind; their eyes operate correctly so they can clearly look at the painting. The issue is that they are stuck at the concrete level.

Fortunately, people do not live exclusively at the concrete level; meaninglessness is not our perpetual condition. Our experience is larger than that of other creatures who, like us, have empirical sensors. Our liberation comes from the fact that we also inhabit the abstract level of experience. Abstract experience occurs in your mind. It is the world of imagination, of fantasy.

The abstract level is composed of concepts which, when linked together, form propositions. Concepts are abstract terms for organizing sense experience. Place six pens in front of you. Examine the pens. You will likely experience each one differently—as objects of different length, diameter, and colour. Some may contain teeth marks, while others are unflawed. Perhaps they even have distinctive smells. Notice, however, that you refer to each of these concretely different objects by the same abstract concept. Each of them, to your mind, is a pen.

---

[2]Familicide occurs when one spouse kills the other and at least one of their children. Neil Websdale has carefully studied 211 cases of familicide and provides an insightful account of the common motivations of these killers in his (2010) book *Familicidal Hearts: The Emotional Styles of 211 Killers*.

Through this naming process, called conceptualization, you organize concrete experience by placing the different objects into a single, meaningful category.

Your mind is full of concepts. They let you organize and give meaning to concrete experience. Your mind also relates concepts to one another and, in doing so, forms propositions. Propositions are abstract statements that express the relationship between two or more concepts. Imagine someone says, "Watch, table, education, income." After hearing this list of concepts, it is doubtful that you will say, "Now that's a good idea!" Concept lists are not propositions. Propositions emerge when your mind connects concepts in a meaningful way. "The watch is on the table" and "More educated people receive more income" are propositions.

A key consideration regarding the abstract level of experience is that this is where the meaning in human life resides. How much or little meaning some specific experience has for you, how much or little meaning your life has for you, is a function of the number and nature of the concepts and propositions that your mind can bring to experience. Persons who appreciate terms like *role conflict*, *Oedipal complex*, *recombinant DNA*, or *sticky wicket* have a different understanding of the world than those whose minds do not appreciate these concepts. Similarly, individuals whose minds contain propositions that characterize the relationship between things (e.g. "An armadillo is larger than a possum") or people (e.g. "I love you") or symbols (e.g. $F = m \times a$) have a different understanding than those who do not.

Figure 1.1 illustrates these two basic levels of experience and their components.

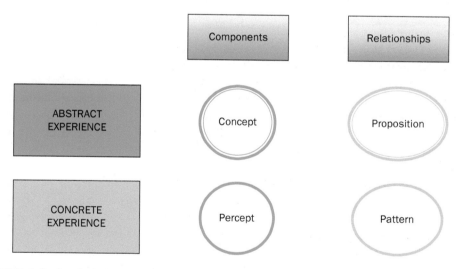

**FIGURE 1.1** Basic Levels of Experience

An appreciation of this model includes the following key consideration: *experience* (at the concrete level) is different than *understanding* (at the abstract level). Understanding requires *interpreting* experience. Imagine watching a beautiful sunset on a warm summer evening. The fact that you had this concrete experience is quite different from understanding the experience in terms of concepts and propositions relating the rotation of the earth on its axis relative to the sun. Understanding requires interpretation, which occurs through the application of abstract concepts and principles. The problem of making sense of empirical experience, of trying to understand what is occurring to and around you, is a fundamental problem of human life. Think of the person who struggles to discern whether his partner's actions are really acts of love. Or the president who wonders whether the actions of some other nation constitute an act of war. Or the radiologist who tries to discern whether the X-ray actually portrays cancer.

## On the Social Construction of Reality

The meaningful understanding of any event is a function of the concepts and propositions an individual *brings to* her concrete experience. In this way an individual's experience of reality literally is "constructed."[3] Such construction occurs through the merger of the concrete and abstract levels of experience. Figure 1.2 provides a visual representation of reality construction.

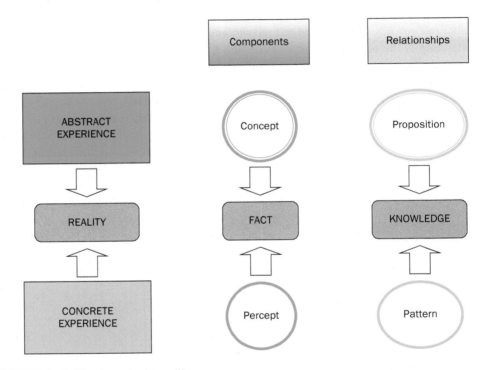

**FIGURE 1.2**  The Level of Reality

Constructed reality, like the other levels of experience, is composed of components and their relationships. The components of reality are **facts**, which are appropriately defined empirical experiences. If there is an orange piece of fruit on your desk and you label it a *blizzard*, the result is not a fact. It is a fact, however, that the piece of fruit is an orange. Of course, you may protest at being corrected and persist in calling the object a blizzard. When you order breakfast at a restaurant, besides your sausage and eggs, you may request a glass of "blizzard

---

[3]This is the basis for the famous Thomas Theorem, which states that if an individual defines a situation as real, it is real in its consequences. See Berger and Luckmann's (1967) classic *The Social Construction of Reality*.

juice." But if you persist in such behaviour you risk being labelled *abnormal* (or worse). That is why *appropriately defined* is part of the definition of a fact. In general, whether a label is appropriate or not is a matter of the level of public agreement. In most cases, an appropriate level of public agreement comes ready-made in the form of existing social legitimations. Most parents, teachers, preachers, coaches, and other leaders define their task as passing along conventional definitions of reality. In other cases, however, obtaining an appropriate level of public agreement is a challenge. Is your friend's new partner really an idiot? Are the uprisings in developing nations really bringing about democratic rule?

Each of us is welcome to construct our private realities, but if we veer too far from conventional labelling, we risk others challenging our grasp of reality.[4] Notice, however, that within limits, each of us constructs individual reality. Few of us have escaped genuinely fretting over health or social concerns that had limited basis in reality. Or consider the number of persons who actually believe they are smart, beautiful, or charming when they are not. Such delusions are commonplace.[5]

As Figure 1.2 indicates, the merger of concept and percept creates a fact. Concepts can, of course, exist independently of percepts. In this case, the concepts refer to imaginary things, not real things. Mermaids, hobbits, and ghosts qualify on this account.[6] What about the case where percepts occur in the absence of concepts? In such instances the experiences pass by the individual unobserved. Shame, for instance, is a ubiquitous reality in social interaction;[7] yet it is hardly recognized. Such a reality is barely noticed since very few persons have a clear concept of what shame is. Lacking the concept, the empirical experience of shame goes either unnoticed or is vaguely experienced. One benefit of a liberal education is its ability to liberate persons to experience a wider reality by enlarging their command of concepts and percepts.

As with the concrete and abstract levels of experience, reality contains relationships between its components. In the case of reality, **knowledge** is the term that describes these relationships. Knowledge refers to statements about reality in the form of empirical rules. You know something about the real world when you can state a rule connecting facts. In this sense, if you can only recite the facts of the matter, you are not knowledgeable. Knowledge is the kind of understanding that comes from being able to *connect facts in a ruleful way*. It is a factual matter to be able to identify the earth and the sun. Being able to knowledgeably specify how the movement of the earth is *related to* the sun is another matter. Similarly, it is a factual matter to be able to identify a person's social class and their mental health. It is quite another domain to know how social class is related to mental health.

Clarifying that having knowledge is more than being able to identify facts is not to denigrate the latter. Obtaining facts is a necessary condition for acquiring knowledge. And it is no simple task to be confident of factual matters about nations (How many Canadians actually are there?), organizations (Is your university really free of gender bias?), or individuals (What

---

[4]See Milton Rokeach's classic study (1964) *The Three Christs of Ypsilanti* for a powerful illustration of maintaining a realistic identity label in the face of competing interpretations.

[5]For a classic demonstration of how marriages are "constructed" see Berger and Kellner's (1964) "Marriage and the Social Construction of Reality."

[6]Of course, there are those who believe ghosts are real. Estimates suggest this is about a third of the population. For these individuals, the reality of ghosts is sustainable because they have matched some empirical experience with the term.

[7]Thomas Scheff (1990) convincingly argues that shame is the "master emotion" of everyday life.

is your level of neurosis?). Knowledge is built on specifying ruleful connections between facts. The aphorism "knowledge is power" is based on this understanding of knowledge. If you know how some sector of the universe actually operates, then you can make reliable predictions of what will occur and, within the limits of available technology, effectively intervene to influence outcomes. Absent knowledge, predictions and interventions are haphazard. You could ask your professor to fix your car or replace a valve in your heart, but you are advised to seek the counsel of a qualified mechanic or heart surgeon. The latter know more about the operation of vehicles and hearts than your professor does.

Figure 1.2 also provides another way of appreciating what knowledge means. It can be conceived of as the merger of a proposition with an empirical pattern. In other words, knowledge requires that an idea (proposition) be supported by concrete evidence. As with the concepts/percepts connection, it is possible to have one of these components without the other. Speculation is the term for a proposition (idea) for which empirical evidence has yet to be collected. If a proposition has been tested against empirical patterns and found to be wanting, the label *bad idea* is appropriate. Alternatively, it is possible that there are empirical patterns which we have not yet conceived of. This is knowledge that remains to be constructed.

## Theory and Ideology

We have empirical experiences at the concrete level, but gain understanding at the abstract level. The propositions that link concepts at the abstract level are ideas. By virtue of being at the abstract level, all ideas are acts of the imagination.

Our imaginations are useful for all kinds of interesting purposes. One purpose is intrinsic; we simply enjoy ideas for themselves. Those who experience pleasure from fantasizing about anything from exotic vacations to the person next door are using their imagination for intrinsic benefit. The imagined vacation or the interactions with the neighbour are not real; they are enjoyable fantasies. Reading works of fiction or poetry are similarly so.

Beyond intrinsic benefit, ideas can have the practical purpose of helping us live more efficiently and effectively. Ideas that serve this purpose, however, must be more than fantasy; they must be valid and reliable maps of how the empirical world operates. In other words, ideas that serve instrumental purposes need to describe empirical patterns; they must be a form of knowledge.

At this point it is worth saying something about theory. In popular usage, the term *theory* is often used as a synonym for terms like *idea*, *opinion*, or "random thought I just made up." Such common usage points in the right direction to the extent that it indicates that theory exists at the abstract level of experience. However, it is incomplete.

A proposition or idea does not constitute a theory. Imagine a child says, "The cow jumped over the moon." The child is stating a proposition, since the utterance specifies a connection (jumped over) between two concepts (cow and moon). But this statement is not a theory. A **theory** requires the construction of a coherently interconnected set of propositions. In other words, a theory is composed of a set of ideas that are logically knit together to tell a story. If the child expressed a coherent story about why the cow jumped over the moon and what happened when it landed on the other side, then she would, to our amazement, be sharing a theory.

All sciences have theories. Psychology has Freudian and behaviouristic ones. Sociology has conflict and interactionist ones. Biology has evolutionary ones. Since science is aimed at generating knowledge about the empirical world, its definition of *theory* is restricted to a special kind of story. A **scientific theory** is a set of interrelated propositions, some of which are empirically testable. The proviso that scientific theories contain propositions that can be empirically tested is an important one. This is an important consideration since it implies that a defining characteristic of scientific theory is that it be falsifiable. The falsifiability criterion ensures that scientists must be able to imagine some kind of empirical evidence (patterns) that would indicate that the theory is wrong. In this way, scientific theories constitute open systems of thought, where *open* means the story could be modified or discarded if it is not aligned with empirical patterns. Scientific theories are sustained, or not, based on the collection of evidence. If a theoretical story is such that it cannot be falsified, then it does not qualify as scientific theory.

In contrast to scientific theories, many systems of ideas are closed systems. Such closed systems are called ideologies. Like scientific theories, ideologies are composed of interrelated sets of propositions; they tell a coherent story. However, the acceptance of ideologies does not hinge on an assessment of the preponderance of empirical evidence. Instead, the central propositions of an ideology are accepted (or rejected) on the basis of faith. Most systems of religious and political thought are of this type. One doesn't become a committed Catholic or Communist on the basis of empirical evidence. A person enters (or leaves) the system of belief through faith. As closed systems, ideologies are important because they provide subscribers with a powerful sense of certainty.[8] Ideological stories provide coherent explanations without the nagging doubt that future facts may collapse the system of thought.[9] Subscribers to ideologies "know" the truth, while scientific theorists are perpetually doubtful.

## Induction, Deduction, and Research Methods

For the way of knowing called science, the test of credible propositions is the extent to which they account for observed empirical patterns. A credible theory in astronomy helps us predict when and where the next solar eclipse will occur. A credible economic theory predicts when the next recession will emerge. A credible sociological theory explains why the divorce rate is increasing. The question then becomes: How do concrete empirical patterns become attached to propositions so that the statements about reality called knowledge emerge? (See Figure 1.2.) The construction of knowledge occurs through one of two basic processes, induction or deduction.

Induction is the process of going from the specific case to the general case. With reference to Figure 1.2, induction begins when observations are made of empirical instances or patterns. These observations constitute specific cases. If you go on a blind date and enjoy the company of your brown-haired partner, you have made one observation. If your next 50 blind dates all reveal that brown-haired persons are more fun than others, then a pattern is emerging. From your experiences you might generalize about the connection between hair colour and fun: Brown-hairs are more fun. Such knowledge might become your rule for understanding the

---

[8] Hence Daniel Bell's famous definition of an ideologue as a person shouting "I've got an answer, I've got an answer; who's got a question?"

[9] The flip side of such certainty is their danger. When competing ideologies clash, disagreements cannot be resolved through reasoned consideration of empirical evidence. Subscribers of conflicting ideological systems display the courage of their convictions. The historical record of much interpersonal violence and bloody international conflict demonstrates that, when ideologies collide, there are only two options for resolution: force or divorce. See Harris (2005), *The End of Faith: Religion, Terror, and the Future of Reason.*

reality of the dating world. You may generalize that it is prudent to pass over those with black hair, blondes, redheads, and others if you want to have fun. Stated more formally, inductive reasoning looks for commonalities among different specific cases, and then generalizes from these shared characteristics. With reference to Figure 1.2, induction begins at the concrete level of experience, and after observing a requisite number of cases, formulates a general proposition (at the abstract level).

By contrast, **deduction** begins with some general idea and then tests whether the idea is credible. The original propositions used in deduction come from theoretical reasoning which, as noted above, occurs at the abstract level. For example, before making any empirical observations, there may be good theoretical reasons for thinking that contemporary students are less empathetic than those of previous generations, or that the European Union will collapse by 2025. By themselves, these abstract assertions are speculations—but they are propositions that can be tested. If the empirical testing of these abstract propositions confirms them, then an addition to knowledge results. In this way, deduction moves from the abstract to the concrete levels of experience.

Induction and deduction are the two principal ways for bridging the gap between the abstract world of imagination and the concrete world of empirical experience. They underlie the two principal means for formulating predictions about the real world. In cases of **actuarial prediction**, the researcher searches for common characteristics among empirical patterns and generalizes from them. For instance, through actuarial prediction, insurance companies know that smokers, skydivers, and motorcyclists are at greater risk of injury and death than those who practise different less dangerous diversions. Hence, all persons in these categories experience higher premiums. By contrast, practitioners of **clinical prediction** begin with theorizing and then apply their abstract reasoning to specific cases. For anyone interested in gaining insight into the real world that is more than mere haphazard guessing, speculation (grand theorists), or confirmation of preconceived ideas (ideologists), then methods based on either induction or deduction must be employed. Figure 1.3 locates induction, deduction, and research methods.

The term *research* literally means "to look again." As Figure 1.3 illustrates, *research methods* refers to procedures and techniques aimed at constructing facts and knowledge about reality. Realistic understanding comes from merging empirical experience (percepts and patterns) with thoughtful labels (concepts and propositions). There are a wide range of research methods in different disciplines. However, all research methods utilize either induction or deduction procedures to forge connections between concrete and abstract experience.

## Narratives and Numbers

Like other social science disciplines, sociology contains a wide range of specific research methods. The discipline has several different tools and techniques for "looking again" to gain insights into the real world. Commonly, however, the discipline's methods are categorized into two complementary groups, including qualitative and quantitative approaches.

**Qualitative methods** include techniques such as participant observation, in-depth interviewing, and textual analysis. While these techniques vary widely, they share a common purpose, which is to gain insight into how people actively construct and give meaning to their worlds. Generally, qualitative methods examine a small number of cases in an effort to obtain

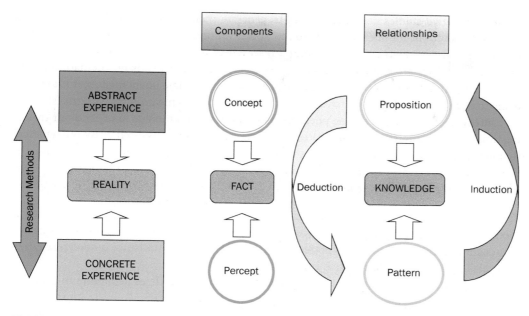

**FIGURE 1.3** Induction, Deduction, and Research Methods

extensive detail about the lives lived.[10] Qualitative studies generally begin with some empirical evidence (cases) of interest, and then study the details of these cases in an effort to make sense of what is occurring. In this way qualitative studies tend to rely on induction. They begin with specific cases and study them in order to generate an understanding of what is occurring in this general class of events. Qualitative studies are rooted in theory-independent empirical observations. Practitioners of participant observation, for instance, often employ a protocol called analytic induction to make sense of the cases. The result of the best qualitative studies are stories or narratives (at the abstract level) that provide an intuitively appealing understanding of the class of cases studied (at the concrete level).

Quantitative methods, by contrast, take a different approach to connecting abstract and concrete levels of experience. Large-scale surveys exemplify the quantitative approach. The idea behind quantitative methods is to generalize to a large number of cases and, in doing so, the methods must necessarily focus on a limited number of considerations. Quantitative approaches generally begin at the abstract level with propositions that the researcher has good (theoretical) reason to believe describes some aspect of reality.[11] The research then goes on to construct a concrete test of these propositions. In other words, quantitative approaches rely on deduction. The formal protocol employed in quantitative research is the hypothetico-deductive method. Since quantitative studies collect observations from a large number of cases, some techniques must be used to see whether there are patterns in the large data sets. Enter statistics.

---

[10]Some classic illustrations include: Braroe (1975), *Indian and White: Self Image and Interaction in a Canadian Plains Community;* Ellis (1986), *Fisher Folk: Two Communities on Chesapeake Bay;* Lofland and Stark (1965), "Becoming a World-Saver: A Theory of Conversion to a Deviant Perspective."

[11]One of the cardinal sins of survey research is to ask a question only because it is "interesting." Such questions generally yield useless results. Survey research, as a method employed in a scientific discipline, strives to ask theoretically informed questions.

While qualitative studies construct stories to generate their narratives, quantitative studies apply statistics to the numbers assigned to their observations.

The narratives constructed by qualitative researchers and the statistics generated by quantitative researchers are complementary approaches. Their use depends on the nature of the question at hand. If a researcher has little understanding of some phenomenon, it is sensible to begin by looking in depth at a few cases and trying to discern meaningful patterns (stories). By contrast, as knowledge of some phenomenon grows, it is reasonable to expand our thinking by testing the empirical plausibility of new ideas. This testing is assisted by tools (statistics) that help summarize and reveal patterns within large numbers of empirical observations.

## Quantitative Model

This book is about understanding social statistics. So far, the discussion has provided a model for understanding how knowledge is generated, and where qualitative and quantitative approaches fit into this model. To better understand the place of statistics, however, requires some elaboration of the quantitative approach.

The quantitative approach begins at the abstract level and utilizes deduction to generate new ideas worthy of empirical testing. We can use the following propositions to illustrate this hypothetico-deductive process:

> Proposition 1: People living in nations with greater inequality experience more stress (Wilkinson and Pickett, 2009).

> Proposition 2: People who experience more stress have poorer health (Kiecolt-Glaser, 2005).

The references cited after each proposition document that these ideas are established in the research literature; in other words, these statements are backed by empirical research results. If we think about it, we can use these ideas to generate the following new idea:

> Proposition 3: People living in nations with greater inequality have poorer health.

Proposition 3 is a new idea, derived from linking Propositions 1 and 2 through deductive reasoning. Note that, like the first two propositions, the third one exists at the abstract level and states how two concepts (inequality and health) are connected. Like all abstract propositions, this new idea is speculation.

In order to translate this abstract speculation about how inequality and health are related into knowledge about the relationship of these terms in the real world, the proposition must be tested against actual empirical patterns. This requires making observations at the concrete level. And, at this point, serious problems emerge. Have you ever seen, tasted, touched, smelled, or heard national inequality? What does it mean to concretely experience health? Abstract concepts exist in your mind and, by themselves, have no concrete content. The quantitative model addresses this problem by translating abstract propositions into a researchable form through a process called **empirical deduction**.

Figure 1.4 uses numbers to identify various relationships. Number 1 in the diagram refers to an abstract proposition (idea) that a researcher thinks is plausible and wants to test empirically. In our example, 1 refers to the statement "People living in circumstances of greater inequality will have poorer health."

The two arrows labelled *2* in the diagram refer to a process called **operationalization**, which involves identifying variables used to measure the concept under consideration.

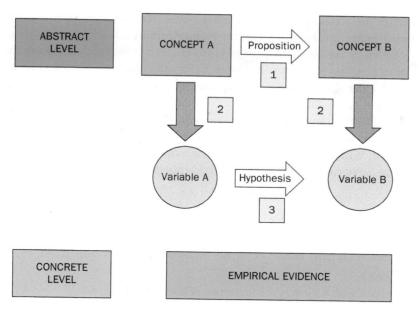

**FIGURE 1.4** Empirical Deduction: Translating Propositions into Hypotheses

**Variables** are properties of objects that can change. Variables are composed of attributes (sometimes called values) that specify the possible scores the variable can take. For example, social class is a variable. It specifies a property of individuals (i.e. their relative social location) and can take on different attributes (i.e. some people are upper class, some middle class, others working class). Variables play a crucial role in quantitative research since they detect empirical differences between cases.

In our example, the challenge of operationalization is to find a variable (or variables) that adequately captures what we have in mind by the abstract terms *inequality* and *health*. There are many alternatives. Here are some choices a researcher could make, along with the rationale.

Concept: National Inequality

Variable: The proportion of wealth controlled by the richest fifth of the population compared to the poorest fifth of the population.

Rationale: Dividing a population into fifths creates quintiles. An equalitarian society would have the same proportion of wealth in the top and bottom quintiles, and would score a 1 on this variable. In more unequal societies the rich will control greater proportions of the wealth, so the variable will have larger scores. For example, in a society where the top quintile controls 60 per cent of the wealth and the bottom quintile 10 per cent, the variable score is 6.

Concept: Health

Variable: Incidence of heart attacks

Rationale: A key to health is a functional heart. As the number of heart attacks per 1,000 population increases, it is a sign that a nation's population is less healthy.

For each of these concepts, other variables might have been plausible alternatives. For instance, health might have been measured by *Infant mortality rate* or *Obesity rate* or *Life expectancy*. In quantitative methods, the issue of epistemic correspondence speaks to the validity of the choice of variables to measure concepts. If, for example, the health of a population was measured by *Number of moose per capita*, the epistemic correspondence of the measure would be questionable.

With the concepts translated into variables (the arrows labelled *2*), an empirical deduction can occur. Empirical deduction is the thinking process that transforms a proposition into a hypothesis. With reference to Figure 1.4, empirical deduction takes arrows 1 and 2, and generates arrow 3. Here is how empirical deduction applies to our example:

IF

1.  People living in nations with greater inequality will have poorer health.

    AND

2.  Greater inequality occurs as the top quintile's share of wealth grows.

    AND

3.  Poorer health occurs as the incidence of heart attack increases.

    THEN

4.  As the top quintile increases their share of the wealth, the incidence of heart attack will increase.

Notice that 3 expresses the logical consequence (deduction) of 1 and both parts of 2. Deduction 3 is a hypothesis, which is a statement about the expected relationship between variables. In grade school, students often learn that a hypothesis is an educated guess. Figure 1.4 lets you appreciate what this statement means. A hypothesis is not a wild guess; it is an educated one. The educated part of a hypothesis stems from the fact that a hypothesis is logically derived from an abstract proposition (idea). The abstract proposition underlying the hypothesis is, in turn, derived from a theory. In short, there is a lot of thinking (education) embedded in a hypothesis. However, a hypothesis is still a guess about concrete patterns. It is not knowledge (see Figure 1.3) since it has not been tested to see how well its prediction aligns with empirical experience.

In the quantitative tradition, successfully translating a proposition into a hypothesis turns the theoretical idea into a testable form. The hypothesis is testable because the researcher can measure the variables and their relationships at the concrete level. The variables are the empirical referents of the concepts and, as such, move the original idea away from the abstract level toward the concrete.

### Instrumentation and Measurement

Testing a hypothesis involves examining the actual empirical patterns among the variables. With reference to Figure 1.4, the gap between the hypothesis and the empirical evidence is filled by two processes, instrumentation and measurement. Instrumentation is the process of creating a tool for measuring the variable under consideration. For example, the variable *Temperature* is measured by the instrument called a thermometer. Four centuries ago, thermometers did not exist, so researchers had to create one. Doing so was an act of instrumentation. Some tool is required to measure any variable. Sometimes tools exist and researchers can just use them (e.g. thermometers and scales measuring self-esteem). On other occasions, either appropriate instruments do not exist or existing ones are deficient. In these cases, researchers have to construct instruments.

With instruments in hand, researchers can take the last step in acquiring empirical evidence, which occurs through the process of measurement. **Measurement** involves applying an instrument to an object in a ruleful way. Through measurement a researcher applies numbers to objects in a reliable way. You measure a person's temperature using a thermometer by following certain rules. If it is an old-fashioned mercury thermometer you must shake the liquid down the tube to reset the thermometer, and then place it under the subject's tongue for a certain period of time. If you do not follow the rules, the measurements may be inaccurate. The same is true of taking all measurements. A tape measure allows you to measure length accurately, but only if you follow its rule for proper use. Figure 1.5 diagrams the instrumentation and measurement processes.

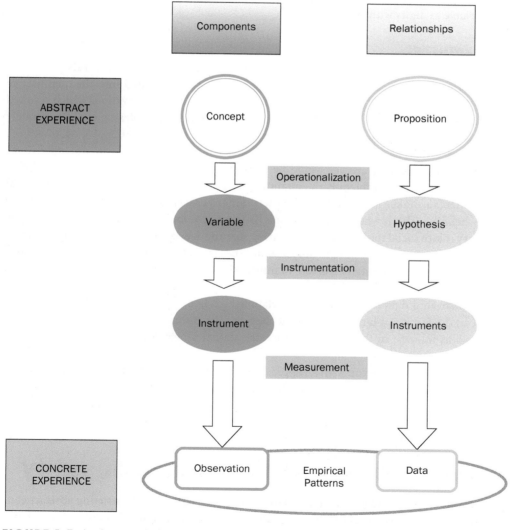

**FIGURE 1.5** Instrumentation and Measurement

Note that when an instrument is applied to objects through the measurement process, the result is an observation. In the quantitative tradition, an observation is the empirical measurement of an object in terms of some variable. If you apply the thermometer correctly, you may observe that a person's temperature is 35°C. If you apply the tape measure correctly, you may observe that the desk is 96 centimetres wide. If you apply Rosenberg's self-esteem scale correctly, you may observe that your friend has low self-esteem. When observations are collected across several cases on a variable or variables, the result is data. Data refers to the aggregation of quantitative measurements (observations) across several variables and/or cases.

Remember that, in the quantitative tradition, observations (the result of measurement) are in the form of numbers. Therefore, a data set is actually a large set of numbers. In most quantitative studies, the data set is in the form of a data matrix, which includes the variables as columns and the cases as rows, and reports numbers which measure each object's score on each variable as digits within the matrix.

### Enter Statistics

Imagine that, a week after a class test, a student in a large class of 175 students raises his hand and asks the professor, "How did the class do on the test?" Translated into the quantitative model, the student is really asking about the data set of empirical observations which resulted from measuring each student on the instrument (the test). The professor could accurately answer the student's question by reciting each student's test score: 75, 59, 36, 93, 64, 67, 88, 55, 51, 48, 99, 56, 67, 77, 58 . . . until she had recited all 175 observations.

While the professor's answer would be accurate, it is not that helpful. Such unhelpfulness captures a central problem is the quantitative approach. Simply stated, when numerical measurements are taken for several variables across many cases (objects), the volume of evidence is overwhelming. This is where statistical tools like those covered in this book enter the picture. A primary role of social statistics is to summarize quantitative data. Instead of reciting 175 numbers, if the professor in our example simply said, "The class average on the test was 78 per cent," he would have employed a statistical tool to summarize the observations and the student would have received a meaningful answer to her question.

Summarizing data for variables and relationships is not all that the tools of social statistics can do, but it is a large part of what they do. The summaries that statistics provide fit Figure 1.5 as follows: In the quantitative model, statistics help researchers detect empirical patterns in large data sets.

This section provided you with an appreciation of where statistics fits into the quantitative tradition of research and where this tradition, in turn, fits into the larger quest for understanding our experience. The next section introduces some key concepts and considerations employed by quantitative researchers.

# 2. Types of Variables and Causality

Variables play a central role in the quantitative model of research. By translating abstract concepts into variables, ideas become researchable. Figure 1.5 illustrates this point. Hypotheses are the objects that quantitative researchers test. Hypotheses, in turn, state the expected relationship between variables. Therefore, variables are the key to quantitative research.

## Goals

In quantitative research various schemes are used to classify variables. This section introduces you to one classification scheme. Based on this classification scheme, the idea of causality is introduced.

## Types of Variables: Independent, Dependent, Control

Earlier you learned that statistics are used to summarize data, which takes the form of empirical evidence. A common procedure is to *sequentially* summarize data regarding individual variables. In this practice, variables are considered in isolation. For example, imagine an opinion poll that collected data from 1,500 respondents. In the poll, information was collected from respondents on a number of variables such as age, sex, social class, attitudes toward the prime minister, recreational activities, etc. One use of statistics on this polling information is to summarize the variables sequentially, which is what you see reported in newspapers when the headline reads "Prime Minister's Popularity in Free Fall." The contents of the newspaper article will report what percentage of the poll respondents rated the prime minister unfavourably. The article may also report what proportion of the sample were young people or what percentage were middle class. In each of these illustrations variables (political attitudes, age, social class) are being reported individually. When this occurs, the statistics are being used to perform **univariate analysis**.

Univariate analysis is important, but it is not the only form of analysis. Rather than focusing on single variables and their distribution in some sample, sometimes we are interested in the *relationship between two variables*. In this case, statistics will be used to perform **bivariate analysis**. Later you will learn various bivariate analysis procedures, but at this point it is important to understand the types of variables that emerge in bivariate analysis.

Earlier you learned that variables are properties of objects that can change. The possible changes (alternatives) for any variable are defined by its attributes. For instance, *Sex* is a variable because it is a property of objects (people) that can differ on the attributes *male* and *female*. Univariate analysis focuses on summarizing information about single variables. In bivariate analysis the focus shifts from summarizing the variables to *summarizing the relationship* between two variables. This difference is illustrated in Figure 1.6.

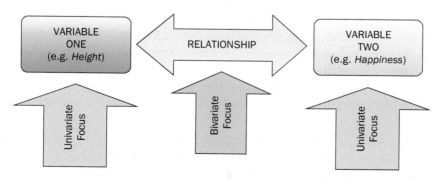

**FIGURE 1.6** Univariate and Bivariate Focus

Note that the univariate focus is on the single variables, asking questions like "What is the average height in the sample?" or "Does the sample contain more happy than unhappy people?" The bivariate focus in not on the variables but on the *relationship (connection) between the variables.* A bivariate question about the connection between two variables might ask, for example, "Are taller people happier than shorter people?" Notice that this bivariate question cannot be answered by knowing just about height (a univariate question) or just about happiness (another univariate question). Answering the bivariate question requires knowing about the *connection* between the variables.

Bivariate analysis yields a first type of difference between variables, which is the distinction between independent variables and dependent variables. To appreciate this distinction, ask yourself the following questions: Are you basically an independent or a dependent type of person? How do you know? If you answered that you are an independent type, you probably made reference to the fact that you like to take charge of situations, control your fate, make things happen, etc. Alternatively, if you claim you are a dependent type, then you likely gave evidence about going with the flow, being easygoing and accommodating, etc. This same classification applies to types of variables. **Independent variables** are those that initiate change, while **dependent variables** express outcomes. So, in the example in Figure 1.6, *Height* would be the independent variable, and *Happiness* the dependent variable.[12]

It is important to note that there is *nothing about the variable itself* that leads it to be either the independent or dependent type. Take the variable *Education* for example. Is this an independent or a dependent variable? You cannot tell just by looking at the variable. Instead, you need to look at the hypothesized relationship in which the variable is used. If the hypothesis is "People with more education obtain better jobs," then *Education* is an independent variable. Alternatively, if the hypothesis is "Children from lower class families obtain less education," then *Education* is a dependent variable. The designation of a variable as independent or dependent depends on the specific research issue under consideration.

To recap, the quantitative research tradition tests abstract ideas by translating them into hypotheses, using the process of operationalization. Hypotheses, in turn, express the anticipated relationship between variables. The variables that are connected in a hypothesis can be of the independent or dependent type. The patterns expected by a hypothesis are tested against actual concrete empirical patterns through the processes of instrumentation and measurement. Social statistics are tools used to detect patterns in empirical data about the variables collected through measurement.

In most cases, however, the actual situation is more complex. You can appreciate this complexity by thinking about your relationship with somebody close to you. By way of analogy, you can think of both yourself and the other person as variables (i.e. you both change) who have some kind of relationship (e.g. you love one another). You know from experience that the love which characterizes your relationship is not fixed. Sometimes it is more intense, at other times less intense. Sometimes the love disappears for a period and is replaced by bitterness or some other less attractive state. This fluctuation in the state of human relationships is also evident in the relationship between variables. In short, relationships are *context dependent.*

---

[12]In this example the distinction between independent and dependent is clear. Obviously, changes in *Happiness* are not going to affect *Height.* In other cases, the distinction is less clear. Which is the independent and which the dependent variable among the following two variables: *Dropping out of high school*; *Drug abuse?*

The context which surrounds a relationship can affect its character. People who have been drinking and declare their "undying love" often sober up and find the love has disappeared. Just as drug consumption can affect a relationship, so can many other variables, including money, health, intruding parents, patrolling police officers, and the like. Again, what is true in everyday life is mirrored in the relationship between variables.

To take the importance of context into account, researchers specify another type of variable, called control variables. **Control variables** (also called third variables) specify the influence of context on an independent–dependent variable relationship. Figure 1.7 diagrams this situation.

In Figure 1.7 the larger box in which the independent variable, the dependent variable, and their relationship is embedded is the context. This context is specified by identifying control variables which may affect the nature of the independent–dependent relationship. As you shall see in the next section, specifying the effect of control variables on independent–dependent variable relationships plays an important role in quantitative research.

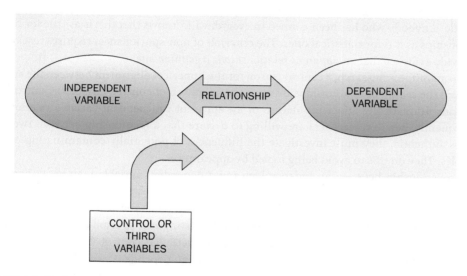

**FIGURE 1.7**  Contextual Control Variables

## Identifying Causal Connections

Like the human experience, relationships between variables come in a wide variety of types. Quantitative researchers have a special interest in one type of relationship; namely, relationships that specify a **causal connection** between an independent variable and a dependent one. Does an increase in studying cause an increase in grades? Does eating carrots cause an improvement in vision? Does raising the minimum wage cause higher unemployment levels?

In practice, quantitative researchers use three criteria (or tests) to determine whether a relationship between an independent variable and a dependent one is a causal connection. These criteria include (1) association, (2) sequence, and (3) non-spuriousness. For the assertion of cause, quantitative researchers need to provide evidence that a specific relationship passes each of these tests. Later we will discuss the nature of the empirical evidence required on each of these accounts. At this point we need only concern ourselves with the root ideas.

**Association:** The criterion of association requires that researchers demonstrate that an empirical relationship is evident between the independent and dependent variables. The test of association requires evidence that a change in the independent variable is connected to a *systematic change* in the dependent variable.

**Sequence:** This criterion concerns time-ordering and asks that evidence be provided that the independent variable changed *prior to* changes in the dependent variable.

**Non-spuriousness:** While obtaining the necessary empirical evidence may be challenging, the first two criteria (association and sequence) are straightforward in their meaning. The third criterion (non-spuriousness) is less so and requires some elaboration.

Understanding non-spuriousness begins with the first criterion, association. When association is demonstrated there is evidence of an apparent connection between the independent and dependent variables. But here is the rub: *The appearance of a relationship between two variables does not always signify that the relationship is real.* Appearances and reality do not always coincide. Anybody who has been conned in everyday life knows that this is as true for human relationships as it is for statistical ones. The criterion of non-spuriousness requires researchers to provide evidence that an *apparent relationship is a genuine one.*

In quantitative research, a central reason for the imperfect alignment between the appearance and the reality of relationships involves the operation of control variables. Control variables can dramatically affect the nature of an apparent independent-variable relationship. Consequently, before researchers are willing to declare that a relationship between two variables is authentic, they must investigate the influences of potentially contaminating control variables. They do this to avoid being fooled by appearances.

A relationship between an independent and a dependent variable may be authentic or phony. Researchers call phony relationships spurious. Non-spuriousness (authenticity) is the third criterion researchers use to determine causality. Later you will be introduced to the statistical tests that quantitative researchers use to distinguish spurious relationships from non-spurious ones. For now, it is sufficient to appreciate that control variables are used to determine the effects of context on apparent independent–dependent variable relationships.

To summarize, researchers conclude there is a causal connection between two variables when they successfully demonstrate that:

1. the variables systematically change together (the association criterion);
2. the independent variable changed before observed changes in the dependent variable (the sequence criterion); and
3. the observed relationship is an authentic one (the non-spuriousness criterion).

# Chapter Summary

This introductory chapter provided a review of the statistical maze, the landscape in which quantitative social statistics (the focus of this book) are embedded. Extended details of this landscape are available in any good research methods book. Our goal is to provide the minimum number of concepts and principles you need to appreciate in order to know where social statistics are located in the larger project of turning experience into understanding. Having completed this review of the landscape at the beginning of our journey, you should understand the following central points:

- Empirical experience must be connected to abstract concepts and principles in order to become meaningful.
- Scientific theory generates new ideas to help explain empirical experience.
- The quantitative research tradition tests new ideas by translating them into hypotheses and then comparing expectations to recorded empirical patterns.
- Statistics provide tools for detecting patterns within empirical data sets.
- Quantitative researchers have a particular interest in establishing causal connections between independent and dependent variables.

With these fundamental ideas in place, the next chapter introduces three key questions that must be answered in order to understand the pattern of the statistical maze.

# 2

# The Logic of Social Statistics

## Overview

The first chapter located the place and function of social statistics, which is really a set of tools for summarizing and interpreting large amounts of quantitative information. A large number of statistical tools are available. Look at the size of this (or any other) text that covers social statistics. All such books are substantial volumes containing many statistical tools and large amounts of technical information.

Different statistical tools are located in different parts of the statistical maze. Given the content, confronting the statistical maze can be intimidating. With so many statistical tools available, how do you know which ones to choose? This chapter provides a way of becoming confident in selecting appropriate social statistics and, in doing so, should reduce your anxiety about going down the path toward mastering specific techniques. Confidence in selecting statistical tools comes from understanding the pattern of the statistical maze.

This chapter provides the grid lines around which the statistical maze is organized. It introduces three questions you need to answer in order to select appropriate statistics from your statistical toolkit. In answering these three questions you will learn about:

- the difference between descriptive statistics and inferential statistics;
- the difference between univariate, bivariate, and multivariate analysis; and
- the difference between nominal, ordinal, interval, and ratio levels of measurement.

Although this is a short chapter, it is a very important one. The blessing of this chapter can also be its curse. The blessing is that when you thoroughly understand and can answer the three basic questions this chapter poses, you will be able to select appropriate statistical tools for practical application. The curse comes from not thoroughly understanding the chapter's contents, since it will impede your confidence in selecting statistical tools. Before beginning on any journey it is always better to be blessed than cursed, so please pay careful attention to what follows!

## Identifying the Kind of Problem

Research literally encourages us to "look again" at some situation. The goal of research is to add to our understanding, our knowledge, of some topic of interest. Research always begins with some problem.

Research problems are typically stated in the form of a question: "Why did she divorce me?" "What is causing our crime rate to rise?" "What percentage of the population is illiterate?" Problems are a source of anxiety, and this anxiety is rooted in uncertainty. The most common reason people are anxious is that they are uncertain about something.[1]

Experiencing uncertainty is another way of saying you "don't know." Student test anxiety is remarkably reduced if the professor provides the questions in advance and guarantees they will be the only questions on the exam. In this situation students know (are certain of) what is on the exam. Knowledge reduces uncertainty. Since research aims to add to our knowledge about some topic, research activity can be considered an uncertainty-reduction activity.

Choosing appropriate statistics from a wide range of tools is potentially a generator of uncertainty. Fortunately, the selection of appropriate statistical tools need not be an anxiety-laden activity. There is a logic to the way statistical techniques are organized. Different statistical tools are designed for different kinds of problems. If you are able to identify the kind of problem you face, then there are usually only a few statistical tools to choose from. The selection of an appropriate technique then becomes straightforward. Alternatively, if you do not know what kind of problem you face, then selecting an appropriate statistical tool is overwhelming. Since it is preferable to have easier choices than overwhelming ones, it is important that, at the outset, you are able to identify the kind of problem you face.

Before explaining how to distinguish between particular kinds of statistical problems, let's remember the context outlined in Chapter 1. When we speak of a "kind of problem," we are speaking in the context of the quantitative research tradition. Moreover, within the quantitative tradition, we are talking about statistical analysis issues, not issues of operationalization, instrumentation, or measurement. Different research traditions and different issues within the quantitative tradition will have different ways of defining problems. Here we are restricting ourselves to problems of choosing an appropriate statistical technique.

That said, there are three different questions you need to answer in order to narrow the selection of an appropriate statistical technique. These problems include:

1. Does your problem centre on a descriptive or an inferential issue?
2. How many variables are being analyzed simultaneously?
3. What is the level of measurement of each variable?

Each of these questions has different possible answers. Therefore, combinations of answers among the questions yield a large number of alternatives—which is why there is such a large number of available statistical tools! However, as you answer each of these questions, the statistical alternatives become more restricted. After you answer all three questions, the choice of statistics is down to one or two.

The goal of this chapter is to elaborate on each of these three questions so that you know what each question means and how to answer it.

---

[1]Universities are organizations that continually create uncertainty. For this very reason, most university students experience considerable anxiety. Courses are filled with new content that must be mastered. Students are unsure about the points made in class. You never know for sure what is going to appear on the upcoming tests, or how your term paper or project will be received. All this on top of uncertainty about how you are going to meet your next tuition payment, whether you will be able to find a job after graduation, or if you are ever going be able to find a date.

## Question 1: Does your problem centre on a descriptive or an inferential issue?

As noted in Chapter 1, quantitative data analysis is performed on variables. Information on variables is collected from cases which, in social analysis, range from individuals (micro) through organizations (meso) to states and other global aggregates (macro). The question "Does your problem centre on a descriptive or an inferential issue?" stems from the following key consideration: *In almost all circumstances a researcher never has data on all the relevant cases.*

Think about the variable *Blood type*. Blood types are measured on the basis of antigens. Based on the presence and type of antigens, blood is classified (grouped) into four types: A, B, A/B, and O. In order to measure your blood type, researchers have to collect some of your blood. Thankfully, when researchers collect your blood for analysis, they do not hang you by your ankles, slit your throat, and drain all the blood from your system! In other words, they do not collect all the relevant "cases" of your blood. Instead, they collect only a few drops for analysis.

The same holds true in social analysis. When you write a test in sociology, the professor does not ask you every possible test question. When researchers conduct a public opinion poll, they don't ask every Canadian what their opinion about the current government is. Even when researchers try to gather data from every case, which is the goal of a census, they are not successful.

The point is that researchers can analyze only the data they have actually collected and, almost always, the collected cases are fewer than all possible cases. When researchers analyze and interpret the cases they have in hand, they are using descriptive statistics. **Descriptive statistics** are used to summarize patterns in the responses within a data set.

In most cases, descriptive statistics are not the end of the story. After all, you are not really interested in what type of blood characterizes the few drops taken from your fingertip. You want to know the type of all your blood. Similarly, your sociology professor is interested in what you know about the entire content of the course. The pollster is not really interested in the opinions of 1,500 adults but of all adult Canadians. In short, researchers are interested in generalizing from the specific cases of their data to the general case. Inferential statistics are used for this task. **Inferential statistics** are used to estimate how accurately the specific cases summarized in the descriptive statistics can be used to characterize all cases.

With these distinctions in place, we can return to answering the first question: Does your problem centre on a descriptive or an inferential issue? If you are interested in *summarizing the results of the data you have*, then you require descriptive statistics. If your interest is in *generalizing from the specific cases you examined* to the general case, then inferential statistics are what you require.

As we shall see later, researchers often use both descriptive and inferential statistics. However, it is important to note that these different kinds of statistics address very different questions. Being clear on the purpose of the analysis you are performing at any specific time is important, which is why it is important to begin by answering the first question clearly.

## Question 2: How many variables are being analyzed simultaneously?

Quantitative researchers can examine variables in various combinations. The simplest way to examine variables is through univariate analysis. Univariate analysis techniques examine the variables one at a time. In univariate analysis, single variables are analyzed in sequence. For example, a researcher may look at the sex distribution of cases in the sample, and then look at the age distribution, and then examine the responses to a question about attitudes toward capital punishment. In this analysis, since the variables (sex, age, and attitudes toward capital punishment) are being analyzed in sequence, the researcher would be conducting three separate univariate analyses.

Univariate analysis examines the variables separately, asking questions such as, What proportion of the sample is women? Are retired people over-represented in the sample?, or, Do most Canadians favour the reintroduction of the death penalty? While univariate analysis provides useful information, these techniques do not answer all the kinds of questions of interest. Often researchers are interested in the relationships between two variables. For questions about the character of the relationship between two variables, bivariate analysis is required. Bivariate analysis techniques are used when a researcher wants to examine two variables simultaneously. If, for instance, the research question was, "Are there gender differences in the attitudes toward capital punishment?" then bivariate statistical tools are required. These tools allow investigation of the connection (relationship) between the independent (sex) and dependent (attitudes toward capital punishment) variables. Figure 1.6 in Chapter 1 illustrated the difference in focus between univariate and bivariate analyses.

Chapter 1 also introduced the idea of causal analysis. In that discussion you learned that, in order to establish causation, a bivariate analysis between an independent and a dependent variable needs to take account of the context in which the relationship occurs. The context is established by control variables. In causal analysis at least three variables are taken into account, including the independent and dependent variables, and at least one control variable. Causal analysis enters the realm of multivariate analysis.

Multivariate analysis occurs when statistical tools take account of three or more variables simultaneously. The simplest form of multivariate analysis is trivariate analysis, which examines how a control variable affects the relationship between an independent and a dependent variable. However, *multivariate analysis* is the general term which applies when 3, 4, 5, 6, or more variables are considered simultaneously. Multivariate analysis can take many forms. The example of causal connections is one case. Others include a researcher wanting to examine the impact of several independent variables on a dependent variable (e.g. what effect does the combination of studying, gender, social class, and intelligence have on university grades?), or one independent variable on several dependent variables (e.g. what effects does level of education have on income, happiness, divorce rate, and social class?), or any other combinations.

This second question about how many variables are being examined simultaneously is an important one. Just as researchers have different statistical tools at their disposal for answering descriptive or inferential questions, they also have different statistical tools depending on whether the task involves univariate, bivariate, or multivariate analyses.

## Question 3: What is the level of measurement of each variable?

Answering this third question involves somewhat more technical considerations than answering the first two. Coming to terms with this question begins with a brief review of some issues addressed in Chapter 1.

Chapter 1 discussed the processes of operationalization, instrumentation, and measurement by which researchers link abstract concepts and propositions to concrete empirical patterns. Together, these three steps constitute an operational definition. An **operational definition** is the process by which a researcher establishes an empirical understanding of what is meant by a concept. Operational definitions are different than conceptual definitions. A **conceptual definition** tells us what an abstract concept means by expressing it in terms of other abstract concepts. This is what dictionary definitions are. When you look up the word *intelligence* and see it expressed as "the ability to think abstractly," you are being provided with a conceptual definition. Conceptual definitions are useful in clarifying what notions we have in mind when we use a concept. Operational definitions expand on conceptual ones by reporting how to observe (experience) a concept. It is one thing to know what a lemon soufflé is (through a conceptual definition), quite another to experience (eat) one (the product of an operational definition). Similarly, knowing what the term *alienation* means is different from being able to determine an individual's level of alienation.

The final step in an operational definition is measurement. Measurement is the process of *systematically applying* a variable to an object. The key here is in understanding what *application* and *systematic* mean in this context. You will recall that variables are composed of components called attributes. Attributes specify the mutually exclusive and exhaustive set of categories that constitute a variable. The attributes, in other words, identify all the possible scores that a variable can take. In the quantitative tradition, the attributes of a variable are scored with numbers. For example, *Gender* might be scored with the alternatives 1 for *male* and 2 for *female*.

The application part of measurement is now evident. Application requires placing a number (representing the variable's score or attribute) on an object. For *Gender*, 2 is applied to Jane; 1 to James. The application could be literal, but in most cases, it is symbolic. Your instructor could, for example, provide you with your score on the first test by pasting a sticky note with your grade on your forehead (literal), but is more likely to record your score beside your name in the gradebook (symbolic).

For measurement to properly occur it is not enough that scores be applied to objects; they must be applied *systematically*. In other words, the application of scores needs to follow some rules. The haphazard application of numbers does not yield good quality measures. You would rightfully protest the instructor who assigned grades randomly. Instead, you expect grades to be assigned through an answer key, which acts as a template of rules for accumulating points on a test.

The goal of measurement is to *capture information*. More specifically, it is to capture information about a particular variable with respect to a particular object. We want to learn how much after-tax income (the variable) Fredrika (the object) has. In other words, measurement extracts (i.e. takes an imprint of) information about a variable from an object. The goal of a research methods test is to find out what you know about the readings and lecture material. Once extracted, the quantitative information is in the researcher's possession and can be analyzed using statistics.

It is important for a researcher to capture information about objects for two reasons. First, in most cases, the researcher is unable to keep the objects of interest. A psychology professor is no better positioned to keep the subjects of her experiment after their appointments are complete than an astronomer is to keep the comet after its properties have been measured. Secondly, even if researchers could keep the objects of their attention, there is a good chance they will change. Students' aggression levels change, as do the radiation levels comets exhibit. In short, the objects will not be the same later.

As a result, researchers take measures (i.e. readings) of objects and store them. Through the construction of variables that indicate concepts (i.e. operationalization), the construction of appropriate measurement tools (i.e. instrumentation), and the systematic application of instruments to objects (i.e. measurement), researchers capture the information they require for later analysis.

To repeat, the goal of measurement is to capture *information* about *differences* between objects regarding a variable or variables. For example, respondents might express their attitude to the statement "My research methods professor is incompetent" on a five-point scale:

1. strongly disagree
2. disagree
3. neutral
4. agree
5. strongly agree

The differences between two respondents' scores, for example, is identified by assigning one of them a 2 and the other a 5. Long after the respondents are done answering, these numbers allow the researcher to statistically analyze the information about their different attitudes on the subject under consideration.

A problem soon emerges, however; and this problem is related to the fact that the information captured by the measurement numbers *differs in terms of its complexity*. Take the following simple example. Imagine you are asked to rate two folk songs on a 10-point scale ranging from *a terrible folk song* (scored 1) to *a magnificent folk song* (scored 10). Let's say you rate one of the folk songs as 3 and the other as 9. Clearly, these differences measure your ratings of the quality of the different folk songs.

Now let's say you are asked to rate two symphonies on a 10-point scale ranging from *a terrible symphony* (scored 1) to *a magnificent symphony* (scored 10). And, in this case, you rate one of the symphonies as 3 and the other as 9—again expressing your ratings of the different qualities of the symphonies.

You can now easily compare the ratings (measures) of the two folk songs and identify the better one (scored 9). Likewise, you can compare the measurements of the symphonies and identify the one scoring 9 as the superior one. These are called within-class differences because you are measuring differences within a particular class of music (i.e. either "folk song" or "symphony" differences).

The following puzzling question now emerges: Is the 9 given to the folk song the same as (i.e. comparable to) the 9 given to the symphony? After all, they both look the same; they are both 9s. The answer is that these digits are *not* comparable even though they look identical. Even though they are both accurate measures, *they are not measuring the same thing*. One is a "folk song 9"; the other, a "symphony 9."

The differences between folk songs and symphonies are differences *between classes* of music. Folk songs are a relatively simple class of music (e.g. you can play hundreds of folk songs if you know three simple chords), while symphonies are a more complex class of music. The folk song rating is a "simple 9," while the symphony a "complex 9."

The general point here is the following. The properties researchers are interested in measuring differ in the amount and complexity of the information they contain. Sometimes researchers are trying to capture differences in relatively simple information; other times they are trying to capture differences in relatively complex information. The numbers used in a measurement can readily identify within-class differences (i.e. between objects of the same type). However, since digits are one-dimensional, they cannot tell us about between-class differences. Accordingly, researchers developed a system to distinguish the amount and complexity of the information captured by a measurement (i.e. between-class differences). This system is called **levels of measurement**.

Imagine a friend asks for your help in hanging a picture on the wall. A small finishing nail is going to be placed in the wall to hold up the picture and you are asked to hold it between your thumb and forefinger in the correct location. Your friend exits the room to get the tool for driving the small nail into the wall and returns, to your surprise, lugging a sledgehammer! Would you protest? Undoubtedly so—on the grounds that this is the wrong tool for the job. A little nail requires a little hammer; a big job (like demolishing a room) a big hammer.

Levels of measurement can be thought of in a parallel way. The different levels represent different tools. In the levels of measurement case, the tools differ in their ability to capture less or more sophisticated information. The lower levels of measurement are simple tools used to measure properties of objects that are not too sophisticated in their complexity. As you move up the levels of measurement the tools become more sophisticated in that the measures they produce include more sophisticated information.

There are four levels of measurement and, as *levels* implies, they can be ranked. From least to most sophisticated the names of the levels of measurement are nominal, ordinal, interval, and ratio. Here is a summary of the properties of each level:

**Nominal:** This is the lowest level, which means the numbers provided by this level of measurement provide the least information. In fact, the numbers provided by **nominal** measures are not numbers at all—since they do not provide any quantitative information. The numbers used at this level of measurement are really replacements for names (which, by the way, is what *nominal* means). The only meaning the digits convey is that objects with the same number are equivalent and are *qualitatively* different from objects that have a different number. Think of the digits on a hockey player's jersey. The player who wears 66 is not six times as good as the player who wears 11 on her jersey. The number 66 simply acts as a substitute for the player's name. The scores of variables measured at the nominal level work the same way. The digits simply identify different categories of the variable (which is why nominal measures are sometimes called categorical variables).

**Ordinal: Ordinal** variables contain the same information as nominal variables, *plus an additional characteristic.* (This is the case because the levels of measurement are nested in one another—similar to those Russian stacked dolls, where every larger one contains all the smaller ones.) The additional characteristic possessed by ordinal measures is that the attributes of the variable can be *ranked* (i.e. ordered). In other words, the scores can be ranked from highest to lowest with respect to the measured property. The five-point scale introduced earlier rating the statement about your research methods instructor's competence (1, *strongly disagree*; 2, *disagree*; 3, *neutral*; 4, *agree*; 5, *strongly agree*) exemplifies the ordinal level of measurement. The nominal property of this score exists because all respondents who, for example, score 4 all have the same amount of agreement, and this amount is different from all those who score 1, 2, 3, or 5. The additional property which makes the scale ordinal is that respondents' scores can be ranked in terms of the intensity of their agreement with the statement, meaning that those answering 1 to this question are lower (feeling less satisfied) than those scoring 5 (feeling more satisfied).

**Interval:** The **interval** level of measurement starts with an ordinal measure and adds another important characteristic. The additional characteristic required by interval measures is that the distance between each of the categories (i.e. the interval) is *fixed* or standardized. Notice that this feature is absent from ordinal measures. If runners are measured in terms of their place of finish in a race (e.g. first, second, third, etc.) we can rank their performance (which makes it ordinal)—but we *do not know* how much faster one finisher was than another. From the order of their finish we cannot tell whether the second-place finisher was a millisecond behind the winner or a minute behind. Interval measures, which use a fixed unit to indicate differences between categories, can provide us such an answer.

Ratio: The highest level of measurement is ratio. Given the nested nature of levels of measurement, the ratio level is an "interval plus." The "plus" provided by the ratio level of measurement is an **absolute zero** point. This contrasts with interval measures which utilize an arbitrary zero. An arbitrary zero means that the original selection of what *zero* means could have been for any reason. For example, the Fahrenheit and Celsius temperature scales use arbitrary zeros. Anders Celsius decided to identify zero on his scale as the melting point of ice. By contrast, Daniel Fahrenheit chose to identify his zero point as the lowest temperature he could record in his lab by mixing together water, ice, and sea salt. This certainly qualifies as a whimsical choice of zero! By contrast, the absolute zeros, which ratio levels of measurement use, literally mean "no amount" of the property under consideration. When the bank teller informs you that your bank balance is zero, this is not a capricious choice; it literally means you have zero dollars in the account.

Here is an analogy that may help your understanding of levels of measurement and their connection to statistics. When researchers measure an object they "suck" information from it—like drawing fluid into a sponge. The different levels of measurement indicate different amounts of "fluid" (i.e. types and complexity of information). After measurement, the researcher now has the information stored—like having the liquid-filled sponge in hand. At a later time (i.e. the statistical analysis stage), the researcher can wring the information out of the sponge for analysis purposes.

For present purposes there is one final point about levels of measurement that is important to appreciate. First, let's state the point negatively: You *cannot* determine the level of measurement by looking at the name of the variable. For example, knowing that the variable under consideration is *Income* does not tell you what its level of measurement is. To determine a variable's level of measurement you need to look at the attributes of the variable (i.e. how it is actually measured). In other words, to determine a variable's level of measurement, look at how the variable is scored (i.e. its attributes) and ask yourself the following questions:

- Are the categories only names? If so, then the variable is nominal.
- Can the categories be ranked? If so, then the variable is ordinal.
- Is the difference between the categories fixed and does it include an arbitrary zero? If so, then the variable is interval.
- Do the fixed differences between the categories include an absolute zero? If so, then the variable is ratio.

# Statistical Possibility Space

A **possibility space** identifies a range of defined outcomes that emerge from combining different properties. Let's assume you know a lot about genetic mutation (i.e. that is one of your properties). Imagine you babysit your four-year-old niece and want to have an intelligent conversation with her about genetic mutation. Is this going to occur? Hardly. That outcome is not occurring because it is simply not possible. Having an intelligent conversation with a typical four-year-old about genetic mutation is not in the possibility space. Everything is restricted by its properties, and when two or more restricted things come together, the outcomes are necessarily restricted.

Like all mazes, the statistical maze is similarly restricted. You can't wander around everywhere. Your routes, and therefore your potential locations, are limited. The statistical maze represents a possibility space where only some things are possible. Understanding the pattern in the statistical maze comes from understanding how its routes are created.

The pattern in the statistical maze results from different combinations of answers to the three central questions discussed in this chapter. Differing answers to the three questions create a possibility space in which alternative statistical procedures are located. This possibility space is characterized in Figure 2.1.

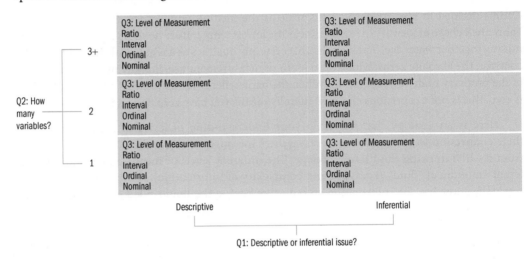

**FIGURE 2.1**  Statistical Possibility Space

**TABLE 2.1**  Statistical Selection Checklist

| Check one item for each question: |
| --- |

1. Does your problem centre on a descriptive or an inferential issue?
   - ❏ Descriptive
   - ❏ Inferential

2. How many variables are being analyzed simultaneously?
   - ❏ One
   - ❏ Two
   - ❏ Three or more

3. What is the level of measurement of each variable?
   - ❏ Nominal
   - ❏ Ordinal
   - ❏ Interval
   - ❏ Ratio

One way to think of this possibility space is as a checklist (Table 2.1). The three questions and their alternative responses are listed and different combinations of checkmarks produce different possibilities. Specific statistical procedures are designed for different possibilities. For example, one kind of statistic is useful for the combined responses: descriptive, univariate, ordinal; a different kind of statistic, for the combined responses: inferential, bivariate, interval.

This checklist will be used repeatedly throughout the upcoming chapters. It is your map to the statistical maze. The checklist guides the selecting of appropriate statistical procedures, so please review the materials in this chapter until you are confident answering each question. The confidence that comes with competence is the blessing that begins your journey.

# Chapter Summary

Selecting appropriate statistical procedures is governed by answers to three questions. The first question asks whether your interest is in describing the data you have collected or generalizing from existing data to broader populations. The second question asks how many variables the analysis is going to simultaneously consider. The third question asks what the levels of measurement are of the variables under consideration. After studying this chapter you should appreciate that:

- Different kinds of problems require different statistical techniques.
- Providing clear answers to each of the three central questions is necessary for selecting appropriate statistical techniques.
- If you can answer the three questions clearly, then appropriate statistical techniques are clarified. If not, then not.

The goal of this book is to teach you a wide variety of quantitative statistical procedures. Chapter 1 situated these techniques within the general context of conducting social research. This chapter clarified what information is required to answer the three central questions that lead to the selection of appropriate statistics. The next chapter outlines the approach that will be used in the remaining chapters to solidify your understanding of the each statistical procedure.

# 3

# Calculations and Computers

## Overview

The first two chapters oriented you to the general landscape surrounding the statisticalmaze and introduced you to the plan directing your journey. You also learned how the statistical maze and the tools it contains are organized. This chapter explains how you will be proceeding down the various paths of the maze, leading toward statistical competence. This chapter begins by emphasizing that:

- optimal understanding of any subject matter combines both knowledge and application; and
- learning specific statistical procedures is optimized by taking five systematic steps toward competency.

After these sections you will have a good idea of how your journey through the maze will proceed.

One important step in your journey is learning how to use computer software to generate statistical results. The last section of this chapter provides an overview of the statistical software and data set used throughout the book. The statistical software we are using for this textbook is IBM SPSS Statistics software ("SPSS"). We assume you know nothing about this software, and the materials in this chapter provide a basic orientation. For this reason, an additional objective of this chapter is to:

- provide a basic orientation to SPSS and the Student Health and Well-Being Survey data set.

There is no need for you to feel anxious or intimidated by the SPSS orientation. It covers a lot a ground but is only there to provide an overview. You will return to this material in later chapters in order to solidify your understanding.

Before getting to the SPSS material, let's begin by discussing the objective and strategic plan for navigating the statistical maze.

# Navigating the Maze
## Knowing and Doing

Here is an aphorism that educators at all levels find mean-spirited: "Those that can, do; those that can't, teach." This aphorism turns on a crucial distinction between *knowing* and *doing*, and contains the value-laden premise that doing is preferable to knowing. But a moment's reflection makes it clear that the story is not so simple; there are varied connections between knowing and doing.

Let's return to the understanding of *knowing* introduced in Chapter 1 (Figure 1.2). There you learned that knowledge refers to statements about reality. Specifically, you know something about some topic when you can state a ruleful connection between facts. Knowledge, then, is a type of understanding that comes from appreciating the ideas that make sense of empirical experience. You know why the sun sets in the west when you appreciate ideas related to the rotation of the earth. Knowledge is rooted in your mind, in acts of the imagination. By contrast, *doing* is rooted in application. Doing is about intervening, about utilizing techniques to create or change something. You "do" golf or welding or psychoanalysis.

Since knowing and doing are largely located in different realms (knowing in your mind, doing outside of it), different kinds of knowing–doing connections are possible. Three clear cases include knowing without doing, doing without knowing, and knowing with doing. Let us briefly consider each case. Knowing without doing occurs among those who have imaginative understanding but refrain from acting on what they know. Examples include parents who "know better" but refrain from providing their children with appropriate discipline, or students who know their community would be improved if they volunteered but never get around to doing so. By contrast, doing without knowing is quite different. Here we have people who are actively engaged in practice, but without an imaginative understanding of why the outcomes are actually occurring. These are ritualistic practitioners who may, or may not, be successful in their interventions. Nonetheless, they try. Many golf games qualify as an illustration; so do many people's efforts at making love to their partners.

The final type results from the joining of knowing and doing. These are the informed practitioners, who have an imaginative (theoretical) understanding of the principles governing some realm of experience and successfully work at applying these principles to concrete situations. Here, for example, we have educated mechanics and informed teachers.

For practical purposes, of these three types, the optimal one combines knowing and doing. After all, if you require heart bypass surgery, would you prefer someone who understands what is required but has never done it, a person who is willing to try but has limited understanding of anatomy and physiology, or a seasoned practitioner who is up to date on the latest techniques? It is clear that knowing without doing is idle; and doing without knowing, potentially dangerous. The optimal approach combines understanding (knowing) with practice (doing).

The goal of this text is to give you a "knowing plus doing" base to your understanding and execution of statistical techniques. If you study this book and come away with an understanding of statistics but poor application techniques, we will be disappointed. Similar disappointment will result if you end the book able to generate statistical results but don't understand why. Our goal is informed application of appropriate statistical techniques to social science questions.

## Downside of Computers

Computing capacity continues to increase exponentially, doubling every two years according to Moore's law. All of us benefit both directly and indirectly from this growth. However, most things, including computing speed and sophistication, have costs associated with their benefits. When it comes to learning social statistics the downside of sophisticated computing software is a belief in the "magical black box."

*Black box* is a term that refers to some object whose internal workings are unknown. Such a black box becomes magical when it encourages a belief that the box operates according to inaccessible, mysterious principles. Relating this point to the knowing–doing connection, sophisticated statistical computing software, by encouraging a belief in the magical black box, has reduced the knowing–doing connection. Many statistical users have little interest in or understanding of what is taking place when statistics are calculated. Too often, as a result, they become poorly informed "doers."

The reliance on a mysterious black box carries a serious implication. Namely, if you have little appreciation of what is occurring in the black box, or how it is occurring, then you have limited appreciation of the results. Unless you assume that the black box is foolproof (and what is?), you have no idea whether the results it produces are reasonable. Moreover, without an appreciation of how results are generated, how can you determine what they mean?

## Where You Probably Stand

Our students are unique and interesting characters. The uniqueness and interest comes from the fact that they are enormously varied. Students come in all shapes and sizes, with varied experiences, and wide-ranging competencies. For present purposes, let us hazard a guess about some of the characteristics of a typical student reading this book. First, you have varying levels of something called "math phobia." You didn't do well in math in high school and don't even think about trying to balance your chequing account. You are anxious about math and, since you associate a statistics book with math, you are somewhat fearful of the course you are enrolled in. Second, you are intimately connected to applications that rely on computing technologies, such as Facebook, YouTube, and various apps on your personal technology. Third, you are essentially clueless about how Facebook, apps, or other modern technologies actually work. All that you care about is that they do work.

Fair enough; if this is where you stand then you are probably a typical reader. The problem with staying in this position is that you are at risk of becoming a ritualistic practitioner of social statistics. If you seek to avoid doing calculations and prefer to rely on the mysterious black box, this is probably the best you can expect.

As we noted, this outcome is not that desirable. Being able to competently analyze and interpret quantitative data is one of the most marketable skills you are likely to obtain in your undergraduate studies. To exploit this advantage, the key is competence and this, in turn, requires, knowing *plus* doing. "Doing" that relies primarily on the mysterious black box is insufficient.

## How We Will Proceed

To repeat, our goal is to help you become a knowledgeable practitioner (knowing plus doing) in using social statistics. The position of most students, however, biases them toward ritualistic practice (doing with limited knowing). A gap typically exists between where most students are and where it is optimal for them to be with respect to the subject matter of this book. The question becomes: How can this gap be closed?

Our strategy is to use a five-step approach toward moving students to becoming knowledgeable practitioners. The same five steps are repeated in almost all of the remaining chapters of the book. Since using this book subjects you to this approach, it is worthwhile if you understand what each step entails and why it is useful. Here is a summary of the strategic steps.

### Step 1: Understand the statistical technique in everyday language

We appreciate that many of you are anxious about math and are proceeding with trepidation. Much of this material is foreign, challenging, and produces anxiety. For many students an initial impulse is to avoid the material through procrastination and other common student sins. While understandable, such responses are not helpful to mastery. To induce more favourable initial reactions to each new tool, we shall try to present the basic ideas in everyday language. That way you will be able to make an initial, familiar connection to the material, which provides a base on which to build your understanding.

### Step 2: Learn the steps in the calculation of the statistic

Statistical computing software is very powerful and wonderfully helpful. As we have noted, however, it cultivates a mysterious black-box approach to social statistics that is not helpful to informed understanding. The only way to reduce the mystery of a computing box is to open its lid and learn what is actually going on inside. And what you will see is that the box is only doing calculations that you can do—except with exceptional speed and accuracy. Let us repeat: *It is doing what you can do.* Therefore, the second step is to learn the basic computing that the computer performs as it does its work.

In practice there is no avoiding that this step means learning some equations and doing some hand calculations. Our task is to guide you through both the equations and calculations in ways that let you master them.

### Step 3: Learn to use conventional computer software

While learning to calculate statistics by hand is essential to a thorough understanding, it is impractical in realistic circumstances, which involve at least hundreds, and often thousands, of cases. Doing calculations on large numbers of cases is both tedious and error-laden. Since it serves nobody's interest for you to be bored and making mistakes, we will show you how to use statistical analysis software to generate the statistic(s) under consideration on realistic data sets.

### Step 4: Practise, practise, practise

Learning anything valuable is difficult and challenging. If the task wasn't difficult and challenging then anybody could do it and, by definition, the value of the outcome is reduced. People don't value or pay much attention to your walking down the street, since almost everyone does it. However, if you can walk on a tightrope strung between two office towers in downtown New York, you are likely to receive considerable attention.[1]

---

[1] See, for example, what some have labelled the "artistic crime of the century" captured in the wonderful 2008 documentary *Man on Wire*. The film documents Phillippe Petit's 1974 illegal tightrope walking on a cable strung between the twin towers of New York City's World Trade Center. The documentary examines the difficulty and challenge of his actions, and the attention they garnered.

The bestselling author Malcolm Gladwell suggests that the mastery of valuable skills is subject to the "ten thousand hour rule." Gladwell estimates that this is how much someone has to study and practise in order to gain mastery or expertise over some subject. Gladwell's long list of illustrations includes Bill Gates's mastery of computing and the Beatles's mastery of popular music. Now, although you don't have 10,000 hours to devote to learning social statistics, you need to practise (do) in order to solidify what you are learning (knowing). And the more serious practice you perform, the more competence you will gain. Consequently, after you learn the techniques in each chapter, you will be given opportunities to solidify your understanding through practice.

### Step 5: Learn to interpret the results

Chapter 1 highlighted that statistical techniques are simply ways of identifying patterns in quantitative data. Remember, however, that the reason some researcher took the time, trouble, and expense to collect the quantitative data in the first place was rooted in the quest to answer some important question. In other words, *statistics are means to ends*. Statistics are tools that researchers employ to help answer questions.

It follows that the meaning of statistics comes from their role in answering important questions. By themselves, statistics are meaningless. Since statistics acquire their meaning through answering imaginative questions, then a central task involves interpreting the statistical results. At this point we see the deficiency of relying on the magical black-box approach to statistical calculations. If you don't know how specific statistics are actually calculated, then it is very problematic to know what the results mean.

Computers are helpful in calculating statistical results, but they are useless in informing us what the results mean. Meaning requires interpretation which, in turn, requires relating the findings to the research question of interest. Interpretation is a very valuable practical skill and, for each of the statistics introduced in this book, we will help you learn to interpret what the results mean.

# SPSS Essentials

IBM® SPSS® Statistics software ("SPSS") is a common statistical analysis software package. *SPSS* stands for Statistical Package for the Social Sciences. SPSS produces several full-length guides for using the many sophisticated features of this program.[2] In addition, there are extended manuals available for using the common elements of SPSS.[3]

Our introduction to SPSS is different from that found in many other books. First, we assume that you know *nothing* about SPSS. So there is no reason to be intimidated or scared; everything you need to know is included here. Second, we provide you little shortcuts and tidbits of information to make your experience with SPSS as easy, productive, and gratifying as possible.

---

[2]See, for example: Morgan et al. (2010), *IBM SPSS for Introductory Statistics: Use and Interpretation*. See also: Leech et al. (2011), *IBM SPSS for Intermediate Statistics: Use and Interpretation*.

[3]Our students find the following volume helpful: George and Mallery (2012), *IBM SPSS Statistics 19 Step by Step: A Simple Guide and Reference*.

The detailed steps for producing specific statistical results are included in the chapters that introduce the statistical procedures. In this section we cover the material you need to get set for those operations.

# Starting the SPSS Program

To begin, you need access to the SPSS program, either by purchasing a version or using a resident version on a public computer. There is a student version available from the SPSS website, which is relatively inexpensive and is commonly purchased by our students. At the SPSS website, you can also download a free 14-day trial version. SPSS is available for both Windows-based and Mac computers. Alternatively, you may find SPSS on a public computer lab at your campus.

- *If you purchase SPSS or download a trial version*, you need to load the SPSS program onto your computer following the download instructions.
- Next, *whether you are on your own or a public computer*, you need to find the SPSS program and open it.

# Opening a Data File

Data files are those that contain the results of a researcher's investigation, translated into quantitative (numerical) form. Data files for SPSS use the suffix ".sav." Data files can be accessed from any conventional source (e.g. USB key or an internal drive). The examples and practice questions in this book use evidence from a Student Health and Well-Being Survey. A data file for this evidence is available on the book's website, so you will need to download and save the data file before you begin.

Once the data file is downloaded and saved, go to the File menu and select: Open → Data. This opens the Open File box. Look in the location where you have saved the data file. Select the ".sav" file and click Open. This set of procedures will make the data set available for analysis.

Alternatively, by default, when you open SPSS a start-up window automatically appears, as shown in Figure 3.1. Here, if the "Open an existing data source" radio button is selected, you can find the desired data file.

# Navigation in SPSS

After opening the program, you will find that there are three main windows in SPSS. For now, we will be using two of these windows—(1) Data Editor (which automatically appears once you open SPSS) and (2) Statistics Viewer (which does not open until you analyze some data). Both windows are described in more detail below.

## The Data Editor

The Data Editor is the main window in SPSS and is where you will instruct the program to do what you want it to do. In order to learn more about the Data Editor, we have divided this section into 2 parts: (1) Data Editor Menus, and (2) Data Editor Views.

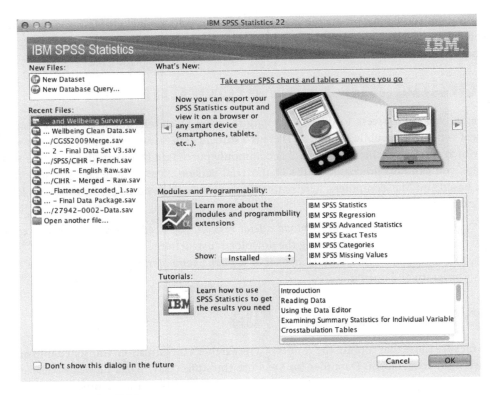

**FIGURE 3.1** Default Menu

Source: Reprint Courtesy of International Business Machines Corporation, © International Business Machines Corporation.

## 1. Data Editor Menus

Most program commands in SPSS are listed in drop-down menus. These program commands are listed *across the top* of the SPSS Data Editor window (in both the Data View and the Variable View). Use a menu item the same way you would in any other program in Windows or on your Mac.

The data editor menu is shown in Figure 3.2. Below are descriptions of important items from the data editor menu. Please do not feel overwhelmed if, as you move down the list, you do not know the terms you are reading. The vast majority of them will be explained to you in this chapter or in subsequent ones. Remember, your first reading of this material is a general overview.

| SPSS Statistics | File | Edit | View | Data | Transform | Analyze | Direct Marketing | Graphs | Utilities | Add-o |
|---|---|---|---|---|---|---|---|---|---|---|

Untitled1 [$DataSet] – IBM SPSS Statistics Data Editor

|  | var | var | var | var | var | var | var | var | var | var |
|---|---|---|---|---|---|---|---|---|---|---|
| 1 | | | | | | | | | | |
| 2 | | | | | | | | | | |

Data Editor Menu

**FIGURE 3.2** Data Editor

Source: Reprint Courtesy of International Business Machines Corporation, © International Business Machines Corporation.

**File**: Used to create a new SPSS file, open or close an existing data set or output file, save files, export output files, print files, and exit the program.

**Edit**: Used to modify or copy text from the Viewer/Output or Data Editor windows. This is where you can use the Undo button to reverse an editing action and set SPSS options (see Setting Options in SPSS section below for more instructions).

**View**: Used to change fonts and view toolbars, status bars, and search for variables (which is useful when using extra large data sets).

**Data**: Used to make global changes to SPSS data files. These changes do not affect the permanent data files unless you save the file on top of the original data. Popular changes made under the Data menu are Sort cases, Merge files, Split file, Select cases, and Weight data. If you are working with items under the Data menu, it is wise to first save your data set under a different name (File → Save As—and rename your file). If you mistakenly save your original data set after making changes, the original data is permanently lost!

**Transform**: Used to make changes to selected variables in the data file and to compute new variables based on the values of existing ones.

**Analyze**: Used to run statistical procedures (such as frequencies, crosstabulations, *t*-tests, and chi-square tests which we will learn about later in this book).

**Graphs**: Used to create pie charts, bar charts, histograms, scatterplots, and other graphs.

**Utilities**: Used to display information on the contents of SPSS data files or open an index of SPSS commands. This menu is used to display information on the definition (i.e. coding scheme) of specific variables (described in "2. Examining Codes of a Specific Variable under Interpreting Scores," on p. 41).

**Window**: Similar to Microsoft Word, under Window, you can switch back and forth between multiple files (e.g. switch from the Data Editor to your Viewer/Output file).

**Help**: Every window has a Help menu on the menu bar. The Topics option provides access to the Contents, Index, and Find tabs, which you can use to find specific help topics. Tutorial provides access to the introductory tutorial. The Statistics Coach asks you a set of questions about your proposed analysis and then suggests a statistical procedure to follow.

> **Computer Tip**: There is no need to feel intimidated by all of this information. In fact, as the remainder of the book will demonstrate, you will be using only a few of the wide-ranging capabilities of SPSS. Once you become more familiar with SPSS, we encourage you to play around in order to become more comfortable with the program. As with all computer programs, the more you practise, the more you discover!

## 2. Data Editor Views

In the Data Editor window at the bottom left-hand corner, you can toggle between Data View (the default, which allows you to view and enter/alter actual data) and Variable View (which allows you to specify and view the variables) by using the appropriate tabs, as shown in Figure 3.3.

### A. Data View

Below are key terms that will help you understand the function of the Data View.

**Data Matrix**: The information presented in the Data View window is called a *data matrix*, and is a "flat file" similar to those found in Microsoft Excel.

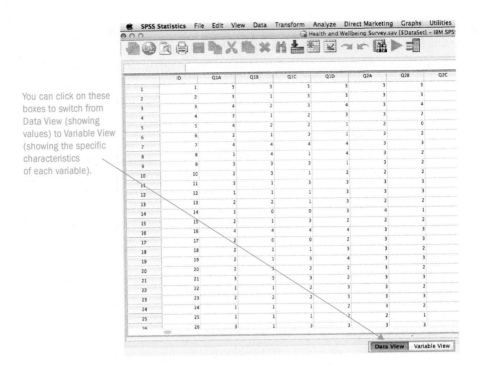

You can click on these boxes to switch from Data View (showing values) to Variable View (showing the specific characteristics of each variable).

**FIGURE 3.3** Data View & Variable View

Source: Reprint Courtesy of International Business Machines Corporation, © International Business Machines Corporation.

**Row**: Each **row** is a single **case** (observation). For example, the data set displayed in Figure 3.3 shows cases 1 through 11, and then 39 through 41 (we cut out the cases in between to save space). Case 1 corresponds to respondent 1, and so on. In the Student Health and Well-Being Survey, if you scroll to the bottom of the data file, there should be 1,245 cases (the sample size).

**Column**: Each **column** in the data set corresponds to a single *variable*. For example, in Figure 3.3, there are five variables displayed: ID, Q1A, Q1B, Q1C, and Q1D (the labels for these variables will be explained in the Variable View section).

**Variable**: A characteristic or property of an object that expresses differences or can be changed (i.e. it varies). Examples of variables are *Respondent sex*, *Sexual identity*, and *Ethnic identity*. Variables are represented in the data set as columns.

**Value**: The *responses* for each variable are called **values**, attributes, or scores. For example, the values for *Respondent sex* are listed in the *cells* as 1 (for *female*) and 2 (for *male*). Values are most often numbers, but not always.

**Cells**: **Cells** are the little boxes visible in the Data View window. The boxes are the intersection of a particular row and column. The number recorded in each cell expresses the value of a particular variable (the column) for a specific case (the row). For example, in Figure 3.3 the first respondent in the sample (row 1) has a score of 3 for variable Q1A (second column). In the Variable View (described next), we will see that variable Q1A asks participants to rate their satisfaction with their overall physical health, and a value of 3 means that respondent 1 is *satisfied* with his/her physical health.

**Interpreting Scores**

By themselves, the numbers in the cells of a matrix are often meaningless. For example, in our example in the previous section for Q1A, a value of 3 is meaningless and requires interpretation. Similarly, imagine that the cells of a column labelled Ethnicity contained numbers such as 1, 4, 6, 3, etc. What does it mean to say that an individual's ethnicity is 3?

Interpreting the digits in a matrix requires a codebook. Sometimes data sets come with a hard or digital copy of a codebook that allows you to interpret the scores for each variable. More commonly, you will need to retrieve this information from the data set. There are two common ways of doing this: creating a full codebook and examining codes of a specific variable.

1.  *Creating a Full Codebook:* You can have SPSS generate a codebook in the Output/ Viewer window (described later in this chapter), which will list all of the variables and their corresponding values for the entire data set. To do this, go to the File menu and select Display Data File Information, and then select Display Working File. The results will take the form of the example in Figure 3.4.

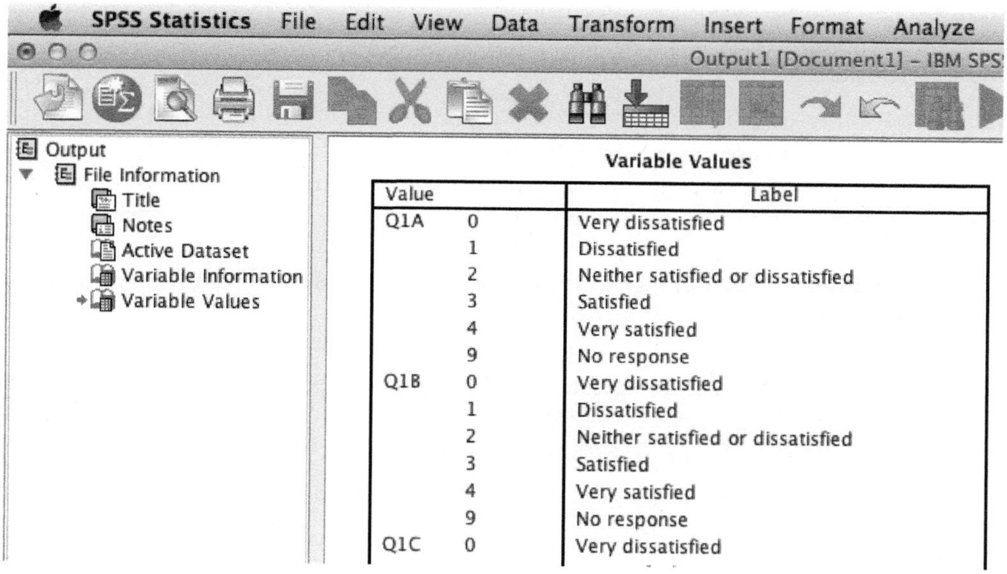

**FIGURE 3.4** Codebook

Source: Reprint Courtesy of International Business Machines Corporation, © International Business Machines Corporation.

2.  *Examining Codes of a Specific Variable:* If you want the *coding scheme* of a particular variable, proceed as follows:

    a.  Go to the Utilities menu → select Variables.
    b.  On the left side, scroll down the list of variables until you reach the variable of interest. Highlight the variable name; its coding scheme (Value Labels) will be displayed on the right. For example, in Figure 3.5 we see that for the variable Q7, the Label is Q7. *In the past year, have you consumed alcohol*, and a value of 1 denotes that the respondent has consumed alcohol (i.e. *yes*).

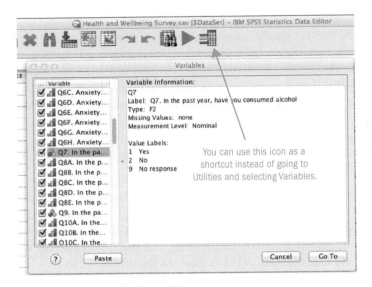

**FIGURE 3.5**  Specific Variable Information

Source: Reprint Courtesy of International Business Machines Corporation, © International Business Machines Corporation.

### B. Variable View

The Variable View tab contains more detailed information about your data set. It is in this *view* where you can give your variables descriptions, label values/attributes, set levels of measurement, and set values to *missing*. Unlike in the Data View, each row represents a variable (in the Data View, each column represents a variable).

Variable View is shown in Figure 3.6. Below are descriptions of relevant columns in the Variable View.

| | Name | Type | Width | Decimals | Label | Values | Missing | Columns | Align | Measure |
|---|---|---|---|---|---|---|---|---|---|---|
| 42 | Q13D | Numeric | 2 | 0 | Q13D. In the p... | {1, Yes}... | None | 11 | Right | Nominal |
| 43 | Q13E | Numeric | 2 | 0 | Q13E. In the p... | {1, Yes}... | None | 11 | Right | Nominal |
| 44 | Q14A | Numeric | 2 | 0 | Q14A. Coping... | {0, Does no... | None | 10 | Right | Scale |
| 45 | Q14B | Numeric | 2 | 0 | Q14B. Coping... | {0, Does no... | None | 10 | Right | Scale |
| 46 | Q14C | Numeric | 2 | 0 | Q14C. Coping... | {0, Does no... | None | 10 | Right | Scale |
| 47 | Q15A | Numeric | 2 | 0 | Q15A. Forgive... | {0, Not at al... | None | 11 | Right | Ordinal |
| 48 | Q15B | Numeric | 2 | 0 | Q15B. Forgiven... | {0, Not at al... | None | 11 | Right | Ordinal |

**FIGURE 3.6**  Variable View

Source: Reprint Courtesy of International Business Machines Corporation, © International Business Machines Corporation.

**Name:** The variable name. It is the same as the columns in the Data View.

**Type:** By default is set to Numeric. SPSS can analyze non-numeric data such as Strings (i.e. text) or Dates; however, for our purposes, all variables should be left as numeric (i.e. numbers, which can be seen in the cells in the Data View).

**Width:** The default is set to eight. The width represents the number of spaces on the left side of a decimal. For the vast majority of variables, the default of eight is sufficient; however, if you have an income variable, you may need to increase the width to accommodate billionaires (they would require a width of 10—1,000,000,000!). You can increase or decrease the width by clicking on the width cell for the variable you want to change; you will see an up/down arrow.

**Decimals:** The number of places on the right side of a decimal, and related to width. The default is two (e.g. an hourly wage of $17.82). Most variables, however, do not require decimals (e.g. Likert scales where 1 is *strongly disagree*, 2 is *somewhat disagree*, and so on). If you wish, variables that do not need decimals can be changed to zero. If left unchanged (i.e. the default of 2), your output will simply show two decimals in the output (e.g. 1.00 *strongly disagree*, 2.00 *somewhat disagree*, etc.).

**Label:** Most variable names are short (e.g. Q1) so they can be analyzed without long descriptions. In the Label column you can describe what variables mean, which is important for recodes (described in a later section). You can enter or edit a variable label by clicking on the cell. The variable label will appear in your Output/Viewer window if you run analyses on the variable.

**Values:** Similar to Label, Values displays the responses/attributes for a variable. You can create or edit a value by selecting a cell and clicking the three dots on the right side of the cell. A new window will appear. Here you can add/create, change/edit, or remove/delete values and/or labels. Make sure you select OK to save your work.

**Missing:** As will be discussed in the next chapter, not all values/attributes should be included in your analysis. For example, often we wish to exclude respondents who did not answer a question. Typically, we give those participants a specific value and label it *no response*. In order to exclude them from statistical analysis, you need to set them to *missing*. As for Values, click the three dots. Select the "Discrete missing values" radio button, and in the box below, type the value you wish to exclude (e.g. 9). You can exclude up to three values (e.g. *don't know* and *no response*). If you want to include all values in your analysis, select the "No missing values" radio button. Remember to click OK to save your work.

**Measure:** Each variable has a level of measurement (i.e. nominal, ordinal, interval, and ratio). SPSS refers to an interval or a ratio variable as a scale (i.e. continuous or quantitative variable).

## The SPSS Statistics Viewer

Although not immediately shown, the Viewer is where you see statistics and graphics—often referred to as the output—from the analysis work you performed using SPSS (e.g. frequency tables, pie charts, bar charts, measures of central tendency, crosstabulations, etc.). An example of a pie chart in the Output/Viewer window is shown in Figure 3.7.

When you run a statistical procedure, the default Viewer file is called *Output1*. Newer versions of SPSS have a file suffix of ".spv" while older versions have a file suffix of ".spo". Unfortunately, the older Viewer files with ".spo" extensions are not compatible with the newer versions of SPSS.

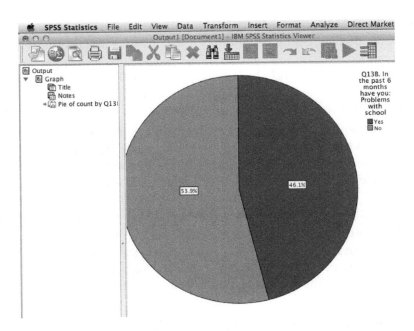

**FIGURE 3.7** Output/Viewer Window

Source: Reprint Courtesy of International Business Machines Corporation, © International Business Machines Corporation.

The Viewer contains many useful features so that you can edit your output, much like you would do with a word-processing file. As shown in Figure 3.8, an underused tool in the Viewer is the Output contents located on the left side of the Viewer. Here you can move (i.e. cut/copy and paste) and delete various statistics that are displayed in full on the right side of the Viewer. Many students are afraid to delete anything from their Viewer window, and after several runs they find it difficult to scroll up and down the statistics displayed on the right side. A shortcut is to scroll up and down on the left side. Once you find what you are looking for, click it on the left side and the statistic will automatically appear on the right side. Finally, double-clicking items on the left side will suppress the statistical output on the right side. The best way to master both sides of the Viewer is to practise—don't be afraid to play around as it will make your SPSS experience more organized and, therefore, more enjoyable!

Another feature found in the Viewer is Pivot Tables. By double-clicking on an output (i.e. a frequency distribution or a pie chart), you activate the Pivot Table. You will know when the Pivot Table is activated because there will be small dashed lines around the outside of whatever statistic you selected. Once the Pivot Table is activated, you have even more editing options at your disposal. For example, you can delete rows or columns from a table or change the table look (to do this, go to the Menu and click Format, then TableLooks, where you can find plenty of ready-made Output styles, or create your own!).

A common frustration for students occurs when they try to print tables with many columns (for example, a crosstabulation that has several values on the independent variable). You can overcome this problem by using the Pivot Table. To do so, activate the Pivot Table (i.e. double-click to get the small dashed lines) and under Format click Table Properties. From

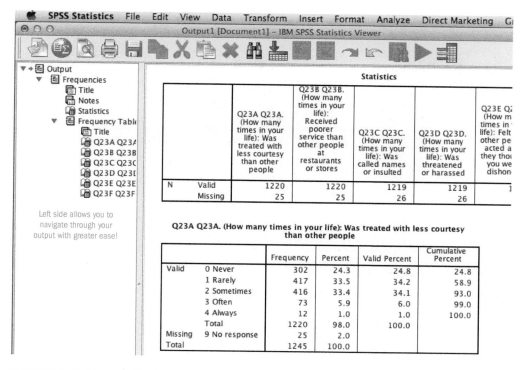

FIGURE 3.8 Viewer Tools

Source: Reprint Courtesy of International Business Machines Corporation, © International Business Machines Corporation.

Table Properties, go to the Printing tab and check "Rescale wide table to fit the page" (and click OK). Now when you print, the font will be smaller, but it will all fit across on one page. (Of course, in extreme cases, the font is so small, this option is useless!)

Many students also find it frustrating when they want to print from the Viewer. Here are some simple tips to ease your frustration. If you go to File → Print and select OK, you will print all of the information located on the right side of the Viewer. However, if you want to print only one table or one graph, just click the one you want to print. You will see a solid black or yellow line (depending on what version of SPSS you are using) around your selection. If you go to File → Print you will see under Print Range that the Selected Output radio button is highlighted. Now if you click Print, only the table/graph you selected will print. Similarly, if you hold down the Control key on your computer and click multiple selections (either on the right or left side of the Viewer), only what you have chosen will print.

Many students use SPSS on a public-use computer and want to save their output to view at a later date on their personal computer. Unfortunately, you need to have SPSS installed on your own computer in order to view your output files; however, there is a simple way to export your Viewer to a PDF. You can do this by going to Menu → File → Export. Under File Name click Browse where you can save your output file to a USB key or some other destination. Under Type you can also export your output file to an HTML, a PowerPoint, or a Word/RTF file, if you prefer one of those formats instead of a PDF. Figure 3.9 provides an example.

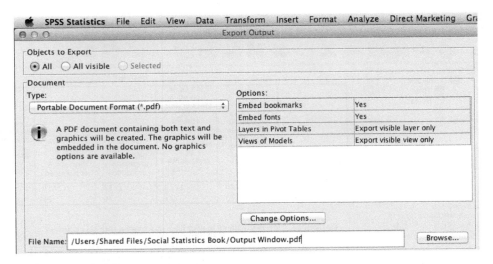

**FIGURE 3.9** Exporting Files

Source: Reprint Courtesy of International Business Machines Corporation, © International Business Machines Corporation.

## Setting Options in SPSS

Before beginning some analysis, it is advisable to set some display options in SPSS.

1.  Go to Edit menu → Options.
2.  The Options dialogue box has a series of tabs at the top of the screen. The General tab is the default; if not, select it.
3.  Under the Variable Lists option, check Display Name as well as File. This will display the short names of the variables and arrange them in alphabetical order.
4.  You also want to set some options for displaying your output. Click the Output Labels tab (see Figure 3.10), and look at the setting under Outline Labeling. Using the down arrow, select Names and Labels and Values and Labels. Under Pivot Table Labels, also select Name and Labels and Values and Labels. This will allow you to see both the names and labels of your variables and the codes (i.e. attributes) and labels for your variables in all of your output

## A Note on Entering Primary Data

All of the examples and practice questions in this book involve secondary analysis from the Student Health and Well-Being Survey data file. In **secondary research**, the data are originally collected by some other source, and you have obtained them for your own (secondary) research purposes. In many research situations, however, you collect your own data. In other words, you are conducting **primary research**. In these cases, the data matrix is not provided to you; *you have to create a data matrix.*

To create a data matrix, you first need to open SPSS. This can be done one of two ways, although both start with first opening the SPSS program. First, by default, SPSS opens with the following box (Figure 3.11):

The second option, especially if you have checked the "Don't show this dialog in the future" box, is to go to File → New → Data (assuming you already have an existing data set open).

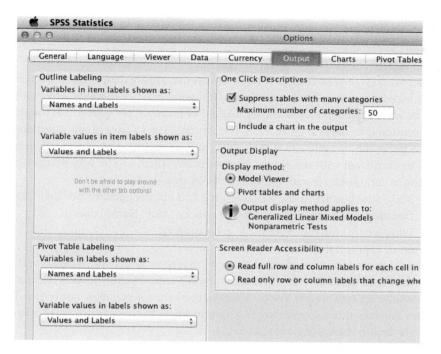

**FIGURE 3.10** Output Labels

Source: Reprint Courtesy of International Business Machines Corporation, © International Business Machines Corporation.

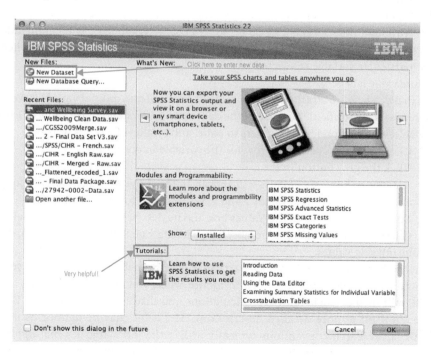

**FIGURE 3.11** Entering New Data

Source: Reprint Courtesy of International Business Machines Corporation, © International Business Machines Corporation.

Now that you have a blank data set, entering data is a fairly simple five-step process.

1.  Name the variable.
    The first step in defining your variables is to give the variable a name, with no spaces. Older versions of SPSS restrict variable names to no more than eight characters; however, newer versions allow longer names. It is good practice to follow a format similar to the survey questionnaire you are using. For example, for question 1, the standard convention is to name it Q1 in SPSS. If question 2 has a series of sub-questions, most users name the variables Q2A, Q2B, Q2C, and so on. This way, the variables in SPSS will be consistent with the survey questionnaire, which can be very useful if you are trying to communicate with a colleague, professor, or client.
    a.  Make sure you are in the Variable View and not the default Data View tab.
    b.  Double-click the first blank cell that you see in the Name column.
    c.  Enter a name for your variable (e.g. Q1 or *Age*).

2.  Define an appropriate number of decimal places.
    The default is two decimal places. Use the cell's arrows to increase or decrease this stipulation. For nominal or ordinal variables you can enter 0 (zero), because decimal places are usually not appropriate. For continuous interval/ratio variables, one or two decimal places are sometimes appropriate (e.g. hourly wage), but most often these also do not require decimal places (e.g. year of birth).

3.  Define variable labels.
    Although you may have given names to your variables, variable labels help us to better describe and remember exactly what each variable is about.
    a.  Double-click the appropriate cell in the Label column.
    b.  Enter a description up to 256 characters long (e.g. "Q1. Respondent sex"). Although descriptions can be up to 256 characters long, it is recommended to use far fewer—remember your label will appear in all of your output when the particular variable is selected.

4.  Define value labels.
    Normally, labels are attached to each value of nominal and ordinal variables. Without these labels, it is difficult for users of the data set to understand that a value of 1 for the variable "Q1. Respondent sex" refers to *male*, and a value of 2 refers to *female*.
    a.  Click the Values cell in the row for the variable in question.
    b.  Click the grey button in the cell.
    c.  A dialogue box Value Labels will appear. In the above example, we assigned *male* and *female* to numeric values 1 and 2, respectively, for the variable Q1.
    d.  Enter the data value in the Values box and its descriptive label in Value labels.
    e.  Click Add to record the value label for each data value and OK when you are done.

5.  Define missing values.
    When you do not have data for a particular case, you should specify how you will denote values that you want SPSS to ignore when you run your statistics. If not, SPSS will include such values in calculations and distort your output (see B. Variable View for a more detailed description). Any value may be defined as *missing*. However, research conventions typically utilize the following: 7 = *not applicable*; 8 = *don't know*; 9 = *no response/refusal* (or, depending on the range of the variable, 77, 777, 7777; 88, 888, 8888; or 99, 999, 9999, etc.). Make sure to give these values Value Labels as well.

Now that you have set up data entry specifications, entering simple numeric data is easy.

1. Switch from Variable View to Data View by clicking the tab at the bottom of the Data Editor window.
2. Select a cell for your first case (row) and the first variable (column); enter a value.
3. Press Enter to record the value—or you can use the right or down arrow keys or your mouse to move to the next cell. Remember, unless you have specified differently, values entered in the cells should be numeric.

## Saving Files

Output files (i.e. results of your analyses) and the data set can be saved using the File menu. Saving files in SPSS is the same as saving word-processing and spreadsheet files on your computer, although you must save changes made to the data set and the output files separately.

1. Go to File → Save (or Save As if you want to retain an original copy of a secondary data set).
2. In the dialogue box, name the output or data set. It is useful to name versions or use dates to differentiate files, especially when you are repeating actions many times or altering data sets over time.
3. Once you have named the file, save it to your hard drive or USB key.

## Exiting the SPSS Program

1. Go to File → Exit (or click the X [PC] or red dot [Mac] in the upper-right corner of the program window).
2. If you have made changes to the data file and you have not already saved your data file, SPSS will ask something like: "Save contents of data editor?"
3. If you have made changes to the data you wish to retain, click Yes; otherwise click No.

# Description of the Student Health and Well-Being Survey Data Set

The computer examples and practice questions in this text use a data set we collected from students through the Student Health and Well-Being Survey. Students enrolled in Introduction to Sociology and Introduction to Research Methods courses at our university make up the sample of 1,245 undergraduates. Respondent ages ranged from 17 to 45, with an average age of 20. Overall, 90 per cent of the participants were under the age of 24. Consistent with the overall enrolment of Canadian university students,[4] 62.4 per cent of participants were female and 37.6 per cent were male.[5]

We chose to centre the computer analysis in the book around the Student Health and Well-Being Survey data set for two reasons. First, the data set covers a wide range of interesting questions and, second, the undergraduate sample is similar in many respects to readers just like you. We hope you will find your exploration of the data set engaging.

---

[4]Statistics Canada (2006). University enrolment. *The Daily*. www.statscan.gc.ca/daily/061107/d061107a .htm. Retrieved 25 July 2009.

[5]More detailed information about the methodology, questions, response rates, and ethics of the survey are included on the companion website.

# Chapter Summary

This brief exposition has identified our goal (knowing plus doing), hypothesized where most students typically stand (somewhat anxious), and set out a five-step learning plan to move you toward becoming an informed user of social statistics. Just as the checklist of statistical selection questions introduced in Chapter 2 is routinely repeated in upcoming chapters, this five-step mastery plan will also be replicated. The routine application of these two templates will take you down the path toward mastering basic social statistics. As you will see, most of the remaining chapters implement the five-step plan as it relates to various combinations of answers to the three critical questions identified in Chapter 2.

The chapter also introduced you to SPSS data analysis software. This introduction tells you everything you need to know *except how to perform specific statistical analysis procedures*. The details of generating specific statistics with SPSS are covered in the chapters that introduce the procedures. This chapter covered what you need to know and do in order to get started.

This concludes Part 1, your general orientation to navigating the statistical maze. In the next chapter, you enter the statistical maze.

# PART II
Univariate Analysis

# Introducing Univariate Analysis

## Overview

The previous three chapters provided you with:

- an understanding of the place and importance of quantitative statistical analysis in the scientific quest for knowledge (Chapter 1),
- an understanding of the questions that need to be answered to select appropriate statistical tools (Chapter 2), and
- an appreciation of the five-step model this book uses to advance your statistical education (Chapter 3).

You are now ready to enter the statistical maze and begin moving down the path to mastering specific statistical tools.

This chapter takes some initial steps down the path by:

- introducing the part of the landscape called univariate statistics;
- identifying why univariate statistics are important;
- explaining the grouping of variables into categorical and continuous types; and
- teaching you about the most fundamental statistical tool, frequency distributions.

Part II focuses on univariate (one variable) analysis. Remember that selecting appropriate statistical techniques involves answering the following three questions:

1. Does your problem centre on a descriptive or an inferential issue?
2. How many variables are being analyzed simultaneously?
3. What is the level of measurement of each variable?

Part II answers the first question with "descriptive"; in other words, these chapters introduce statistics that summarize features of an existing data set. Part V (Chapters 18 through 21) deals with inferential issues.

Regarding the second question, Part II focuses on one variable. Univariate techniques characterize single variables in a data set. In univariate analysis, variables are considered one at a time, in isolation. Single variables, however, can be measured at any of the four levels of measurement (nominal, ordinal, interval, ratio). Since there are different answers to the third question, there are a variety of different univariate statistics—ones for nominal variables, others for ordinal variables, etc. Chapters 5 through 8 introduce various univariate statistics, how to calculate them, and what they mean.

The following statistical selection checklist identifies the tools covered in this chapter and provides your coordinates in the statistical maze.

## Statistical Selection Checklist

1. Does your problem centre on a descriptive or an inferential issue?

   ☑ Descriptive

   ❑ Inferential

2. How many variables are being analyzed simultaneously?

   ☑ One

   ❑ Two

   ❑ Three or more

3. What is the level of measurement of each variable?

   ☑ Nominal

   ☑ Ordinal

   ❑ Interval

   ❑ Ratio

We are now ready to enter the maze. Following our five-step model, let's begin with understanding the tools.

## STEP 1 Understanding the Tools

Before delving into the different univariate measures, this chapter considers three preliminary issues. First, it discusses the need for univariate statistics. Second, it introduces a common way of grouping the four levels of measurement and associated statistical techniques. Third, it informs you about frequency distributions, the most basic univariate summarization tool.

# The Need for Univariate Statistics

Let's begin by reminding ourselves what descriptive univariate statistics are about. First, the term *descriptive* tells you what the answer to the first decision question is (Does your problem centre on a descriptive or an inferential issue?). Descriptive statistics report on information about a variable in an actual data set. Second, *univariate* tells you that the answer to the second decision question (How many variables are being analyzed simultaneously?) is "one." These statistics examine one variable at a time. Third, *statistics* tells you that the data from a single variable is being summarized in some manner.

Here's an example of a descriptive univariate situation. Imagine you asked 100 of your Facebook friends this question: "Do you think it is a good idea for me to drop out of university?" Their possible answers were as follows:

1. Yes, absolutely.
2. Yes, but it's risky.
3. No, but you might want to reconsider after finishing this course.
4. Definitely not.

A univariate descriptive statistic of this variable would do three things. First, it would *only* report on the responses of your 100 Facebook respondents. Descriptive statistics would not tell you about any persons other than those who actually responded (not all your Facebook friends, not all persons using Facebook, not all Canadians). Second, it would only tell you about responses to this single question. You would not learn about whether males answered this question differently than females, or whether the views of university graduates differed from non-grads. Third, the 100 responses would be summarized in some fashion. For example, the statistics might tell you what most respondents thought you should do.

With these kinds of single variables, descriptive summaries are useful on two accounts. One use is obvious; the other, less so. The obvious utility of univariate descriptions is the summarization they provide. Remember the earlier example where you are enrolled in a large university course of 175 students. The day after the first unit test, a student raises his hand and asks the professor the following question: "How did the class do on the test?" The professor could accurately answer by saying: "75, 59, 36, 93, 64, 67, 88, 55, . . . etc." until she had recited the 175 specific grades that students obtained on the first test. If the professor answered this way, the student would not likely be satisfied. The experience of information overload would only be frustrating. By contrast, a satisfying answer to the student's question might be, "The class average was 78 per cent," or "Most students passed the test." Univariate descriptive statistics are useful because they provide just this kind of overall summary of how a large group of respondents answered a specific question. Our ability to get our minds around a large amount of information is quite restricted, and, in these instances, summaries are obviously helpful.

The other use of univariate descriptive statistics is less obvious and concerns what are called distribution assumptions. The logic of social statistics described in Chapter 2 showed that, depending on how the three decision questions are answered, social statistics can become quite complicated. Imagine that the answers to the three decision questions are "descriptive," "11," and "ratio." In this case you would be examining the multiple relationships between 11 variables that contain sophisticated (ratio) information. Understandably, the statistical tools for constructing such descriptions are going to be quite powerful and complicated. Often, the use of more sophisticated statistical procedures that involve examining complicated relationships between multiple variables requires making assumptions about what the variables are like in isolation. In other words, advanced techniques make distribution assumptions about the univariate character of the variables included in the analysis. If the variables are not of the right type (i.e. they do not meet these distribution assumptions), then the advanced statistical analysis techniques cannot be used, or must be applied with caution.

In summary, univariate descriptive statistics are useful on two accounts. First, they provide useful summaries of large amounts of information about a single variable. Second, through tests of distribution assumptions, they are useful in making decisions about whether more advanced statistical procedures can reasonably be applied.

# Grouping Levels of Measurement

Chapter 2 reviewed the characteristics of the four levels at which variables can be measured: nominal, ordinal, interval, and ratio. Understanding these different types of variables is important since you need this information to answer the third statistical choice question (What is the level of measurement of each variable?). In general, the decision regarding this final question is made easier since the levels of measurement are often grouped into two types. These types include categorical and continuous variables.

Categorical variables are also called discrete or qualitative variables. For most purposes, nominal and ordinal variables are of the categorical type. The various names for this type of variable (categorical, discrete, qualitative) comes from the nature of the values (scores) of the variable. Take the nominal variable *Religion*, which might be, for example, organized into six values: *Christian, Jew, Muslim, Hindu, Buddhist, other*. The values of this variable are only categories. This remains the case even if you assign numbers to the categories such as 1 = *Christian*, 2 = *Jew*, 3 = *Muslim*, 4 = *Hindu*, 5 = *Buddhist*, 6 = *other*. The numbers do not designate anything arithmetic; they are merely replacements for names. The 5 does not mean that Buddhists are five times as religious as Christians! The numbers could have been just as easily assigned as 1 = *Hindu*, 2 = *Jew*, 3 = *Buddhist*, 4 = *Christian*, 5 = *Muslim*, 6 = *other*. In this sense the categorical variable is qualitative, since the categories represent different qualities (types) of the variable *Religion*. The categories of this type of variable are also discrete in that each category is a separate group that cannot be divided into smaller units. For practical purposes either you are a Buddhist or you are not.

While the lower levels of measurement (nominal and ordinal) are of the categorical type, variables measured at the higher levels (interval and ratio) are of the continuous type. Continuous variables are quantitative variables. The numerical values of these variables can be subject to legitimate arithmetic operations. A family with five children actually has five times as many children as a family with a single child.

This distinction between categorical and continuous variables is useful since it makes the organization of statistical choices easier. In general, with respect to the third decision question (What is the level of measurement of each variable?), there are statistical techniques that apply to lower levels of measurement (categorical variables) and ones that apply to higher levels (continuous variables). We will use this distinction in the following discussion about the most basic univariate summarization tool, frequency distributions.

# Frequency Distributions

Frequency distributions provide the simplest descriptive univariate summaries. Imagine you collected information on the variable *Eye colour* from 150 students. If each student's response was written on a small piece of paper, you could hold the 150 responses (data) in your hand. Examining the papers, you would see that some say *blue*, others say *brown, grey, green*, etc. But if you browsed through the responses on each piece of paper, you would be hard pressed to accurately make a statement about the eye colours of the 150 students; there is simply too much information.

Frequency distributions provide a way of organizing and summarizing data so that you can see patterns in the responses. The term *frequency* means "count," so a frequency distribution is really a count of how responses are distributed across the categories of a variable. Here is an illustration using the eye colour example.

**TABLE 4.1** Students' Eye Colour

| Eye Colour | Frequency (f) | Per cent (%) | Cumulative Frequency (F) | Cumulative Per cent |
|---|---|---|---|---|
| Blue | 30 | 20.0 | 150 | 100 |
| Brown | 50 | 33.3 | 120 | 80.0 |
| Grey | 45 | 30.0 | 70 | 46.7 |
| Green | 10 | 6.7 | 25 | 16.7 |
| Other | 15 | 10.0 | 15 | 10.0 |
| Total (N) | 150 | | | |

First, notice how the table is set up. The variable under consideration (eye colour) is listed at the top of the left column and the categories of the variable are listed in the rows of that column (blue, brown, etc). The other columns in the table identify various kinds of summary information about the data collected on the variable under consideration. Here is what each of these additional columns means.

The Frequency column is often identified by a lower-case letter *f*. So if you ever see a column that only contains *f*, interpret it as meaning "frequency." The numbers in each row of this column are the *count* of how many respondents in the data provided that particular response. In Table 4.1, 30 (of the 150 responses) reported they had blue eyes; 50 reported brown eyes; 45, grey eyes; 10, green eyes; and 15, some other colour of eyes. The numbers in the *f* column always have to add up to the total number of respondents in the sample. In this case the total sample size (typically designated by the letter *N*) is 150 cases.

The frequency column is itself a simple statistical summary. Recall, if you held the 150 pieces of paper listing eye colours in your hand, you couldn't say anything precise because there was too much information. Now, with the information on these 150 cases summarized as frequencies, you can precisely state some patterns. For instance, you can see that "brown is the most frequently occurring eye colour." Or you can state that "the sample contains more people with grey eyes than blue eyes." Statistics help you summarize patterns in the data, as these simple illustrations demonstrate. Now let's look at the next column.

The Per cent column expresses the *f* column in a standardized form. While the frequency column in this example was based on 150 (i.e. 150 respondents), the Per cent column is based on 100. The Per cent column tells you that "if there were 100 respondents in the sample, 20 of them would have blue eyes; 33.3, brown eyes; 30, grey eyes, etc." The standardization provided by percentages is very helpful when comparing variables that come from different sample sizes. For example, if a different sample of 50 students reported that 30 had blue eyes, you could not conclude that blue eyes were equally prevalent in both samples. Both samples contained 30 persons with blue eyes, but in the first case this was 30 of 150 respondents, while in the second case this was 30 of 50 respondents. Since these two samples are of different bases (150 and 50), the only way to compare them is to convert them to a common base. Percentages provide such a standardized conversion.

The third column in the table is labelled Cumulative Frequency. Notice that this column is often designated by a capital letter *F*, rather than the name *cumulative frequency*. As the name indicates, this column also contains counts (frequencies), but of a special sort. The term *cumulative* shares the same root as the term *accumulation*, and means "adding up." So, a cumulative distribution begins with a simple frequency distribution (the *f* column) and adds up the frequencies. In Table 4.1, the adding up begins at the bottom of the column, with the Other row. That is why both the *f* and *F* columns contain the number 15 for the Other row. But then the numbers change. For the Green row, the *f* number is 10 but the *F* number is 25. The reason is that the *F* number beside Green includes the adding up (accumulation) of the Other frequency (15) and the Green frequency (10), for a cumulative frequency of 25. The *F* for Grey comes from taking this 25 and adding to it the *f* of 45, for a total of 70. Can you see where the Brown *F* of 120 and the Blue *F* of 150 come from?

Since the cumulative frequencies come from an adding-up process here is how you interpret them. The 25 beside Green means that there were 25 persons in the sample who had green or "other" coloured eyes. The 70 beside Grey means that there were 70 respondents who had grey, green, or "other" coloured eyes. Can you provide the interpretation for the 120 beside Brown and the 150 beside Blue in the *F* column? In general, any particular *F* score is interpreted as the number of respondents who had that particular response or *any of those below it*.

In Table 4.1 the *F* scores were created by accumulating from the bottom up. It is possible to also create an *F* distribution by accumulating from the top down. Imagine that such a column were added to Table 4.1. Can you see that the column would contain the following numbers: Blue 30, Brown 80, Grey 125, Green 135, Other 150? Can you also interpret each of these cumulative totals? Can you see that the 125 beside Grey means that 125 respondents in the sample had grey, brown, or blue eyes?

The final column in Table 4.1 is labelled Cumulative Per cent. Like the one beside it, this column results from an adding-up (accumulation) process. In this case, the items being accumulated are the percentages in the second column of data (%). Again, the accumulation is being done from the bottom up, so the 16.7 beside Green means that 16.7 per cent of the sample had either green or "other" coloured eyes. Similarly, the 46.7 beside Grey means that 46.7 per cent of the sample had one of grey, green, or "other" coloured eyes. For practice, you should interpret the meaning of the Cumulative Per cent numbers beside Brown and Blue.

# Some Additional Items

Before moving on to the applied sections, you should be familiar with a few other items related to frequency distributions. The first of these concerns a synonym for the term *frequency distribution*. You have just learned that the frequency distribution appears in the *f* column and provides a count of responses by categories of the variable. Sometimes frequency distributions are called *marginal distributions* or *marginals*. This is an odd term for the straightforward information contained in the *f* column. In a later chapter we can explain why frequency distributions are often called *marginals*; for now, it is sufficient for you to have seen the term.

A second additional issue related to frequency distributions concerns a real-world constraint. In the hypothetical sample of eye colour in Table 4.1, all 150 respondents reported their eye colour. In realistic data collection situations, it is rarely the case that data is present for all of the cases of a variable. This occurs for a variety of reasons, ranging from respondent resistance to data collection errors. Whatever the reason, the analysis of data often includes cases for which data on a variable are missing. The question becomes, How do frequency distributions handle missing data?

It is important to find a way of reporting on missing data. If a large proportion of the data on a variable are missing, it raises questions about the validity of the evidence and readers deserve to be informed of this issue. Frequency distributions handle the missing data issue as follows. First, a separate row in the table is used to designate both the number (*f*) and Per cent (%) of missing cases. Then the columns in the table that use either frequencies or percentages are *recalculated using only the valid cases*. Valid cases are those for which data exist (i.e. non-missing cases). Table 4.2 illustrates what the earlier example on eye colour might look like if it contained missing cases. You should review this table and note how it takes missing cases into account.

**TABLE 4.2** Students' Eye Colour, Including Missing Cases

| Eye Colour | Frequency (f) | Per cent (%) | Valid Per cent | Cumulative Per cent |
|---|---|---|---|---|
| Blue | 28 | 18.7 | 20.0 | 100 |
| Brown | 48 | 32.0 | 34.2 | 80.0 |
| Grey | 43 | 28.7 | 30.7 | 45.8 |
| Green | 8 | 5.3 | 5.8 | 15.1 |
| Other | 13 | 8.6 | 9.3 | 9.3 |
| Missing | 10 | 6.7 | | |
| Total (N) | 150 | | | |

One final consideration before putting these ideas about frequency distributions into practice: Frequency distributions organize and summarize data from a single variable; they are univariate distributions. Remember, however, that single variables can occur at various levels of measurement—nominal, ordinal, interval, or ratio. The eye-colour example illustrates a frequency distribution analysis for a variable at the nominal level of measurement. You can easily imagine a similar kind of frequency distribution report for a variable at the ordinal level. The earlier example regarding the question "Do you think it is a good idea for me to drop out of university?" is ordinal because the possible responses (Yes, absolutely; Yes, but it's risky; No, but you might want to reconsider after finishing this course; Definitely not) are rank ordered. If you obtained several hundred responses to this question, you could create a frequency table similar to the ones just reviewed.

But how would you proceed if the variable was at the interval or ratio level of measurement? Imagine you obtained income statistics for 1,000 professors at a university. These incomes would be reported in dollars (making the variable ratio) and would be wide-ranging. Moreover, few of the thousand professors would have exactly the same income, so perhaps the 1,000 cases would occur in 950 categories. In a frequency distribution table each category of the variable is identified as a separate row. Therefore, if you made a frequency table of this variable it would have more than 950 rows! And, to make matters worse, since few professors share exactly the same income, most of the frequency counts ($f$) in the table would be 1. Clearly, this is not a sensible situation.

In practice, for variables at the higher levels of measurement, frequency distributions are constructed using grouped categories. A grouped category is constructed by organizing an interval or ratio variable into ranges. For instance, professors' salaries could be grouped into categories such as under $39,999; $40,000 to $69,999; $70,000 to $99,999; $100,000 and greater. This process effectively turns an interval or ratio variable into an ordinal one, since, in this case, the four salary categories can only be rank ordered. For purposes of constructing frequency tables this grouping process is very handy. As Table 4.3 illustrates, the frequencies,

**TABLE 4.3** Professors' Salaries, Grouped Distribution

| Salary | Frequency (f) | Per cent (%) | Valid Per cent |
|---|---|---|---|
| Under $39,999 | 200 | 20 | 21.1 |
| $40,000 to $69,999 | 300 | 30 | 31.6 |
| $70,000 to $99,999 | 300 | 30 | 31.6 |
| $100,000 and over | 150 | 15 | 15.7 |
| Missing values | 50 | 5 | |
| Total (N) | 1,000 | | |

percentages, and accumulations can now be reported using four rows—one for each of the salary groups.

In general, when the information is from categorical (nominal or ordinal) variables, creating frequency tables is straightforward. When the information is from continuous (interval or ratio) variables, the cases need to be grouped into fewer categories.

These are the basic ideas for preliminary univariate descriptive data analysis. You have learned why univariate analysis is useful and how the four levels of measurement can be categorized into two basic types. In addition, you have been introduced to the fundamentals of frequency tables. You will now have a chance to solidify your understanding through the following sections that provide detailed illustrations of frequency table construction, practice examples, and steps for creating these products on a computer.

## MATH TIPS

### How to Calculate a Percentage

The calculation of a percentage requires you to have two pieces of information—the total number of cases and the number of cases with the value of interest. We call the first number the denominator. The **denominator** refers to the total number of cases in the sample. Let's say we had a class of 161 students taking introductory sociology; 161 would be the denominator—the total number of students in the class.

Now say that the professor was interested in finding out the number of students who are English majors as a percentage of the class. The professor consults her class list and discovers there are 47 students who are English majors enrolled in her class; 47 would be the numerator. The **numerator** refers to the number of cases in the category of interest. In this case, the category of interest is the people declaring English as their major. We now have two numbers: the denominator which is 161 (the total number of students in the class) and the numerator which is 47 (the number of people who major in English). The formula for calculating a percentage is to divide the denominator by the numerator and multiply by 100. The following formula is used:

$$(\text{Numerator} \div \text{Denominator}) \times 100$$

Let's perform this calculation in two steps. First, take the numerator (47) and enter it into the calculator, press the ÷ or / key, and then enter the denominator (161). We get 0.291925. If we multiply that number by 100, we get 29.1925 which we would express as a percentage: 29.1925 per cent. It means that 29.1925 per cent of the students in the professor's introductory sociology class are English majors.

Notice there are many numbers behind the decimal place. It doesn't make much sense for us to carry so many decimal places in our answer. How do we determine the numbers we keep? Most social statisticians would save only one number after the decimal place. In this case, we would say that 29.2 per cent of the class are English majors. Notice that we rounded up the number 1 to 2, because there was a 9 following the 1. In general, when the next number is 5 or higher, we round up; when the next number is 4 or lower, we round down (if the result had been 29.1325, we would have rounded down and the final percentage would be 29.1 per cent).

## STEP 2 Learning The Calculations

# Creating Meaningful Categories for Continuous Variables

Sociologists frequently deal with variables that have many, many categories. In the example of the salaries of 1,000 professors described earlier, we could imagine that there are hundreds of unique categories of salary because no two professors earn exactly the same amount of income. Let's work on another example of a variable with many categories.

Let's say that we were hired by a pediatrician to analyze some data on autistic children in her practice. She provides us with the following information about the age of first diagnosis for 155 of her autistic patients.

**TABLE 4.4** Age Distribution of Autistic Pediatric Patients

| Age | Frequency (f) |
| --- | --- |
| 1 year | 0 |
| 2 years | 17 |
| 3 years | 17 |
| 4 years | 32 |
| 5 years | 39 |
| 6 years | 27 |
| 7 years | 7 |
| 8 years | 9 |
| 9 years | 3 |
| 10 years | 0 |
| 11 years | 1 |
| 12 years | 0 |
| Missing values | 3 |
| Total | 155 |

We actually have two problems here. First, there are many rows in this table; far too many to make a sensible conclusion. We need to find a way to reduce the number of rows so we can make sense of the data. Since each row in the table represents a value (attribute) of the variable, the issue here involves collapsing (reducing) the number of categories of the variable. The second problem involves calculating the Per cent, Valid Per cent, and Cumulative Per cent. Let's deal with the first problem, the large number of rows.

How do we determine the appropriate number of rows for a table? For continuous (i.e. interval or ratio) variables, the appropriate way to deal with the problem is to reduce the number of categories in a way that makes each new category reflect a similar number. For example, we could collapse Table 4.4 into six rows, with each new category containing two ages. Row one would include 1- and 2-year-olds, row two would include 3- and 4-year-olds, row three would include 5- and 6-year-olds, and so on. Since there are 12 rows (not including the missing cases), we would have six categories (we derived this by dividing 12 rows by 2—the number of years per category).

Perhaps we want to make the table even smaller. Let's say we want to have four rows. In this case, each row would contain three ages (12 rows divided by 4 equals 3). In this case, the first row would have children ages 1 to 3, the second row would have children ages 4 to 6, the third row would have children ages 7 to 9, and the fourth row would have children ages 10 to 12. No matter the final number of rows you wish your table to contain, each row must have an equal number of categories. In this case, since we are talking about the age of the child, each row would contain three years.

Let's collapse Table 4.4 into four rows. The new table template would look like Table 4.5.

**TABLE 4.5** Frequency Distribution Table of Autistic Pediatric Patients, Grouped Distribution (Blank)

| Age | Frequency |
|---|---|
| 1 to 3 years | |
| 4 to 6 years | |
| 7 to 9 years | |
| 10 to 12 years | |
| Missing values | |
| Total | 155 |

Next we add the figures from Table 4.4 to match our new categories. In this case, we add zero for 1-year-olds, 17 for 2-year-olds, and 17 for 3-year-olds to obtain a total of 34 cases in the "1 to 3 years" category. We repeat this procedure for each new category. At this point, you should complete these summary calculations and see if you obtain the results in Table 4.6.

**TABLE 4.6** Frequency Distribution Table of Autistic Pediatric Patients, Grouped Distribution

| Age | Frequency |
|---|---|
| 1 to 3 years | 34 |
| 4 to 6 years | 98 |
| 7 to 9 years | 19 |
| 10 to 12 years | 1 |
| Missing values | 3 |
| Total | 155 |

Double-check your work to ensure that the total adds up to 155. If it does not, then you have made an error in calculating one of the cells. Good researchers will always double-check their work!

# Completing Frequency Distribution Tables

Now that we have resolved the large number of rows in the original table (Table 4.4), let's work on the second problem: constructing this information into a frequency distribution form.

Recall that a standard frequency distribution table has five columns: one for the name of the category, one for the number or frequency ($f$), one for Per cent (%), one for Valid Per cent, and another for Cumulative Per cent. The template for a full version of Table 4.6 is found in Table 4.7.

**TABLE 4.7** Template Autistic Pediatric Patients

| Age | Frequency | Per cent (%) | Valid Per cent | Cumulative Per cent |
|---|---|---|---|---|
| 1 to 3 years | 34 | | | |
| 4 to 6 years | 98 | | | |
| 7 to 9 years | 19 | | | |
| 10 to 12 years | 1 | | | |
| Missing values | 3 | | | |
| Total | 155 | | | |

Let's work on calculating the Per cent (%) column first. If you are unsure how to calculate a percentage, please refer to the Math Tips box. For the first cell in the third column (we are looking at the cell immediately beneath the column labelled Per cent) we would calculate the percentage. In this case, using the nominator as 34 and the denominator as 155 and using the formula given in the Math Tips box, we would calculate the following:

$$(34 \div 155) \times 100$$
$$= 0.21935 \times 100$$
$$= 21.9\%$$

For the second cell in the third column, using the numerator of 98 and the denominator as 155 and using the formula, we would calculate the following:

$$(98 \div 155) \times 100$$
$$= 0.63226 \times 100$$
$$= 63.2\%$$

Continue to use this formula until you have completed the calculations for rows 3 and 4. You should have a table that looks like this:

**TABLE 4.8** Percentage Distribution of Pediatric Autistic Patients, Including Missing Cases

| Age | Frequency | Per cent (%) | Valid Per cent | Cumulative Per cent |
|---|---|---|---|---|
| 1 to 3 years | 34 | 21.9 | | |
| 4 to 6 years | 98 | 63.2 | | |
| 7 to 9 years | 19 | 12.3 | | |
| 10 to 12 years | 1 | 0.6 | | |
| Missing values | 3 | 1.9 | | |
| Total | 155 | 100.0 | | |

To ensure you have calculated your percentages correctly, add together the numbers in the third column. In this case, you would get 99.9 per cent; this is due to rounding and is not troublesome. If the number does not add up to 99.9 or 100, then there has been a miscalculation in one of the cells in this column. Double-check your cell calculations.

Now, let's calculate the column for Valid Per cent. Recall that the **valid per cent** is a calculation that excludes the missing cases from the calculation. In this case, instead of using 155 as the denominator, we would use 152 (which is calculated as the total number of cases, 155, minus the number of missing cases—which in this case is 3).

Following the procedure for calculating a percentage, let's calculate the column for Valid Per cent. For the first cell in the fourth column, the calculation would be $(34 \div 152) \times 100$. We should get a result of 22.4 per cent. See if you can do the remaining calculations. The resulting table should look like this:

**TABLE 4.9** Grouped Distribution of Pediatric Autistic Patients, Without Missing Cases

| Age | Frequency | Per cent (%) | Valid Per cent | Cumulative Per cent |
|---|---|---|---|---|
| 1 to 3 years | 34 | 21.9 | 22.4 | |
| 4 to 6 years | 98 | 63.2 | 64.5 | |
| 7 to 9 years | 19 | 12.3 | 12.5 | |
| 10 to 12 years | 1 | 0.6 | 0.7 | |
| Missing values | 3 | 1.9 | | |
| Total | 155 | 100.0 | 100.1 | |

We ought to double-check our calculations like we did for Per cent. In this case, the four rows should total 100 per cent. In our case, we get 100.1 per cent (the 0.1 per cent extra is due to rounding errors; this is permissible).

Our last step is to calculate the Cumulative Per cent column. As described above, this column takes into consideration the percentage each cell increases from the cell above. In this case, the first cell of the fifth column would be calculated by adding 22.4 per cent to zero (because there is no cell above it). The second cell of the fifth column would be calculated by adding 22.4 per cent (the cell above it) to 64.5 per cent to get 86.9 per cent. For the third cell, take 86.9 per cent and add 12.5 per cent to get 99.4 per cent. Finally, we add 0.7 per cent to 99.4 per cent to get 100.1 per cent for the final cell. Note that we do not include the missing values in this calculation. Your table should look like this:

**TABLE 4.10** Pediatric Autistic Patients by Age Group

| Age | Frequency | Per cent (%) | Valid Per cent | Cumulative Per cent |
|---|---|---|---|---|
| 1 to 3 years | 34 | 21.9 | 22.4 | 22.4 |
| 4 to 6 years | 98 | 63.2 | 64.5 | 86.9 |
| 7 to 9 years | 19 | 12.3 | 12.5 | 99.4 |
| 10 to 12 years | 1 | 0.6 | 0.7 | 100.1 |
| Missing values | 3 | 1.9 | | |
| Total | 155 | 100.0 | 100.1 | |

To summarize, here is how you proceed in creating frequency distribution tables:

1. If the variable is continuous and has many attributes, reduce the number of categories by grouping into equal sizes.
2. Complete the Frequency column, ensuring that the total matches the sample size.
3. Complete the Per cent column, based on calculations including all cases (including missing values). Ensure this total is 100.
4. Complete the Valid Per cent column, based on calculations that exclude missing values. Ensure this total is 100.
5. Complete the Cumulative Per cent column by aggregating the Valid Per cent values over the categories. Ensure the cumulative total is 100.

Now let's see how to generate these results using computer software.

## STEP 3 Using Computer Software

You now know a lot about the organization and interpretation of information in the form of frequency distributions. Now it is time to show you how frequency distributions are generated using IBM SPSS Statistics software ("SPSS"). Before you run any frequency distribution, however, you need to make sure that applicable values are set to *missing*. If you do not, the Per cent column and the Valid Per cent column will be identical in the Output/Viewer window. Here is how you proceed to ensure that the Valid Per cent calculations exclude missing values.

## Setting Values to *Missing*

In SPSS, you can set desired values to *missing* by following these instructions:

1. Go to the Data Editor window.
   a. Select the Variable View tab, located in the bottom left-hand corner.
   b. Find your variable (remember in Variable View, each variable represents one row).
   c. Go to the Missing column, and select the cell (e.g. row Q21F and column Missing).
2. Click the three dots; a Missing Values dialogue box will appear (shown in Figure 4.1).
   a. Select the "Discrete missing values" radio button.

      Note that if your data has already been set to *missing*, the "discrete missing values" button will be selected. Just click OK and proceed to the Generating Frequency Distributions section below.
   b. Enter the value you want to set to *missing* (in our Q21F example, the value we want to set to *missing* is 9). You can select up to three values, if need be.
   c. Select OK. The value you set to *missing* will now appear in the cell (instead of *none*).
3. If you want to remove the value you set to *missing*, follow steps 1 through 2a, and click the "No missing values" radio button.

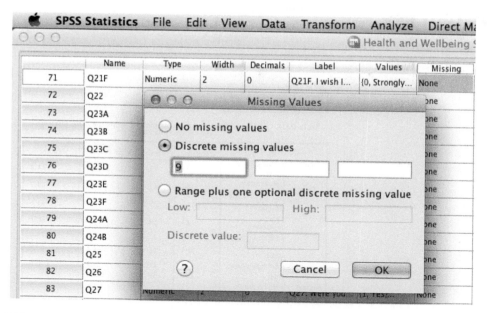

**FIGURE 4.1** Missing Values

Source: Reprint Courtesy of International Business Machines Corporation, © International Business Machines Corporation.

# Generating Frequency Distributions

Now that the missing values are correctly specified, you are in a position to generate a frequency table. Here is how you proceed.

1. Select Analyze from the menu bar.
2. Select Descriptives → Frequencies.
   a. In the left-hand box, scroll to and highlight (click once) the variable you want to analyze (for example, Q21F).
   b. Click the arrow button (alternatively, if you double-click your variable, it will automatically move to the right-hand side).
   c. Repeat the previous two steps if you want to select additional variables to analyze.
3. Select OK to execute the Frequency command.

In the Output/Viewer window, you will find the results of your analysis, an example of which is illustrated in Figure 4.2.

# Recoding Variables

The Recode feature collapses data into categories using existing variables. This is the procedure used for reducing the number of categories of a variable.

In Figure 4.2, we have information based on a Likert agreement scale ranging from *strongly disagree* to *strongly agree*. Suppose, however, that we are only interested in whether or not respondents *agree*, *disagree*, or are *neutral* (i.e. *neither agree nor disagree*). The recode procedure allows you to collapse the original variable into three categories, using the following steps.

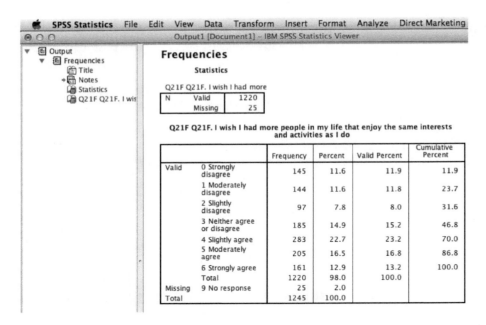

**FIGURE 4.2** Output

Source: Reprint Courtesy of International Business Machines Corporation, © International Business Machines Corporation.

# 1. Run a Frequency Distribution of Your Original Variable

The output allows you to examine the overall distribution, as well as see whether there are any missing values. Often, categories of new (i.e. recoded) variables are based on the number of cases that fall into a specific range; frequencies help you decide whether there are enough cases in each range to do so.

After examining the frequency distribution, we see that there are enough cases to create three separate categories (*agree*, *disagree*, and *neutral*).

# 2. Use the Recode Procedure to Create a New Variable from the Original One

*Procedure*

1.  Go to the Transform menu → Recode into Different Variables.
    a.  Select the variable you want to change (e.g. Q21F) from the list of variables and move it into the Input Variable → Output Variable box.
    b.  Select a name for your new variable (e.g. NQ21F, where *N* signifies that it is a new variable), and type it into the Output Variable box.
    c.  Under Label, you can also add a variable label at this point (e.g. "Q21F recoded"). You can also label the variable by going to the Data Editor and selecting the Variable View (as described in the previous chapter).
    d.  Click the Change button.
    e.  Select the Old and New Values button. This opens the Recode into Different Variables: Old and New Values dialogue box (see Figure 4.3).

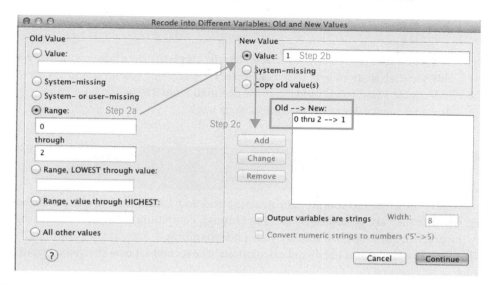

**FIGURE 4.3** Recoding

Source: Reprint Courtesy of International Business Machines Corporation, © International Business Machines Corporation.

2.  Select the Range button under the Old Value dialogue box on the left.
    a.  Enter 0 (zero) in the first box and 2 in the second box.
    b.  Enter the value 1 in the New Value box on the right.
    c.  Click Add. This creates a disagree range (i.e. *strongly disagree, somewhat disagree*, and *slightly disagree*), with a new value of 1.
3.  Next, click the Value button under the Old Value dialogue box on the left.
    a.  Enter 3.
    b.  Enter the value 2 in the New Value box on the right.
    c.  Click Add. This changes our old *neither agree nor disagree* value from 3 to 2 so that it logically follows our first new value 1 or *disagree*.
4.  Select the Range button under the Old Value box on the right.
    a.  Enter 4 in the first box and 6 in the second box.
    b.  Enter the value 3 in the New Value box on the right.
    c.  Click Add. This creates an agree range (i.e. *strongly agree, somewhat agree*, and *slightly agree*), with a new value of 3.
5.  When you finish recoding, select Continue and choose OK.

> **Tip:** Always run a frequency distribution and include your original variable and your new variable to make sure that you did the recoding correctly (new variables will always appear at the bottom of the list in the frequency command). For example, what if you selected the wrong range? In Figure 4.4, we see that 386 respondents disagreed. By totalling the three *disagree* categories in Figure 4.2 (145 + 144 + 97 = 386) you can confirm your work.

**NQ21F Q21F recoded**

|        |            | Frequency | Per cent | Valid Per cent | Cumulative Per cent |
|--------|------------|-----------|----------|----------------|---------------------|
| Valid  | 1 Disagree | 386       | 31.0     | 31.6           | 31.6                |
|        | 2 Neutral  | 185       | 14.9     | 15.2           | 46.8                |
|        | 3 Agree    | 649       | 52.1     | 53.2           | 100.0               |
|        | Total      | 1220      | 98.0     | 100.0          |                     |
| Missing| System     | 25        | 2.0      |                |                     |
| Total  |            | 1245      | 100.0    |                |                     |

**FIGURE 4.4** Recoded Output

## Final Steps

1. You should also give your new variable value labels via the Data Editor window → Variable View tab (bottom left). In our example, 1 = *disagree*, 2 = *neutral*, and 3 = *agree*.
2. In Variable View, you should lower the decimals from 2 to 0.
3. Run a frequency distribution to check your work.

### STEP 4 Practice

You now have the ideas and tools for constructing and understanding frequency distribution tables, whether they are generated by hand or through computing software. Completing the following practice questions will solidify your understanding.

The first set of questions uses hand calculations. The second set uses the SPSS procedures. For each set of questions:

1. Follow the procedural steps and complete the appropriate calculation (Set 1) or software application (Set 2).
2. Check your answers, using the Answer Key in the back of the text.
3. If your answer is incorrect, consult the Solutions section on the book's website. The Solutions provide a complete step-by-step analysis of how the answers are derived.

After you have completed the next section (Interpreting the Results), return to your calculations or output and *provide complete, written interpretations of each of the statistics you have generated.*

# Set 1: Hand-Calculation Practice Questions

1. A hypothetical survey of an Introduction to Sociology class asked the students whether or not they used Facebook. Out of 184 students who responded, 174 said "yes." What percentage of students reported being Facebook users?
2. What percentage of respondents were not Facebook users?
3. Table 4.11 shows the frequency distribution for Facebook users in the Introduction to Sociology class. Reduce the number of rows by grouping the data into four categories.

**TABLE 4.11**   Hours Per Week Spent on Facebook by Introduction to Sociology Students

| Hours | f |
|-------|-----|
| 2 | 5 |
| 3 | 7 |
| 4 | 9 |
| 5 | 15 |
| 6 | 16 |

*(Continued)*

**TABLE 4.11** *(Continued)*

| Hours | f |
|-------|-----|
| 7 | 20 |
| 8 | 19 |
| 9 | 18 |
| 10 | 15 |
| 11 | 10 |
| 12 | 9 |
| 13 | 7 |
| 14 | 7 |
| 15 | 5 |
| 16 | 4 |
| 17 | 2 |
| Missing values | 6 |
| Total | 174 |

4. What number would you use as the denominator to calculate the Valid Per cents for your table of grouped data?

5. Use the table of grouped data you created in question 3 to create a percentage distribution table. This will require you to add three additional columns.

6. What is the Valid Per cent of student Facebook users who reported spending between 6 and 9 hours a week on Facebook?

7. What percentage of students reported spending 13 hours or less a week on Facebook?

8. What percentage of students reported spending 10 hours or more a week on Facebook?

9. Twenty-one third- and fourth-year students attended a workshop offered by the Faculty of Arts focusing on improvement of students' academic writing and research skills. They were asked to fill in a brief survey that would help estimate characteristics of students typically interested in attending such workshops. Among other things, the students were asked about their declared major and their cumulative GPA (rounded to one decimal place), as shown in Table 4.12.

**TABLE 4.12** Student's Declared Major and Their Cumulative GPA (grade-point average)

| Student | Declared Major | GPA |
|---------|----------------|-----|
| 1 | psychology | 2.9 |
| 2 | political science | 3.5 |
| 3 | psychology | 3.4 |
| 4 | sociology | 4.0 |
| 5 | anthropology | 3.1 |
| 6 | sociology | 3.7 |
| 7 | psychology | 3.6 |
| 8 | anthropology | 3.5 |

*(Continued)*

Tips for constructing frequency distribution: The actual frequency distribution will have as many rows as there are categories (attributes) of the variable in question (in our example, the variable *Declared Major* has six categories). Then, for each category (for example, psychology) count the number of people associated with that category (those who indicated that psychology was their declared major). Do this for each of the variable's six attributes (rows); the resulting column is your frequency column with the number in each row of the frequency column representing the frequency of that category of major.

**TABLE 4.12** *(Continued)*

| Student | Declared Major | GPA |
| --- | --- | --- |
| 9 | political science | 3.8 |
| 10 | political science | No answer |
| 11 | sociology | 4.3 |
| 12 | psychology | 4.1 |
| 13 | psychology | 3.7 |
| 14 | psychology | 3.2 |
| 15 | anthropology | No answer |
| 16 | sociology | 3.5 |
| 17 | anthropology | 3.2 |
| 18 | women's studies | 4.0 |
| 19 | sociology | 3.8 |
| 20 | psychology | 3.8 |
| 21 | psychology | 3.9 |

a. Construct a frequency distribution of the variable *Declared Major*, including the percentage column.

b. Construct a frequency distribution of the variable measuring students' GPA. Reduce the distribution into four categories (four equal intervals) and add all the appropriate columns.

# Set 2: SPSS Practice Questions

The following questions use the Student Health and Well-Being Survey data set discussed in Chapter 3. You can access this data set from either your professor or the book's website. Apply the techniques covered in Step 3: Using Computer Software to perform the following operations.

1. Set missing values to 9 for question Q20C of the Student Health and Well-Being Survey data set and then create a frequency table.

2. Recode Q20C into a three-category nominal variable (i.e. collapse the six categories into three as follows: 1 = *disagree*, 2 = *neutral*, 3 = *agree*). Remember to set missing values and to run frequencies on the original and recoded variables.
   Note: When recoding a variable, compare frequency distributions of the original variable and the recoded variable to ensure that the recode worked.

3. Recode Q35 into a two-category (dichotomous) nominal variable by collapsing categories 1 and 2 into a single *yes* category (1 = *yes*), and collapsing categories 3, 4, and 5 into a single *no* category (0 = *no*). Remember to set missing values and run frequencies.

4. Recode Q30 into a new three-category variable (NQ30 *Social class*) by collapsing categories 1 to 5 into a new category 1 (*lower*), collapsing categories 6 to 8 into a new category 2 (*middle*), and collapsing categories 9 and 10 into a new category 3 (*upper*). Remember to set missing values.

5. Set missing values to 9' for question Q1C of the Health and Wellness dataset and then create a frequency table.

6. Recode Q1C into nQ1C, a three-category nominal variable (i.e. collapse the six categories into three as follows: 1 = *disagree*, 2 = *neutral*, 3 = *agree*). Remember to set missing values.

7. Recode Q8E into nQ8E, a two-category (dichotomous) nominal variable by collapsing categories 0 and 1 into a single *no* category (0 = *no*), and collapsing categories 2, 3, and 5 into a single *yes* category (1 = *yes*). Remember to set missing values.

8. Recode Q17 into nQ17, a new three-category variable (nQ17 "Level of Self-Esteem— Ordinal") by collapsing categories 1 to 20 into a new category 1 (*low*), collapsing categories 21 to 40 into a new category 2 (*medium*), and collapsing categories 41 and 60 into a new category 3 (*high*). Remember to set missing values.

9. Create a frequency table for the variable *Satisfaction* with level of exercise (Q1B). Are more people *very dissatisfied* or *very satisfied* with their level of exercise?

10. Recode variable Q15C into a nominal variable, where categories 0 and 1 become 0 (*no*) and categories 2, 3, and 4 become 1 (*yes*).

## STEP 5 Interpreting the Results

Frequency distribution tables provide basic information that requires little technical interpretation. Nonetheless, here are some useful reminders about describing the information contained in frequency distributions.

• Check to see how large the percentage of missing cases is. This is an indicator of the credibility of the variable under consideration. If a variable has a larger proportion of missing cases, treat the results with caution. If, for example, 40 per cent of respondents' data are missing, there is probably something the matter with how the question was presented or how the data were collected. Be skeptical.[1]

• The Valid Per cent figures are the most commonly reported results from the table because they have two advantages. First, since they are standardized (percentages), the results have intuitive appeal. Second, since they exclude missing data, the findings speak to the reported cases.

• When examining Cumulative Per cents, pay attention to which way the percentages were accumulated. Cumulative Per cents can be constructed by adding the Valid Per cents either up or down. This affects the interpretation. There is a lot of difference between saying that 80 per cent of students writing a test achieved grades of "B or better" (cumulating down) or saying achieved grades of "B or worse" (cumulating up).

• When continuous variables are collapsed (recoded) into fewer categories, check to see how this was done. Make sure you are confident that the reconstruction grouped categories in a meaningful way.

---

[1] As a rule, we are not concerned when missing values are under 5 per cent. There are methods we use to deal with missing data, but they are beyond the scope of this book..

# Chapter Summary

This is the first chapter in which you entered the actual statistical maze, journeyed down some paths, and added some basic statistical tools to your toolkit. In this chapter you learned how to summarize data from a single variable using frequency distributions. Having completed this chapter you should appreciate that:

- Univariate analyses are useful for both describing individual variables and checking distribution assumptions.
- The four levels of measurement are often organized into two groups, including categorical and continuous variables.
- Categorical variables include those at nominal and ordinal levels, while continuous ones are at interval and ratio levels.
- Frequency distribution tables provide basic descriptive information about single variables.
- Standard frequency distributions include information about category counts (frequencies), category percentages, valid case adjustments, and cumulative percentages.
- Continuous variables with many categories are often revised (i.e. grouped or collapsed) into frequency distribution tables with fewer attributes.

With these basic understandings of univariate distributions in place, you are now ready to proceed to the next chapter, which introduces statistical tools for determining where the typical cases in a distribution reside.

# 5

# Measures of Central Tendency

## Overview

Your journey down the paths of the statistical maze is now well under way. The last chapter introduced frequency distributions, the most basic way of describing data on single variables. You learned that frequency distributions are used for variables at the nominal and ordinal levels of measurement and, when categories are grouped, for interval and ratio variables as well.[1]

This chapter continues along a path connected to the previous one. In this chapter you will learn about:

- the two features of any univariate distribution that can be summarized with statistics; and
- the three statistics that measure the central tendency of single variables.

Here is what's covered in this chapter in checklist form:

## Statistical Selection Checklist

1. Does your problem centre on a descriptive or an inferential issue?

   ☑ Descriptive

   ❑ Inferential

2. How many variables are being analyzed simultaneously?

   ☑ One

   ❑ Two

   ❑ Three or more

3. What is the level of measurement of each variable?

   ☑ Nominal → Mode

   ☑ Ordinal → Median

---

[1]Since the act of "grouping" categories of an interval or ratio variable transforms it into an ordinal one.

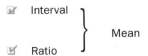

The statistics covered in this chapter provide summaries that are even more compact than frequency distributions, since they summarize a data set in a single number. Here is what you need to know to understand these new tools.

## STEP 1  Understanding the Tools

# Two Features of a Single Variable

Variables are properties of objects that can change. The possible variation in a variable is identified by its scores or values. For example, the variable *Age* has variation that ranges from zero to about 115 years. If you collect information on the ages of a sample of 500 persons, you will have a data set that contains 500 scores: 1, 93, 26, 55, 82, 77, 62, 34, 51, 36, 6, 13, 11, 44, 55, 71, etc. These scores represent the actual variation in ages in the sample under consideration.

The actual scores from a data set on a single variable have two features. One feature is *central tendency*; the other is *dispersion*. **Dispersion** refers to how spread out the actual scores are. Samples of data differ in their dispersion. If one sample included persons whose ages ranged from 78 to 92 years (elderly sample), it would be quite different from one where the ages ranged from 13 to 19 years (teenage sample). There are several statistics that summarize dispersion in a single measure, and these are considered in the next chapter.

This chapter focuses on the other feature of a single variable that can be summarized, central tendency. As the label suggests, central tendency refers to where the centre of a distribution of scores is. In other words, **central tendency** measures inform us about where the scores of a variable are clustered or "hang together." Such clustering tells us what the typical or common scores of a variable are. If a typical age in one sample was 13 years, the sample would be quite different from one in which the typical age was 83 years.

In short, any set of data on a single variable can be summarized in terms of its two basic features. Central tendency provides a summary of what a typical score in a distribution is like. Dispersion informs us how spread out the scores are around the centre. You can appreciate these two features in the following example. Imagine students' scores on the first two tests in an Introduction to Sociology course. On the first test, a typical student score was 44 per cent (central tendency) and the scores included a range (dispersion) of 5 per cent. On the second test, the central tendency was 77 per cent and the dispersion was 20 per cent. Even before the individual grades were distributed, if you were in this class you would be expecting bad news on the first test—since students typically failed and their scores were tightly clustered. Regarding the second test, you would have grounds for optimism, since students typically did very well and the range, although wider, was not extreme. On the basis of these general summaries, you could reasonably conclude you probably did poorly on the first test and well on the second.

# Measures of Central Tendency

You are going to learn about how to calculate and interpret three measures of central tendency. Before discussing these measures, it is worth reviewing where they are situated in the statistical maze. First, these are *descriptive* measures: they inform us about the data actually collected. Second, these are *univariate* measures: they inform us about the evidence regarding a single variable. Third, as we shall see, different measures of central tendency are applicable to variables at different levels of measurement. This third feature will become clear as we discuss each central tendency measure.

## Averages

The term *average* refers to a value that is in the middle or centre of a distribution. If you speak of a person being of average height, you are saying their height is common or typical. In this way, the three measures of central tendency are different types of average. These summary statistics are different ways of calculating where the centre of a distribution of scores is and reporting on a typical score. The three measures of central tendency include mode, median, and mean.

### Mode

The mode is symbolized by a capital *M* and lower-case *o*, as in *Mo*. The **mode** is the simplest measure of central tendency and refers to the most frequently occurring value of a variable. So, to identify the mode, look at the simple frequency distribution of a variable and select the *category of the variable that occurs most frequently*. Note that the mode is *not* the frequency that occurs most commonly; the mode is not a frequency count. The mode is a *category* of the variable.

Table 5.1 presents a simple frequency distribution for the variable *How much do you like chocolate ice cream?*

The mode (*Mo*) of this distribution is *little*. In other words, *little* is the category of the variable *chocolate ice cream preference* that occurs most frequently.

### Median

The median is symbolized by a capital *M* and lower-case *d*, as in *Md*. The **median** is the score that is at the *middle of a rank-ordered distribution*. Imagine you asked all the children in a grade one classroom to line up from the shortest to the tallest. After a predictable amount of

**TABLE 5.1** Chocolate Ice Cream Preference

| Like Chocolate Ice Cream | Frequency (f) |
| --- | --- |
| Very much | 20 |
| Somewhat | 30 |
| Little | 40 |
| Not at all | 10 |
| Total (N) | 100 |

chaos and shuffling about, the students would be rank ordered in terms of the variable *Height*. The height of the student at the middle of that distribution is the median.

Notice that the median separates the rank-ordered distribution into two equal-sized groups. In our example, half of the students would have heights above the median score, and half would be below. The following sections will walk you through the steps in computing the median and provide you practice opportunities. For now, it is sufficient that you appreciate that the median measures where the centre of a rank-ordered variable is located.

### *Mean*

The mean is symbolized as $\overline{X}$, and is the summary statistic that most people have in mind when they say "the average." As you probably learned somewhere in your schooling, the form of average called the **mean** is computed by adding up all the individual scores of a variable and dividing by the number of scores. So, if a sample of five students had percentage grades of 43, 56, 67, 78, and 93 on a test then the mean would be 67.4 per cent.

Now that you have been introduced to these three statistics measuring central tendency, the following sections will give you detailed examples of their calculation and opportunities for practice.

## STEP 2 Learning the Calculations

The previous section provided you with a basic understanding of the three statistics used to measure central tendency. This section shows you the steps in computing each of these statistics.

# Calculating the Mode

The mode is the value (attribute or category) of the variable that occurs most frequently. Determining the mode is not a matter of calculation; it is a matter of inspection. To determine the mode you either look at the scores (if the sample is small) or look at a frequency distribution (if the sample size is larger), using the procedures illustrated in the following examples.

**Example 5.1a**  Hair Colour in an Introduction to Sociology Class

| | |
|---|---|
| Bobby: | red |
| Sam: | blond |
| Laura: | blond |
| Helen: | brown |
| Doug: | black |
| Katie: | blond |

This data set includes only six cases of the variable *Hair Colour*. The mode for this distribution will be the most frequently occurring hair colour. Since the sample contains only a small number of cases, you can simply examine the distribution to determine the mode. In this example, the mode of the variable hair colour is *blond*, since this is the attribute of the variable that occurs most frequently. In the sample of six students, there are three blondes and one person each with red, black, and brown hair.

Notice that, if you wanted to, you could transform the data in Example 5.1 into the following frequency distribution. Although the small sample size makes this unnecessary, it does clarify how determining the mode simply involves selecting the category of the variable that is most common.

**Example 5.1b** Hair Colour in an Introduction to Sociology Class

| Hair Colour | f |
|---|---|
| Blond | 3 |
| Red | 1 |
| Black | 1 |
| Brown | 1 |
| Total (N) | 6 |

Here is another example.

**Example 5.2** Average Income for Towns in Manitoba (in dollars)

| | |
|---|---|
| Altona: | 67,000 |
| Brandon: | 55,000 |
| Gimli: | 45,000 |
| Thompson: | 67,000 |
| The Pas: | 45,000 |
| Winnipeg: | 67,000 |

Answer the following questions/steps and complete the table below.

1. What is the name of the variable in these data? Place your answer in the upper left-hand cell of Table 5.2.
2. What are the specific values (attributes) of the variable? List these scores in the rows in the left column of the table.
3. How many times does the lowest score occur in the data? Place this total in the appropriate f-score cell.
4. How many times does the next lowest score occur? Place this total in the appropriate cell.
5. Continue step 4 until all values have frequency scores.
6. Add up the total number of cases in the f column. Place this number in the cell beside N.
7. Double-check to make sure that the N equals the actual number of cases in Example 5.2a.
8. Look down the completed f column of the table and identify the highest frequency.
9. The label in the row to the left of this (highest) frequency is the mode.

If you have followed the steps correctly, then you should have successfully identified the mode and constructed a basic frequency table (without percentages) for Table 5.2.[2]

One other point. The mode identifies the most commonly occurring score of a variable. In the examples, there was clearly one category of the variable that occurred most frequently. In these cases the distribution is said to be *unimodal*; it contains only a single mode. Distributions are commonly unimodal, but they don't have to be. Examine Table 5.3.

---

[2]The correct answer is $67,000, since this value occurs three times in the distribution.

**TABLE 5.2**  Frequency Distribution for Example 5.2a

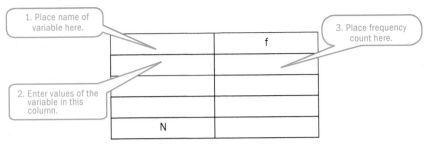

**TABLE 5.3**  Final Math Grades of High School Students

| Grade | f |
| --- | --- |
| A+ | 15 |
| A | 20 |
| B+ | 45 |
| B | 40 |
| C+ | 45 |
| C | 35 |
| D | 25 |
| F | 20 |
| Total (N) | 245 |

If you scan down the *f* column in this table you see that the highest frequency is 45. But this frequency occurs twice. There are 45 students in the sample who received grades of B+, and 45 who were awarded C+. In this example there are two categories of the variable (Grade) that occur most frequency—C+ and B+. In this case, the distribution is *bimodal*; it has two modes.

# Calculating the Median

The *median* is a second measure of central tendency. The median represents *the middle score in a rank-ordered distribution.* In this case, *middle* means that half of the observations fall above this score, and half fall below it. Example 5.3 illustrates how to determine the median.

**Example 5.3**  Test Scores on an Introduction to Sociology Test

| | |
| --- | --- |
| Bobby: | 67 |
| Sam: | 96 |
| Laura: | 75 |
| Helen: | 76 |
| Doug: | 34 |
| Katie: | 85 |
| George: | 73 |

1.  Rank-order the scores in the data set. This step involves putting the scores in order from highest to lowest, or vice versa. Here is the result of this step for the data in Example 5.3.

| Sam:    | 96 |
| Katie:  | 85 |
| Helen:  | 76 |
| Laura:  | 75 |
| George: | 73 |
| Bobby:  | 67 |
| Doug:   | 34 |

2. Determine where the centre of the rank-ordered distribution is. To determine the middle value, use the following formula: $(N + 1) \div 2$. Remember from the earlier discussion of frequency distributions that $N$ refers to sample size (the total number of cases we are observing). In effect, this formula tells you to add one to the sample size and divide this total by 2.

   Applied to the data in Example 5.3, the $N$ is 7, so the middle score (applying the formula) is $(7 + 1) \div 2 = 4$. This result means that the median is the fourth score from the bottom of the rank-ordered distribution.

3. Count off from the bottom of the rank-ordered distribution to determine the median score. Here is the rank-ordered distribution of scores for Example 5.3 (from step 1).

| Sam:    | 96 |
| Katie:  | 85 |
| Helen:  | 76 |
| Laura:  | 75 |
| George: | 73 |
| Bobby:  | 67 |
| Doug:   | 34 |

Laura's score of 75 is the fourth on the list—so the median is 75.

   Counting from the lowest score, the fourth score is 75, which is the median.

In summary, there are three closing points worth noting about the median. First, the median is the actual score in the data set. So, in Example 5.3, the median is the score 75; it is not the name of the person (Laura) who had that score. (In fact, in realistic situations you will only have the data set scores of a variable. You will not know which persons are associated with particular scores because the data will typically be anonymous.)

Second, in some rank-ordered distributions there will be more than one case with the same median score. Look at the following distribution of scores: 3, 4, 6, 6, 6, 7, 8. Applying the formula (step 2), the median is the fourth score from the bottom—which is 6. It doesn't matter that the distribution contains the individual cases with the score 6. Just rank-order the distribution, apply the formula, count from the bottom, and be confident that whatever score appears is the median.

Finally, sometimes the median turns out to be a number that is not one of the actual scores in the distribution. This often occurs when the sample contains an even number of cases. Examine the following set of scores which contains 8 cases: 3, 4, 5, 6, 7, 8, 9, 10. Applying the formula from step 2, the median will be the 4.5th case from the bottom of the rank-ordered distribution [$(8 + 1) \div 2$]. Counting from the bottom, where is the 4.5th case? It falls between 6 and 7. Where this occurs, the median is the value half-way between the two scores. In this case, the median is 6.5.

To convince yourself that you understand how to identify the median among an even number of cases, apply the steps to the following distribution. What is the median among this set of scores?[3]

| | |
|---|---|
| Sam: | 96 |
| Katie: | 85 |
| Helen: | 76 |
| Laura: | 75 |
| George: | 73 |
| Bobby: | 67 |

# Calculating the Mean

As noted earlier, the mean is the measure of central tendency that most students are used to calling the average. Calculating the mean involves adding up all the scores for a variable in a data set and dividing by the number of scores.

Here is how this simple arithmetic is expressed in statistical terms:

$$\overline{X} = \frac{\Sigma x}{N}$$

It is worth emphasizing here how statisticians can take a simple idea and make it look foreign and intimidating in a formula. The lesson is that you should not be intimidated by formulas that look complicated. If you follow the calculation steps, the result will unfold. Here are the steps for determining the mean $\overline{X}$.

1. Make sure you understand how to interpret each of the symbols in the formula. The calculation formula has the symbol for the mean $\overline{X}$ on the left side of the equal sign and a set of squiggles on the right side. Here is how to interpret each squiggle:[4]
   - $\Sigma$ is the summation sign and it means "add up."
   - $x$ represents each individual score on the variable in the data set.
   - $N$ refers to sample size.

   Taken together, the formula tells us to calculate the mean by adding up all the individual scores in the data set and divide this total by the sample size.
2. Solve the $\Sigma x$ part of the formula. You do this by taking all the individual scores on a variable and adding them.
3. Divide the total from step 2 ($\Sigma x$) by the sample size ($N$), which results in the mean $\overline{X}$.

Here is an example of these steps applied to the following data set.

**Example 5.4** Test Scores of an Introduction to Sociology Class

| | |
|---|---|
| Bobby: | 67 |
| Sam: | 96 |
| Laura: | 75 |
| Helen: | 76 |
| Doug: | 34 |
| Katie: | 85 |

---

[3]In this case the median occurs at the 3.5th value from the bottom of the rank ordered distribution. The 3.5th observation occurs between 75 and 76. Therefore, the correct answer is 75.5.

[4]The squiggles are, for the most part, actually Greek symbols.

1. Review the formula to make sure you understand its components.

$$\overline{X} = \frac{\Sigma x}{N}$$

2. The $\Sigma x$ is 433 because $67 + 96 + 75 + 76 + 34 + 85 = 433$.

3. The $N$ is 6 because there are 6 students in the data set. Dividing the step 2 total (433) by $N$ (6) (i.e. 433/6) results in a mean of 72.17.

   To summarize, these steps resulted from solving the formula for the mean.

$$\overline{X} = \frac{\Sigma x}{N}$$

$$72.17 = \frac{433}{6}$$

You now have the basic ideas and calculation methods for the three measures of central tendency—mode, median, and mean. The calculation examples have used small sample sizes ($N$) to keep the arithmetic simple. In practice, realistic research situations involve much larger sample sizes, and doing the calculations by hand involves both considerable time and the risk of mistakes. For this reason, larger data set calculations are done using computer software. The following section shows you how to apply IBM SPSS Statistics software ("SPSS") to obtain these measures of central tendency.

## STEP 3 Using Computer Software

In SPSS, measures of central tendency are selected through the Frequency Distribution procedure, as discussed in the previous chapter. Before you run a frequency distribution, you need to make sure that applicable values in the data are set to *missing*. If you do not, the measures of central tendency will include the value of the *missing* cases (i.e. 9 = *no response*) in its calculations, thereby giving you an incorrect measure of central tendency. See the previous chapter for instruction on how to identify appropriate values as *missing* (under the section Set Values to *Missing*).

Once you have confirmed that the proper values are set to *missing*, you can run a frequency distribution with measures of central tendency by following these steps:

1. Select Analyze from the menu bar.
2. Select Descriptives → Frequencies.
   a. In the left-hand box, scroll to and highlight the variable (or variables) you want to analyze.
   b. Press the arrow button to move the variable to the right-hand side.
   c. Go to the Statistics box.
   d. In this menu, you can select your preferred measure of central tendency (i.e. mean, median, mode). You can select more than one.
3. Select Continue and OK to execute the Frequency command.

In the Output/Viewer window, you will find the results of your analysis, an example of which is illustrated in Figure 5.1.

**Statistics**

Q14A Q14A. Coping with stress by seeking social support

| N | Valid | 1228 |
|---|---|---|
| | Missing | 17 |
| Mean | | 11.16 |
| Median | | 12.00 |
| Mode | | 14 |

Measures of central tendency

**Q14A Q14A. Coping with stress by seeking social support**

| | | Frequently | Per cent | Valid Per cent |
|---|---|---|---|---|
| Valid | 0 Does not use this coping strategy at all | 15 | 1.2 | 1.2 |
| | 1 | 10 | .8 | .8 |
| | 2 | 6 | .5 | .5 |
| | 3 | 14 | 1.1 | 1.1 |
| | 4 | 26 | 2.1 | 2.1 |
| | 5 | 41 | 3.3 | 3.3 |
| | 6 | 63 | 5.1 | 5.1 |
| | 7 | 52 | 4.2 | 4.2 |
| | 8 | 78 | 6.3 | 6.4 |
| | 9 | 81 | 6.5 | 6.6 |
| | 10 | 105 | 8.4 | 8.6 |
| | 11 | 115 | 9.2 | 9.4 |
| | 12 | 130 | 10.4 | 10.6 |
| | 13 | 113 | 9.1 | 9.2 |
| | 14 | 133 | 10.7 | 10.8 |
| | 15 | 94 | 7.6 | 7.7 |
| | 16 | 56 | 4.5 | 4.6 |
| | 17 | 40 | 3.2 | 3.3 |
| | 18 | 31 | 2.5 | 2.5 |
| | 19 | 16 | 1.3 | 1.3 |
| | 20 Uses this coping strategy all of the time | 9 | .7 | .7 |
| | Total | 1,228 | 98.6 | 100.0 |
| Missing | 99 No response | 17 | 1.4 | |
| Total | | 1,245 | 100.0 | |

**FIGURE 5.1**  SPSS Central Tendency and Frequency Distribution Output

In Figure 5.1, "Q14A. Coping with stress by seeking social support," has 17 respondents set to *missing*, because they did not answer the question. The Coping variable ranges from zero (Does not use this coping strategy at all) to 20 (Uses this coping strategy all of the time). In terms of measures of central tendency, we see that that mode is equal to 14, the median is equal to 12, and the mean is equal to 11.16. Based on your understanding of the three measures of central tendency, you should be able to provide written interpretations for each of these statistics.

## STEP 4 Practice

You have been introduced to the meaning of the three measures of central tendency as well as the procedures for calculating each of these statistics, either through hand calculations (if the sample size is small) or computer software (for large sample sizes). To solidify your understanding you need to practise applying the statistical and software procedures.

This section provides you opportunities to practise what you have learned about measures of central tendency. The first set of questions uses hand calculations. The second set uses the SPSS procedures. For each set of questions:

1. Follow the procedural steps and complete the appropriate calculation (Set 1) or software application (Set 2).
2. Check your answers, using the Answer Key in the back of the text.
3. If your answer is incorrect, consult the Solutions section on the book's website. The Solutions provide a complete step-by-step analysis of how the answers are derived.

After you have completed the next section (Interpreting the Results), return to your calculations or output and *provide complete, written interpretations of each of the statistics you have generated.*

# Set 1: Hand-Calculation Practice Questions

1. What is the mode for the distribution in Table 5.4?

**TABLE 5.4** Make of Cars in a Tim Hortons Parking Lot

| |
| --- |
| Dodge |
| Ford |
| Toyota |
| GM |
| Kia |
| Ford |
| Nissan |
| Toyota |
| Ford |
| GM |
| Toyota |
| Mazda |
| Dodge |
| Ford |
| VW |
| Toyota |

2.  Table 5.5 shows the number of students per class at Happy Valley Elementary School.

**TABLE 5.5**  Students per Class at Happy Valley Elementary School

| Grade | Number of Students |
|-------|--------------------|
| Kindergarten | 22 |
| 1 | 25 |
| 2 | 20 |
| 3 | 27 |
| 4 | 24 |
| 5 | 23 |
| 6 | 19 |

   a.  What is the median number of students per class?

   b.  If we exclude Kindergarten, what is the median number of students per class in grades 1 to 6 in Happy Valley?

   c.  What is the mean number of students per class in the school (K–6)?

3.  Calculate the mode, median, and mean for the set of IQ scores in Table 5.6.

**TABLE 5.6**  IQ Scores

| |
|---|
| 119 |
| 98 |
| 91 |
| 109 |
| 105 |
| 127 |
| 89 |
| 113 |
| 98 |
| 103 |
| 98 |
| 109 |

4.  In Chapter 4 we introduced a table with data for 21 third- and fourth-year students who attended a workshop offered by the Faculty of Arts focusing on improvement of students' academic writing and research skills. The students were asked to fill in a brief survey that would help estimate characteristics of students typically interested in attending such workshops. In addition to data about their declared major and cumulative GPA, the students were also asked to indicate, on a scale from one to five (1 = *highly unlikely*; 5 = *highly likely*), whether they think they will apply for graduate school right after completing their undergraduate degree. The data are shown in Table 5.7.

   a.  What is the appropriate measure of central tendency for students' *Declared Major*? What is its value?

   b.  What is the appropriate measure of central tendency for students' GPA? Calculate its value.

**TABLE 5.7** Students' Declared Major, Cumulative GPA (grade-point average), and Their Likelihood of Applying for a Graduate School Directly after Completion of Their Undergraduate Degree, Measured on a Five-Point Scale (1 = *highly unlikely*; 5 = *highly likely*)

| Student | Declared Major | GPA | Apply to Grad School? |
|---|---|---|---|
| 1 | psychology | 2.9 | 1 |
| 2 | political science | 3.5 | 3 |
| 3 | psychology | 3.4 | 2 |
| 4 | sociology | 4.0 | 5 |
| 5 | anthropology | 3.1 | 1 |
| 6 | sociology | 3.7 | 4 |
| 7 | psychology | 3.6 | 3 |
| 8 | anthropology | 3.5 | 2 |
| 9 | political science | 3.8 | 4 |
| 10 | political science | No answer | 2 |
| 11 | sociology | 4.3 | 5 |
| 12 | psychology | 4.1 | 4 |
| 13 | psychology | 3.7 | 4 |
| 14 | psychology | 3.2 | 1 |
| 15 | anthropology | No answer | 3 |
| 16 | sociology | 3.5 | 3 |
| 17 | anthropology | 3.2 | 2 |
| 18 | women's studies | 4.0 | 5 |
| 19 | sociology | 3.8 | 4 |
| 20 | psychology | 3.8 | 4 |
| 21 | psychology | 3.9 | 5 |

   c. What is the appropriate measure of central tendency for students' reported likelihood of applying for graduate school right after completion of their undergraduate degree?

     i. Calculate the value of this measure.

     ii. How does this value change if we eliminate the first student's score?

   d. What is the average GPA of sociology students who attended the workshop?

   e. Are those students whose GPA is 3.8 or higher more likely than all workshop participants in general to say they would like to apply for graduate school right after their undergraduate degree?

# Set 2: SPSS Practice Questions

1. What are the mode, median, and mean values for the *Self-esteem* variable (Q17)?
2. What are the mode, median, and mean ages of Student Health and Well-Being Survey respondents?

3.  What are the mode, median, and mean values for the *Depression* variable (Q4)?

4.  What are the mode, median, and mean values for the *Reason for living—Family or friends variable* (Q24A)?

5.  What are the mode, median, and mean for the *Satisfaction with life* variable (Q22)?

6.  What are the mode, median, and mean values for the *Strength of religious faith* variable (Q18)?

7.  What is the mode for *Reason for living—One's self or one's future* (Q24B)?

8.  What is the mean for *Psychological well-being—Purpose in life* (Q16E)?

## STEP 5 Interpreting the Results

# Use and Selection

This chapter discussed the three statistics that are used to identify the central tendency among a set of scores for a single variable. In addition to being introduced to the mode, median, and mean, detailed examples have shown you how each is calculated and you have examples on which to practise these calculations. Furthermore, you have instructions for how to use computer software to generate each of these statistics.

The closing section of this chapter shares some general points about the use of measures of central tendency and provides direction for interpreting each of these statistics.

## Appropriateness of Using Central Tendency Measures

Measures of central tendency are descriptive statistics. In a summary form, they tell us what a typical or common score is among a sample of scores on a single variable. Central tendency measures are very useful statistics employed all the time. However, two points about the appropriateness of central tendency measures are worth noting.

The first point to note about central tendency statistics is that, like all statistics, they are summaries. The summarization process takes lot of information and distills it into a digestible form. In this distillation process, *information is lost*. Imagine you attended an NHL hockey game. Your experience of carefully watching provides you with a rich, detailed appreciation of the game. On the radio driving home you hear a sportscaster provide the following summary of the game: "Toronto lost again." Your watching the game is analogous to looking at the specific details of a variable's scores in a data set; the sportscaster's summary is analogous to a measure of central tendency. Summaries accurately capture what occurred, but they leave out a lot of interesting detail. So, when using measures of central tendency, remember that they are not the whole story about what is going on in the data, but a summary of it.

The second point about the appropriateness of central tendency measures concerns their usefulness. Central tendency measures inform us what a typical score in a distribution is. *Central tendency measures are more useful as summaries when the data are clustered*. When the data are more dispersed across the range of a variable, then the summaries provided by central tendency measures are less appropriate. Imagine a sample of 10 students whose intelligence scores on a 10-point scale are: 1, 2, 3, 4, 5, 6, 7, 8, 9, 10. These scores are completely dispersed;

they do not "hang together" around any particular intelligence score. You can, of course, calculate various measures of central tendency on this sample of intelligence scores. The point is, however, that the summaries these statistics provide are not very useful descriptions. By contrast, if the intelligence scores of the 10 students were 7, 7, 7, 8, 8, 8, 8, 9, 9, 9 then central tendency statistics would provide very good summaries of this clustered set of scores.

## Selection of Central Tendency Measures

The mode, median, and mean are different measures of central tendency. How do you know which one to use to summarize a particular variable? The answer relates to levels of measurement. In short, the mode is designed for nominal variables; the median, for ordinal variables; and the mean, for interval/ratio variables. Here's why.

The mode is the category of the variable that occurs most frequently. The categories of nominal variables represent names; they have no numerical properties. So, when summarizing nominal distributions, there is not much you can say other than to identify which category of the variable is most common.

Ordinal variables are nominal variables with the additional characteristic that the categories can be rank-ordered. Remember that the first step in the computation of the median is to rank-order the cases on the variable under consideration. This can only occur if the variable is measured at the ordinal level, which is why the median is appropriate for this level of measurement.

Interval and ratio variables move beyond ordinal variables by including the additional characteristics of equal intervals between categories and an absolute zero point. These additional characteristics make it legitimate to do arithmetic operations like addition and division. The calculation of a mean involves both addition and division, which is why the mean is an appropriate measure for variables at higher (interval and ratio) levels of measurement.

Recall, however, that the levels of measurement are nested; each higher level of measurement begins with the level below it and adds some extra characteristic. Ordinal variables are nominal variables whose categories can be rank-ordered; interval variables are ordinal ones with categories that have equal distances; ratio variables are interval ones with an absolute zero point. This nesting has the following implication for selecting measures of central tendency: *Each higher level can appropriately use the central tendency measure(s) designed for levels of measurement below it.* In practice, this means that while the mode is the only central tendency measure appropriate for nominal variables, ordinal variables can be summarized using the median and the mode. Similarly, interval and ratio variables can use any one of the mode, median, or mean.[5]

---

[5]Although the mean is most commonly used for interval and ratio variables, it is important to note that its result is highly sensitive to even a few extreme scores (i.e. outliers). Imagine a test in which every student except one got a grade of exactly 65 per cent. If the one exceptional student got a 95 per cent, the value of the mean would be pulled up remarkably. Likewise, if the exceptional student got 35 per cent, the mean result would move sharply downward. In circumstances where extreme results are part of the distribution, the median is often used, since it is not affected by outliers.

# Interpreting Measures of Central Tendency

You have learned how to compute the three measures of central tendency through either hand calculations or the use of computing software. Either way, the results generate some number. It is important to remember that computing such statistical numbers is a means to an end. The goal is not to compute the statistic, but to use the statistic to tell you something useful about the data set. In the case of central tendency statistics, the specific goal is to learn which values in the distribution of scores are most typical. Achieving this goal requires interpreting the statistical results.

Interpretation requires stating what the statistical results mean in plain language. Interpretation is the most important part of statistical analysis, and often the most difficult. Here is a guide to how to interpret the three measures of central tendency.

## Mode

The mode is the most frequently occurring value in a distribution of scores. Its interpretation is straightforward and takes the following general form:

> The most frequently occurring score on variable $X$ [the name of the variable] is $Y$ [the label of the category which occurs most frequently].

Imagine a data set that contains twice as many females as males. In this case, interpret the mode of this distribution as follows: "The most frequently occurring score on the variable *Sex* is *female*." Of course, you can say this same thing in different ways, such as "Females are the most frequently occurring sex in the data set."

## Median

The median is a useful summary of central tendency when the scores on a variable can be rank-ordered. Remember that the median divides the distribution of scores into two equal-sized groups. Half of rank-ordered cases fall below the median and half occur above it. Interpreting the median uses this feature and takes the following general form:

> Fifty per cent of the cases on variable $X$ [the name of the variable] fall below a score of $Y$ [the value of the median].

If the median happiness score (on a 10-point scale) in a sample of students was 6, the interpretation would be: "Fifty per cent of the cases on the variable *Happiness* fall below a score of 6." Again, you can make this same point in different words, such as "Half of the students had happiness scores above 6."

## Mean

The mean is the most commonly reported form of average, so much so that many people use the terms *mean* and *average* as synonyms. Oddly, even though the mean is the most familiar central tendency measure, it is not the easiest to interpret. Appreciating this fact requires a review of how the mean is computed.

Computing the mean involves a two-step process. First, all the scores are added together, and then this total is divided by the number of cases. If you wanted to compute the mean income of Canadians who pay taxes and had copies of everyone's tax filings, you would add up each taxpayer's net income and divide by the total number of taxpayers. Think about what this process yields. From the first step, you have computed the total amount of net income among all taxpayers (a very large number). Then, in the second step, you have divided this total by the number of contributors. What result have you created? The outcome is what the net income of taxpayer's would be *if everyone had the same income*. You have created a big pie (step one) and divided it into equal pieces (step two).

The mean, then, is a hypothetical sort of typicality. It states what the score on a variable would be *if* everyone had the same score. In reality, of course, it is almost never the case that all individuals have the same scores on an interval or ratio variable. For this reason, the mean, while the most common central tendency statistic, is the most distant from realistic conditions. Interpreting the mean takes the following general form:

> If everyone had the same score on variable $X$ [insert the name of the variable] it would be $Y$ [insert the value of the mean].

For example, you could interpret the net income result as "If all taxpayers had the same net income, it would be $32,800.63." Notice that, given the hypothetical nature of the mean, it is possible that no actual case in the sample actually has the value of mean.

As you can see from the interpretations, the measures of central tendency express different ways of thinking about what is a typical or common case. The mode expresses what value actually occurs most commonly. The median identifies where the middle of the scores occurs. The mean speaks to what every case's score would be if all cases had the same score. These different views of typicality arise because variables measured at different levels have different features. In later chapters you will see that, because variables at higher levels of measurement can employ more than one measure of central tendency, comparing the results of different measures can provide useful insights.

# Chapter Summary

The last chapter showed you how to summarize data from a single variable using frequency distributions. This chapter introduced three statistics that compute a single score to summarize where the centre of a distribution is. Having completed this chapter you should appreciate that:

- Any single variable has two features, including central tendency and dispersion.
- The three measures of central tendency (mode, median, mean) are all forms of averages.
- Selecting an appropriate measure of central tendency is related to the level of measurement of the variable.
- The mode involves the most frequently occurring value of a variable; the median identifies the middle of a rank-ordered distribution; the mean identifies what score all cases would have if they had the same score.

- Each of the measures of central tendency can be computed through both hand calculations and computing software.
- Using central tendency measures involves trading off information loss for summary insight.
- Central tendency measures are more useful when the actual scores are more clustered.
- Because levels of measurement are nested, variables at higher levels of measurement can employ more than one central tendency measure.
- Interpreting the central tendency computations in simple language is a key goal of statistical analysis.
- Because they are based on different views of typicality, different measures of central tendency have specific interpretations.

Data on single variables (univariate distributions) can be summarized with respect to two characteristics, central tendency and dispersion. This chapter has introduced you to the three statistics that measure central tendency. The next chapter introduces you to statistics that measure the dispersion in a distribution.

# Measures of Dispersion

## Overview

Chapter 4 introduced frequency distributions. These tools are the fundamental way of summarizing information collected on a single variable from a sample of respondents. In the last chapter you learned that univariate data are summarized in terms of two characteristics—central tendency (related to the clustering of scores) and dispersion (related to how widely scattered scores are). You know about the measures of central tendency (mode, median, mean). Your statistical toolkit is getting progressively fuller.

This chapter introduces several common measures of dispersion. In this chapter you will learn:

- why dispersion is important;
- statistical tools for measuring dispersion among categorical variables; and
- statistical tools for measuring dispersion among continuous variables.

Before we begin down this path, let's remind ourselves of where we are in the statistical maze. Knowing where we are in this possibility space for selecting statistical tools keeps us oriented. In checklist form, the statistical measures in this chapter are situated as follows:

## Statistical Selection Checklist

1. Does your problem centre on a descriptive or an inferential issue?

   ☑ Descriptive

   ❑ Inferential

2. How many variables are being analyzed simultaneously?

   ☑ One

   ❑ Two

   ❑ Three or more

3.   What is the level of measurement of each variable?

   ☑  Nominal ⎫  for categorical variables: index of qualitative variation

   ☑  Ordinal ⎬  for ordinal–ratio variables: range

   ☑  Interval ⎭  for continuous variables: variance & standard deviation

   ☑  Ratio

As usual, we will begin by presenting the ideas behind these statistical measures.

## STEP 1  Understanding the Tools

# Variables and Dispersion

Chapter 1 explained how to translate theoretical ideas into a researchable form. This translation involves transforming a proposition (which states a relationship between concepts) into a hypothesis, which forwards an expected relationship between variables. Variables, then, are the working material of research. Hypotheses express how a change in one (independent) variable is expected to change another (dependent) variable.

Variables are properties of objects that can change. Objects are measured and described in terms of variables. You can describe the person sitting next to you in class, for example, in terms of an extensive set of variables such as height, weight, hair colour, race, religion, social class, intelligence, extroversion, etc. Similarly, you can describe this textbook in terms of variables such as weight, difficulty of content, number of authors, etc.

A variable is composed of attributes, sometimes called values or scores. Attributes identify the range of possible change in a variable. For example, the opinion survey statement "Alberta is the most progressive province in Canada" is a variable whose range of possible responses (attributes) includes *strongly disagree, disagree, undecided, agree, strongly agree.*

When a variable is applied to a particular object, then that object is given a specific score on the variable under consideration. The first test you take in a university course is a variable (Test 1 Scores) and you (the object) will obtain a specific score (e.g. 78 per cent) on that variable. When a set of objects (a sample) are measured on a variable, then the scores will almost always be different. When students in a university course take the first test, different students will obtain different scores. In short, variables show variation; different cases (objects) have different scores.

Dispersion is the concept that refers to how spread out a variable's scores are in a sample. Dispersion is an important feature of variables. Here is an example showing how the male and female groups in a class performed on the first test.

Males     15  56  85  89  97  98  99

Females   77  77  77  77  77  77  77

Both these groups have the same typical score, which is a mean of 77. So, in terms of central tendency, the male and female groups in the class are equivalent. Clearly, however, the test performance of these two groups is not the same. Females are very tightly clustered in terms of their test performance, while males display extensive differences. Measures of dispersion summarize how widely scores on a variable actually differ in a sample.

# Measures of Dispersion

Like all descriptive statistics, measures of dispersion are summaries; they capture information about how scattered the scores on a variable are and present it in a single number. As always, this simplification trades off an appreciation of detail for ease of understanding. This section introduces you to four common measures of dispersion, beginning with the range.

## Range

It is common for researchers to identify statistics with some short symbol. In the last chapter the measures of central tendency were symbolized by *Mo*, *Md*, and $\overline{X}$. In keeping with this practice, the dispersion measure Range is symbolized by a capital letter R.

The **range** is the simplest measure of dispersion and identifies the *difference between the highest and lowest scores in a distribution*. Note that the range does not report what the highest and lowest scores are; it reports the *difference* between these extreme scores. For example, if on a test the best-performing student received 87 per cent and the lowest-performing student 17 per cent, then the R is 70 per cent (87 minus 17). Also note that the range is determined from the actual scores on a variable in a data set. The range does not attend to theoretical differences, but actual (observed) differences. In the test example, theoretically the worst-performing student could obtain a zero, while the best-performing student could achieve 100 per cent. The size of this theoretical (possible) maximum difference is set when the variable is constructed. The range is determined using calculations from scores obtained when the variable is applied to actual cases through the measurement process.

The range is a measure of dispersion that can be applied to variables at any of the three higher levels of measurement (ordinal, interval, ratio), although it is most commonly applied to variables at lower levels, since more sophisticated summaries of dispersion are available for variables at higher measurement levels. The simplicity of the range is an advantage that comes with some limitations. First, the R value does not provide much information. If the R for a test is 15 per cent, without examining the data further, you do not know whether the differences in test scores vary between zero and 15, or between 85 and 100. Second, the range is highly affected by even a single extreme score. Imagine if all but one student writing a test received a score in the 90s, but the one exception achieved a test score of 20 per cent. In this case the range would be a large number (indicating extensive variation), even though almost all the students were tightly clustered as top performers. In short, although R provides a quick summary of dispersion, it is prudent to also examine the frequency distribution of the variable from which it is derived.

## Index of Qualitative Variation

A measure of dispersion for qualitative variables that overcomes some of the limitations of the Range is the Index of Qualitative Variation. This measure is symbolized by the capital letters IQV.

The section in step 2 will show you the steps for calculating the IQV. The goal of this section is to provide you with an understanding of this dispersion measure. Understanding the

IQV requires an appreciation of how qualitative differences are identified and counted, and how these counts are used in computation. We will consider these points in turn.

### 1. Recognizing Qualitative Differences

Take a simple, nominal variable like *Sex*, whose values are *male* and *female*. Imagine measuring this variable in two grade 1 classrooms. One classroom is a conventional one in a suburban public school. The other classroom is at a private, all-girls academy. Here are illustrative frequency distributions of the *Sex* variable in these two grade 1 classrooms

Remember that the IQV is a measure of dispersion, so what we are interested in is how much variation there is in the *Sex* variable in each of these classrooms. Examining the frequency table (Table 6.1), you can clearly see that the public school has more variation in this variable than the all-girls private school.

**TABLE 6.1**  Sex Distribution in Two Grade 1 Classrooms

| Sex | f Public School | f Private All-Girls School |
| --- | --- | --- |
| Male | 15 | 0 |
| Female | 13 | 28 |
| Total | 28 | 28 |

The all-girls school contains no variation on this variable. The grade 1 classroom in this school is **homogeneous** with respect to sex. Comparatively, the public school contains considerable variation on the *Sex* variable. Its distribution is quite **heterogeneous** since, among the 28 students, there are a lot of sex differences.

The first point in understanding the IQV measure of dispersion is to appreciate that this measure *captures how heterogeneous the actual scores are on a qualitative variable*. In other words, the IQV will be lower as the scores become clustered around a single value of the variable (as they are in the all-girls school). Alternatively, the IQV will be larger as the scores are more spread out across whatever values the variable has (as they are in the public-school case).

With this appreciation in mind, the next question centres on how heterogeneity in scores of a qualitative variable are counted.

### 2. Counting Qualitative Differences

Imagine a classroom that includes only three students: Dexter, Mary, Sally. How can we proceed to count the amount of sex heterogeneity in this classroom? In this simple case, we can see that Dexter's sex is different from Mary's (one difference), and from Sally's (another difference). And Mary's sex is no different than Sally's. So, in this case, there are two sex differences (Dexter–Mary; Dexter–Sally). If we set up this example in a standard frequency distribution form, we can create a general rule.

The general rule for counting the number of actual differences within a qualitative variable is: *Multiply the frequency of each value by every other frequency and (if necessary) sum the products*. Applying this rule to Table 6.2, we find that the actual differences are $1 \times 2 = 2$.

**TABLE 6.2** Classroom Sex Distribution

| Sex | f |
| --- | --- |
| Male | 1 |
| Female | 2 |
| Total | 3 |

Here is an example of this rule applied to a qualitative variable with more values and more cases.

**TABLE 6.3** Religious Distribution

| Religion | f |
| --- | --- |
| Christian | 6 |
| Jew | 2 |
| Muslim | 3 |
| Buddhist | 1 |
| Total | 12 |

Each Christian is different from each Jew ($6 \times 2 = 12$ differences).

Each Christian is different from each Muslim ($6 \times 3 = 18$ differences).

Each Christian is different from each Buddhist ($6 \times 1 = 6$ differences).

Each Jew is different from each Muslim ($2 \times 3 = 6$ differences).

Each Jew is different from each Buddhist ($2 \times 1 = 2$ differences).

Each Muslim is different from each Buddhist ($3 \times 1 = 3$ differences).

To this point we have multiplied the frequency of each value of religion by every other frequency; now the rule states we should sum the products. In other words, we total these numbers: $12 + 18 + 6 + 6 + 2 + 3 = 47$ differences. In this example, there are 47 differences in religion.

## 3. Standardizing the Count

You have seen that the IQV measures the amount of dispersion (heterogeneity) in the cases of a variable. In addition, you know how to count the number of differences. One final idea is necessary for understanding the IQV. The relevance of this idea is evident in the following simple example.

One classroom has 20 students, composed of 10 boys and 10 girls. Another classroom has 30 students, composed of 15 boys and 15 girls. Are these two classrooms different in terms of diversity in the *Sex* variable? If you follow the rule, you calculate that the first classroom contains 100 *Sex* differences ($10 \times 10$), while the second one contains 225 *Sex* differences ($15 \times 15$). It looks as though the second classroom has a lot more diversity (225 compared to 100 differences), but is this true? Both classrooms are composed of equal proportions of boys and girls (50 per cent each), so, in fact, they are equally diverse in terms of sex. If you were a student in either classroom you would experience the same level of diversity on this variable. Half of the students around you would be of a different sex.

The differences in the computed counts arise from the unequal sample sizes. The first class contains 20 students, while the second one has 30. For the IQV to be a meaningful measure of diversity for comparing samples of different sizes, the counts must be standardized. The IQV does this by comparing the actual differences to the maximum possible differences.

The term *maximum possible differences* refers to a hypothetical situation in which, for a given variable and sample size, heterogeneity is greatest. This situation occurs *when the sample size is* equally distributed *across the categories of the variable*. Refer to the all-girls school example in Table 6.1. In this example, there is no sex diversity, since all 28 students are girls. In this homogeneous situation, there are *minimum* possible differences. If there was one boy in this classroom (and 27 girls) it would be slightly more heterogeneous; if there were 2 boys, slightly more so, and so on. As noted, the maximum possible *Sex* differences occur when the sample size (28) is equally distributed across the categories (2) of the variable (*Sex*). In short, for a sample of 28 students *Sex* differences are maximized when there are equal numbers (14) of boys and girls.

To compute the maximum possible differences for any qualitative table here is how to proceed. Take the sample size (*N*) and distribute it equally across the categories of the variable. (It's okay if each category has some fraction of cases, such as 2.4 people.) Then follow the earlier rule:

Multiply the frequency of each value by every other frequency and (if necessary) sum the products.

So, for the religious differences example in Table 6.3, here is how the maximum possible differences value is obtained. First take the sample size (*N* = 12) and equally distribute it across the 4 categories of the variable *Religion*. This results in each religion having 3 cases (12 ÷ 4), as in Table 6.4.

Table 6.4 shows that, for a sample of 12 individuals and 4 values of the variable *Religion*, religious diversity occurs when there are 3 persons of each religion. To compute how many dif-

**TABLE 6.4** Religious Distribution for Maximum Possible Differences

| Religion | f |
|---|---|
| Christian | 3 |
| Jew | 3 |
| Muslim | 3 |
| Buddhist | 3 |
| Total | 12 |

ferences there are in Table 6.4, follow the rule of multiplying each frequency by every other and summing the products. This results in 54 differences. In other words, for a sample of 12 persons with 4 possible religious categories, you could not have more than 54 religious differences.

### 4. Compute IQV

The index of qualitative variation (IQV) is a measure of dispersion (diversity) for qualitative variables. The components of its calculation include the number of actual differences in a sample and the maximum possible differences for a sample. You know how to determine each of these components, so the final step involves merging them to derive the IQV statistic.

The IQV is calculated by dividing the actual differences by the maximum possible differences.

$$IQV = \frac{\text{Actual differences}}{\text{Maximum possible differences}}$$

For the example using the variable *Religion*, the actual differences (using Table 6.3) was 47 and the maximum possible differences (using Table 6.3) was 54. Therefore, the IQV for the variable is the ratio of actual to maximum differences, or 47 ÷ 54 = 0.87.

These are the basic ideas behind the IQV, which is used for measuring dispersion in qualitative variables (nominal and ordinal). Later sections will provide you with specific calculation steps and opportunities for practice. For now, we turn our attention to two measures of dispersion for variables measured at higher levels.

## Variance

The IQV measure of dispersion is appropriate for qualitative (nominal and ordinal) variables. Quantitative variables (interval and ratio) because they are at higher levels of measurement, contain more information. Therefore, statistics measuring dispersion for quantitative variables can be more sophisticated. The variance is one of these more sophisticated measures.

Just looking at the name of the statistic (*variance*) clarifies that it is a measure of dispersion (i.e. how spread out scores in a single distribution are). It is helpful to think of variance as a measure of *deviance*, or "differentness." The more spread out the scores are on a variable, the more differences there are. You are familiar with this basic idea from your understanding of the IQV, which counted dispersion in terms of differences.

If someone asked you how deviant your level of alcohol consumption is, before you could answer you would have to ask: Compared to what? Compared to elementary school children? Other university students? Catholic nuns? Severe alcoholics? The point is that any measurement of deviance occurs with respect to some reference point. Since the variance statistic measures dispersion for variables at interval and ratio levels, it uses a clear point of reference for determining deviance—which is its measure of central tendency.

In Chapter 5 you learned about statistics that measure central tendency or typicality. A typical case is one that is common. For interval and ratio levels, the mean ($\overline{X}$) was used to measure central tendency. The mean is a form of average, so it can be used as a reference point to measure deviance. If you knew that the mean ($\overline{X}$) alcohol consumption of Canadian university students was 5 drinks per week, you could answer how deviant your drinking pattern is. If you typically consume 15 drinks per week, your deviance is 10 drinks (i.e. your score minus the mean).

In short, for interval and ratio variables, a key idea for measuring dispersion is to calculate how different any score is from the mean. This basic idea can be expressed arithmetically as $x - \overline{X}$, where $x$ is any person's score on a variable (e.g. your 15 drinks per week) and $\overline{X}$ is the mean for the variable (e.g. typical students' 5 drinks per week).

This $x - \overline{X}$ arithmetic determines the deviance of any particular individual on a specific variable. If you want to determine the total amount of deviance in a sample then you would repeat this operation for each case and add up all the differences. The arithmetic for determining this total deviance is expressed as $\Sigma(x - \overline{X})$. The only difference between this expression and the last one is the summation sign ($\Sigma$), which simply means "add up the scores." So, if your alcohol consumption deviation is 10 drinks and those of the other students in the sample are 0, 1, 2, 3, 4, 5, 6, 7, 8, 9 drinks, then if you perform the arithmetic $[\Sigma(x - \overline{X})]$ you can determine the total amount of drinking deviance in this sample.

Now, one further idea. If you take the total amount of deviance in a sample $[\Sigma(x - \overline{X})]$ and divide this total by the number of cases, you will get the mean (average) amount of deviance. In arithmetic terms, this now starts to look fancy:

$$\frac{\Sigma(x - \overline{X})}{N}$$

But it is really just the formula for the mean which you have known since elementary school and reviewed in the last chapter. This time it is just a special type of mean. Instead of being the mean of a set of scores on a variable, it is the mean amount of deviance (dispersion) among a set of scores on a variable.

Here is the set of 11 student scores on alcohol consumption noted earlier: 0, 1, 2, 3, 4, 5, 6, 7, 8, 9, 10. And here is the how the average deviance formula is applied to this set of scores:

| Alcohol Consumption Scores (x) | Deviation $(x - \overline{X})$, where $\overline{X} = 5$ |
|:---:|:---:|
| 0 | $(0 - 5) = -5$ |
| 1 | $(1 - 5) = -4$ |
| 2 | $(2 - 5) = -3$ |
| 3 | $(3 - 5) = -2$ |
| 4 | $(4 - 5) = -1$ |
| 5 | $(5 - 5) = 0$ |
| 6 | $(6 - 5) = 1$ |
| 7 | $(7 - 5) = 2$ |
| 8 | $(8 - 5) = 3$ |
| 9 | $(9 - 5) = 4$ |
| 10 | $(10 - 5) = 5$ |
| N = 11 | |
| $\Sigma x = 55$ | $\Sigma(x - \overline{X}) = 0$ |

$$\Sigma x \div N = 5 = \overline{X}$$

$$= \frac{\Sigma(x - \overline{X})}{N}$$

$$= \frac{0}{11}$$

$$= 0$$

The arithmetic tells you that there is no deviation among the scores in this data set. But that clearly isn't correct; the scores do vary, from zero to 10 drinks. Something's wrong here.

In fact, there is something wrong, and it concerns the numerator $\Sigma(x - \overline{X})$. The mean $(\overline{X})$ is the place in a distribution of scores where there is exactly as much deviance above this centre point as there is below it. So, whenever you calculate deviation from the mean for a set of scores, it will always be zero! And there is not much use having a measure of dispersion that is always zero, no matter how spread out the scores are.

Computing a mean amount of deviation is so intuitively appealing that we don't want to give up on the basic idea, so a little fix is added to overcome this "always zero" problem. This fix involves squaring the differences before adding them up. The squaring operation involves multiplying the number by itself and has the virtue of always creating positive results.

The square of both negative and positive numbers is always positive (e.g. $-4 \times -4 = +16$, and $4 \times 4 = +16$). The squaring operation is designated by a superscript 2, as in $s^2$.

If we apply this fix to the previous equation we obtain the measure of dispersion called the variance, which (for samples) is designated by $s^2$. So here is the variance equation:

$$s^2 = \frac{\Sigma(x - \overline{X})^2}{N}$$

The **variance** is a measure of dispersion applicable to variables at the interval and ratio levels of measurement. Basically, it is the mean amount of deviation from the mean (with the added twist that the deviations are squared).

Here is the computation of the variance for our earlier sample of 11 undergraduates' drinking habits:

| Alcohol Consumption Scores ($x$) | Deviation $x - \overline{X}$ | Squared Deviation $(x - \overline{X})^2$ |
|---|---|---|
| 0 | $(0 - 5) = -5$ | 25 |
| 1 | $(1 - 5) = -4$ | 16 |
| 2 | $(2 - 5) = -3$ | 9 |
| 3 | $(3 - 5) = -2$ | 4 |
| 4 | $(4 - 5) = -1$ | 1 |
| 5 | $(5 - 5) = 0$ | 0 |
| 6 | $(6 - 5) = 1$ | 1 |
| 7 | $(7 - 5) = 2$ | 4 |
| 8 | $(8 - 5) = 3$ | 9 |
| 9 | $(9 - 5) = 4$ | 16 |
| 10 | $(10 - 5) = 5$ | 25 |

N = 11

$\Sigma x = 55$ $\Sigma(x - \overline{X})^2 = 110$

$\Sigma x \div N = 5 = \overline{X}$

$$s^2 = \frac{\Sigma(x - \overline{X})^2}{N}$$
$$= \frac{110}{11}$$
$$= 10$$

For many students, looking at the equation for variance is daunting. But, as noted previously, complex-looking equations are simple if decomposed into their little parts. In the case of variance, the equation amounts to the two-step process we discussed. First, the amount of deviance of the cases is computed relative to the mean central tendency measure. Then the mean (average) of these differences (squared) is computed.

## Standard Deviation

The standard deviation is the most commonly reported measure of dispersion. Like the variance, the standard deviation is computed on variables at the interval and ratio levels of measurement. The following definition of the standard deviation makes it sound very sophisticated: The **standard deviation** is the square root of the mean of the squared differences from the mean. Whew!

As promised, however, what appears complex can be understood in steps. The central clue to understanding the standard deviation comes from examining its symbol. For samples, the symbol for the standard deviation is a lower-case $s$. If you look again at the symbol for the variance, you will see that it is symbolized as $s^2$. It is clear that these two measures of dispersion are related. In fact, the standard deviation ($s$) is computed by taking the square root ($\sqrt{}$) of the variance ($s^2$): $s = \sqrt{s^2}$.

In short, to obtain the standard deviation, we simply take the square root of the number computed through the variance formula. Expressed arithmetically,

$$s = \sqrt{\frac{\Sigma(x - \overline{X})^2}{N}}$$

For the 11-student sample of alcohol consumption scores, the standard deviation is 3.16, since this is the square root of 10.

You now have a basic understanding of four common measures of dispersion. These statistics describe and summarize an important feature of univariate distributions: namely, how scattered the scores are across the values of the variable. As always, the appropriateness of different dispersion measures depends on the level of measurement of each variable. In your toolkit you now have measures of dispersion for both qualitative (nominal and ordinal) and quantitative (interval and ratio) variables. With this basic understanding in place, the following sections will specify the steps for calculating each of these dispersion measures, detail how to generate these statistics using computing software, and give you opportunities to master these measures through practice. The final part of the chapter will show you how to create written interpretations of each of these measures of dispersion.

## STEP 2 Learning the Calculations

The previous section provided you with a basic understanding of four statistics measuring dispersion. This section shows you the steps in computing each of these statistics.

# Calculating the Range

1. Rank-order the scores on the variable in the data set from the lowest to highest values.
2. Subtract the value of the lowest score from the value of the highest score. This difference is the range.

Here is an example of how these steps are applied to a sample of six students' IQ scores, which include the values 108, 100, 130, 118, 105, 111.

1. Rank ordering: 100, 105, 108, 111, 118, 130
2. Difference: $130 - 100 = 30$. The range is 30.

# Calculating the Index of Qualitative Variation

1. Calculate the actual differences by multiplying each attribute frequency by every other attribute frequency, and then summing the products.
2. To calculate the maximum possible differences divide the sample size by the number of categories of the variable. Use this number as the frequency for each category. Then follow step 1.
3. Calculate the Index of Qualitative Variation (IQV) by dividing the result of step 1 by the result of step 2.

Imagine three grade 1 classrooms. Classroom A contains 32 girls and no boys. Classroom B contains 25 girls and 7 boys. Classroom C contains 17 girls and 15 boys. It should be obvious that, with respect to the variable *Sex*, these different classrooms have different levels of dispersion (even though, incidentally, they all have the same mode [i.e. girls] as their measure of central tendency). Classroom A contains no dispersion with respect to *Sex*; it is a *homogeneous* classroom (i.e. all girls). Classroom B has somewhat more dispersion since it contains both girls and some boys. Of the three classrooms, Classroom C is the most *heterogeneous*, since it contains extensive diversity on the variable under consideration.

The following example applies the steps to compute the IQV for each of the three grade 1 classrooms mentioned.

Classroom A:

1. Actual variation: $32 \times 0 = 0$
2. Max. possible variation $= 16 \times 16 = 256$
3. IQV $= 0 \div 256 = 0$

Classroom B:

1. Actual variation: $25 \times 7 = 175$
2. Max. possible variation $= 16 \times 16 = 256$
3. IQV $= 175 \div 256 = 0.684$

Classroom C:

1. Actual variation: $17 \times 15 = 255$
2. Max. possible variation $= 16 \times 16 = 256$
3. IQV $= 255 \div 256 = 0.996$

# Calculating the Variance

1.  Calculate the mean ($\overline{X}$) by adding up all the scores and dividing this total by the number of students.
2.  Calculate the deviation using the formula $(x - \overline{X})$ for each student.
3.  Calculate the squared deviation for each student by squaring the step 2 result for each student.
4.  Sum the squared deviations (from step 3) to complete the numerator for the equation $\Sigma(x - \overline{X})^2$.
5.  Divide the numerator (step 4) by the denominator, which is the number of cases in the sample.

Let's say that we are teaching an Introduction to Sociology class with five students. We gave them a quiz and the students received the following scores:

Student 1    78%

Student 2    80%

Student 3    82%

Student 4    82%

Student 5    88%

The following computations apply the steps for calculating the variance among this set of students.

1.  Add the scores ($78 + 80 + 82 + 82 + 88 = 410$). Divide 410 by 5 (the number of students ($410 \div 5$) = 82.

2.  Calculate each student's deviation.

$$
\begin{array}{ll}
\text{Student 1} & 78 - 82 = -4 \\
\text{Student 2} & 80 - 82 = -2 \\
\text{Student 3} & 82 - 82 = 0 \\
\text{Student 4} & 82 - 82 = 0 \\
\text{Student 5} & 88 - 82 = 6
\end{array}
$$

3.  Calculate each student's squared deviation.

$$
\begin{array}{ll}
\text{Student 1} & (-4) \times (-4) = 16 \\
\text{Student 2} & (-2) \times (-2) = 4 \\
\text{Student 3} & 0 \times 0 = 0 \\
\text{Student 4} & 0 \times 0 = 0 \\
\text{Student 5} & 6 \times 6 = 36
\end{array}
$$

4.  Compute the numerator.

$$16 + 4 + 0 + 0 + 36 = 56$$

5.   Complete the calculation of the variance.

$$56 \div 5 = 11.2$$

# Calculating the Standard Deviation

The standard deviation ($s$) is simply the square root of the variance ($s^2$). Once you have calculated the variance, you are one step away from computing the standard deviation.

1.   Follow the steps to compute the variance and then take the square root of this value. The square root sign appears as $\sqrt{\phantom{x}}$ on your calculator. Simply type in the computed value of the variance and then press the square root sign ($\sqrt{\phantom{x}}$).

The previous example computed the variance of five students' scores as 11.2. To compute the value of the standard deviation for this sample:

2.   On your calculator, type in the value of the variance (11.2) and press the square root sign ($\sqrt{\phantom{x}}$). You will obtain the number 3.35, which is the standard deviation.

## STEP 3 Using Computer Software

The procedure to use IBM® SPSS® statistics software ("SPSS") to generate scores pertaining to measures of dispersion is the same as measures of central tendency. Below are the instructions.

# Procedure

First you should run a frequency distribution to ensure that all applicable values are set to *missing*. See earlier chapters' instructions on how to set desired values to *missing*.

Once you have confirmed that the proper values, if any, are set to *missing*, you can run measures of dispersion by following these steps:

1.   Select Analyze from the menu bar.
2.   Select Descriptives → Frequencies.
     a.   In the left-hand box, scroll to and highlight the variable (or variables) you want to analyze.
     b.   Press the arrow button to move the variable to the right-hand side.
     c.   Go to the Statistics box.
     d.   In this menu, you can select your preferred measure of dispersion (i.e. range, variance, standard deviation—shown in SPSS as *Std. Deviation*), which can be found in the lower left side of the Statistics box. Unfortunately, SPSS does not allow for analysis of the index of qualitative variation (IQV).
3.   Select Continue and OK to execute the Frequency command.

In the Output/Viewer window, you will find the results of your analysis, an example of which is illustrated in Figure 6.1.

In Figure 6.1, "Q17, the Rosenberg self-esteem scale" had 79 respondents set to *missing*, because not all respondents answered the question. In terms of measures of dispersion, the range is equal to 59, the variance is equal to 111.414, and the standard deviation is equal to 10.555.

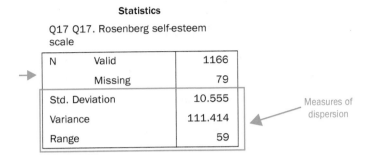

**Statistics**

Q17 Q17. Rosenberg self-esteem scale

| N | Valid | 1166 |
|---|---|---|
|  | Missing | 79 |
| Std. Deviation |  | 10.555 |
| Variance |  | 111.414 |
| Range |  | 59 |

Measures of dispersion

**FIGURE 6.1**  SPSS Output of Dispersion Measures

## STEP 4 Practice

You have been introduced to the meaning of the four measures of dispersion as well as the procedures for calculating each of these statistics, through either hand calculations (if the sample size is small) or computer software (for large sample sizes). To solidify your understanding you need to practise applying the statistical and software procedures.

This section provides you opportunities to practise what you have learned about measures of dispersion. The first set of questions uses hand calculations. The second set uses the SPSS procedures. For each set of questions:

1.  Follow the procedural steps and complete the appropriate calculation (Set 1) or software application (Set 2).
2.  Check your answers, using the Answer Key in the back of the text.
3.  If your answer is incorrect, consult the Solutions section on the book's website. The Solutions provide a complete step-by-step analysis of how the answers are derived.

After you have completed the next section (Interpreting the Results), return to your calculations or output and *provide complete, written interpretations of each of the statistics you have generated.*

# Set 1: Hand-Calculation Practice Questions

1.  When the Johnston and Scribner families get together for the holidays or special occasions, the political discussions are intense and animated. Which of the two families has greater diversity of political affiliations? Calculate and compare the IQV for each family using the data in Table 6.5.

**TABLE 6.5** Family Frequency Distributions

| Johnston Family | |
|---|---|
| Political Affiliation | f |
| NDP | 3 |
| PC | 5 |
| Liberal | 3 |
| Green | 1 |
| Total | 12 |

| Scribner Family | |
|---|---|
| Political Affiliation | f |
| NDP | 8 |
| PC | 0 |
| Liberal | 3 |
| Green | 5 |
| Total | 16 |

2. A team of biologists sampled three neighbouring lakes to assess the diversity of fish species. Calculate the IQV of each lake using the data in Table 6.6. According to this small sample, which lake has the greatest diversity of species and which has the least?

**TABLE 6.6** Fish Diversity in Three Lakes

| Species | Grey Lake F | Lowe Lake f | Crater Lake f |
|---|---|---|---|
| Trout | 2 | 2 | 4 |
| Pike | 5 | 4 | 4 |
| Pickerel | 7 | 8 | 6 |
| Bass | 4 | 3 | 2 |
| Perch | 2 | 3 | 4 |
| Total | 20 | 20 | 20 |

3. Calculate the variance and standard deviation for the set of IQ scores in Table 6.7.

**TABLE 6.7** IQ Scores

| |
|---|
| 119 |
| 98 |
| 91 |
| 109 |
| 105 |
| 127 |
| 89 |
| 113 |
| 98 |
| 103 |
| 98 |
| 109 |

4. Which of the schools listed in Table 6.8 has greater variation in class sizes?

**TABLE 6.8** Class Sizes in Schools

|  | Number of Students | |
|---|---|---|
| | Happy Valley | Colonel Park |
| Grade | | |
| Kindergarten | 22 | 26 |
| 1 | 25 | 24 |
| 2 | 20 | 19 |
| 3 | 27 | 19 |
| 4 | 24 | 25 |
| 5 | 23 | 20 |
| 6 | 19 | 18 |

5. Sanjit is shopping for a new mortgage. Table 6.9 lists the current five-year interest rates for the top 10 banks and credit unions in his city. Calculate the variance and standard deviation of this distribution.

**TABLE 6.9** Interest Rates

| |
|---|
| 2.9 |
| 3.3 |
| 4.0 |
| 3.1 |
| 4.05 |
| 3.9 |
| 3.5 |
| 4.1 |
| 3.3 |
| 3.75 |

6. Shoppers in a pet store were asked to fill in a brief Internet survey at home for a chance to win a $500 gift card. A few of the items on the survey were:
   - What kind of pet do you own?
   - What is your income category? (1 = 0–29,999; 2 = 30,000–59,999; 3 = 60,000–89,900; 4 = 90,000 and above)
   - What is your age?

   The complete data are shown in Table 6.10.

**TABLE 6.10** Pet Ownership by Income and Age

| Respondent Number | Pet | Income Category | Age |
|---|---|---|---|
| 1 | dog | 3 | 36 |
| 2 | fish | 1 | 19 |
| 3 | dog | 2 | 64 |
| 4 | cat | 2 | 32 |
| 5 | ferret | 2 | 25 |
| 6 | cat | 3 | 48 |
| 7 | cat | 4 | 42 |
| 8 | fish | 2 | 31 |
| 9 | dog | 3 | 39 |
| 10 | dog | 2 | 55 |
| 11 | dog | 1 | 21 |
| 12 | shrimp | 2 | 39 |
| 13 | budgie | 1 | 22 |
| 14 | dog | 1 | 18 |
| 15 | cat | 3 | 45 |

a. Calculate the diversity of respondents' answers to what kind of pet they own.
b. Calculate the diversity of respondents' answers about their income category.
c. How diverse are the ages of the respondents?
d. What is the age diversity of those respondents who belong to the second or higher income category?

7. West Town and East Town each decided to hold their own half-marathon and to later compare the results. Table 6.11 shows two randomly selected samples of 10 runners from each half-marathon, indicating how many minutes it took them to complete the race. We now know that runners in the West Town half-marathon were, on average, as fast ($\overline{X}$ = 163.2 min) as the runners who attended the East Town half-marathon ($\overline{X}$ = 163.2 min). But is that the whole story?

**TABLE 6.11** Sample Completion Times in West Town and East Town Half-Marathons

| | West Town Half-Marathon | | East Town Half-Marathon |
|---|---|---|---|
| 1 | 134 | 1 | 85 |
| 2 | 122 | 2 | 195 |
| 3 | 188 | 3 | 220 |
| 4 | 95 | 4 | 180 |
| 5 | 245 | 5 | 69 |
| 6 | 160 | 6 | 155 |
| 7 | 182 | 7 | 248 |
| 8 | 193 | 8 | 77 |
| 9 | 150 | 9 | 312 |
| 10 | 163 | 10 | 91 |

a.  Compute variance and standard deviation for each of the two races.
b.  What does the knowledge of dispersion further tell us about the two races?

# Set 2: SPSS Practice Questions

Apply the appropriate SPSS procedures to the data set to solve the following problems. Remember to set missing values before you begin your calculations!

1.  Use the Frequencies procedure to obtain the variance and standard deviation for the *Age* variable.
2.  Use the Frequencies procedure to obtain the variance and standard deviation for the *Depression* variable (Q4).
3.  Which is more heterogeneous in the Student Health and Well-Being Survey sample: *Strength of religious faith* (Q18) or *Satisfaction with life* (Q22)?
4.  Use the Frequencies procedure to obtain the mean, variance, and standard deviation for the *Psychological well-being* variable (Q16F).
5.  Use the Frequencies procedure to obtain the mean, variance, and standard deviation for the *Feelings of negative affect* variable (Q3B).
6.  Which is more homogenous in the Health and Well-being sample: *Strength of religious faith* (Q18) or *Feelings of positive affect* (Q3A)?
7.  What is the standard deviation for *Coping with stress by avoidant behavior* (Q14C)?
8.  What are the mean, variance, and standard deviation for *Coping with stress by seeking social support* (Q14A)?
9.  Which of the *Psychological well-being* variables (Q16A–F) has the highest mean? The lowest mean? The greatest standard deviation? The smallest?

## STEP 5 Interpreting the Results

Remember that two characteristics describe any set of scores on a single variable. These characteristic descriptions are captured, on one hand, by central tendency statistics and, on the other, by measures of dispersion. This chapter introduced four dispersion measures, including the range ($R$), index of qualitative variation (IQV), variance ($s^2$), and standard deviation ($s$). For each dispersion measure a conceptual understanding, calculation steps, software applications, and practice questions were provided. At this point you should have a solid understanding of how to generate dispersion statistics.

Measures of dispersion, like all statistical results, are numerical facts and, contrary to popular opinion, the facts do not speak for themselves. To understand what any statistic means requires interpretation. Interpretation, in turn, requires stating what the statistical results say about the problem under consideration. The following explains how to interpret each of the four dispersion measures.

# Range

The range ($R$) provides a quick and general notion of how dispersed a set of scores are. The range reports *how much of the possible dispersion in scores actually occurs in a sample for a given variable.* So, if a variable includes 7 values, and the range is 4, we learn that more than half of the possible scores are included in the actual sample. That's all we can tell from the number that is the range. Without additional information, we don't know which of the 7 values are covered, or how the actual scores are distributed across these values.

A standard form of the statement for interpreting the range is:

While scores on the variable $X$ [name of variable] could vary from $a$ to $b$ [theoretical range of variable], in this sample the difference between the highest and lowest scores was $Y$ [calculated value of range].

# Index of Qualitative Variation

The principle of heterogeneity provides the basis for interpreting the results of the IQV statistic. Remember that heterogeneity refers to how much difference there is within the scores of a variable. When all cases have the same score on a variable, the variable is homogeneous. As the cases become more spread out across the variable's values, heterogeneity increases.

The IQV statistic can vary from a low of 0.0 to a high of 1.0. The results tell you *the proportion (or percentage if you multiply the IQV by 100) of the maximum possible variation that is present in the sample.* Take the variable *Religion*, for example. If Classroom A has an IQV of zero, it tells you that this class has none of the possible variations in religion it could have (i.e. it is homogeneous). For Classroom B, an IQV of 0.684 tells you that this classroom contains 68.4 per cent of all the possible variations in religion it could have. If a classroom of 15 students had 4 Catholics, 5 Protestants, and 6 Jews, its IQV for religion would be 0.99. This means that this class contains 99 per cent of the religious variations it could have.

A standard interpretation of the index of qualitative variation takes the following form:

In this sample the variable $X$ [name of the variable] displays $Y$ [IQV score multiplied by 100] per cent of its possible variation.

# Variance and Standard Deviation

Essentially, the variance ($s^2$) is a kind of average (mean) measure of deviance within a sample. Technically, if you review the formula for computing variance and the presentation in step 1, you will see that the mean represents "the mean of the squared differences from the mean." It should be evident, then, that interpreting the result of the calculation of the variance statistic is not so straightforward.

Now, the standard deviation ($s$) is a measure of dispersion computed by taking the square root of the variance ($s = \sqrt{s^2}$). Technically, then, the standard deviation is "the square root of the mean of the squared deviations from the mean." Try to imagine what that is!

The point here is that both the variance and the standard deviation are dispersion measures that, by themselves, have little intuitive appeal. The goal of interpretation is to be able to state what a statistical result means in simple terms. If, among a sample of students, the variable *Income* has a standard deviation value of 1,245, it is not very helpful to state that: "Among these students the square root of the mean deviations from the income mean is $1,245"!

For present purposes, instead of giving a strict interpretation of either the variance or the standard deviation, you need simply appreciate that *the larger the value of the statistical result, the more dispersed the scores are on the variable.* This vague interpretation will become much sharper in Chapter 8, which introduces you to the importance of the normal distribution to social statistics. At that point you will see that the standard deviation is a powerful and useful tool.

# Chapter Summary

The last chapter showed you how to summarize what the typical or common scores on a variable are. These are measures of central tendency. This chapter addressed the other summary feature of single variables—how distributed or scattered the scores on a variable are. You learned the basic ideas, computation procedures, and software applications of four measures of dispersion—range, index of qualitative variation, variance, and standard deviation.

Having completed this chapter you should appreciate that:

- Dispersion is a central feature of every variable.
- Different measures of dispersion are applicable to variables of different levels of measurement.
- The range is the quickest summary of dispersion, but is very limited in the information it provides.
- The index of qualitative variation centres on comparing the actual variation in scores to the maximum possible variation. It provides a useful summary of dispersion for qualitative (nominal and ordinal) variables.
- The dispersion in quantitative (interval and ratio) variables is usefully summarized through variance and standard deviation statistics.
- Both variance and standard deviation are rooted in measuring the average (mean) differences in scores from the mean.
- Larger standard deviation and variance scores indicate more dispersion in the data.
- Although variance and standard deviation are not easily interpreted, they play a key role in later statistical tools.

Data on single variables (univariate distributions) can be summarized with respect to two characteristics, central tendency and dispersion. This chapter has introduced you to the four statistics that measure dispersion. You are clearly moving along the path toward increased statistical competency. You are picking up tools along the way and your toolkit is becoming more sophisticated. More tools at your disposal means more choices. Remember that selecting an appropriate tool is a function of the combination of answers to the three statistical selection questions.

The next chapter adds another useful tool to your set. It shows you how to portray the features of univariate distributions in graphic form.

# 7

# Charts and Graphs

## Overview

Recent chapters have focused on various tools for describing single-variable (univariate) distributions. The appropriate selection and application of these tools depends on correctly identifying a variable's level of measurement. Remember, this is the issue addressed by the third question on the statistical selection checklist. As always, different statistical tools are tailored for the distinctive features of different measurement levels.

Chapter 4 introduced you to the use of frequency distributions for summarizing univariate information. Chapter 5 discussed various measures of central tendency that summarize where the typical scores in a distribution occur. Chapter 6 identified various measures of dispersion that capture how spread out a variable's scores are around the centre of its distribution. This chapter builds upon the previous chapters and explains how univariate distributions are portrayed and described in graphs.

In this chapter you will learn about:

- the importance of form in describing distributions;
- the three dimensions that characterize form in distributions;
- the importance of charts and graphs for understanding evidence; and
- the type of charts and graphs useful for categorical and continuous variables.

As always, it is helpful to orient yourself to the tools introduced in this chapter by locating their place on the statistical maze. The maze, in turn, is organized on a grid defined by the statistical selection checklist. Here is the checklist for the contents of this chapter:

## Statistical Selection Checklist

1.  Does your problem centre on a descriptive or an inferential issue?

    ☑   Descriptive

    ❑   Inferential

2.  How many variables are being analyzed simultaneously?

    ☑   One

    ❑   Two

    ❑   Three or more

3.  What is the level of measurement of each variable?

☑  Nominal  } for categorical variables: pie and bar charts
☑  Ordinal

☑  Interval  } for continuous variables: histograms and polygons
☑  Ratio

As before, our journey in this chapter begins with a conceptual orientation.

## STEP 1  Understanding the Tools

# Form

The first chapter introduced you to the importance of variables to social research. Our minds are filled with ideas (propositions) that express our understanding of how concepts are connected. Some of our ideas are fantasies which speak to imaginary worlds. Other ideas are realistic and describe how empirical worlds operate. Both fantastic and pragmatic ideas are worthwhile, but it is important not to confuse them. You might dream of jumping off a tall building, flapping your arms, and soaring into flight—but reality will ensure you crash. Research is the means by which we learn to distinguish fantasy from reality.

Researching an idea requires translating its concepts into variables. This translation process is called operationalization. The resulting variables act as indicators of abstract concepts. Since variables can be measured, operationalizing concepts makes ideas testable by translating propositions into hypotheses. This entire process was described in Chapter 1, where Figure 1.4 is explained. At this point it might be helpful to look back and review that material.

Variables are the working material of research. They identify properties of objects that can change. By measuring how much of a particular property an object has, we can distinguish objects from one another. Tall people are different than short people in the amount of the variable *Height* they have.

Variables, in turn, are composed of parts called attributes or values. The attributes of a variable capture what its *possible scores* are. By themselves, variables are lifeless abstractions. They come to life when they are applied to objects through the measurement process. When a set of objects is measured on a variable (e.g. when a class of students takes a test), the result is a data set. A data set captures what the *actual scores* of a variable are within a given group.

The previous two chapters introduced statistics for summarizing the central tendency and dispersion of the actual scores in a data set. As this chapter and the next one will show, these characteristics of a data set can be combined to reveal other features. One of these important features is form.

The **form** of an object refers to its shape or configuration. People have body shapes that are described in terms of various forms, including ectomorphic, mesomorphic, and endomorphic. Ectomorphs, for instance, are thin and have small bone structures, while endomorphs are short and stocky types. Cars have forms that let us distinguish sedans from coupes. Forms are everywhere, including in data sets.

A univariate data set includes the scores of a set of objects on a single variable. In Chapter 4 you learned how these data could be presented and summarized in terms of a frequency distribution. The information in a frequency distribution can be expressed in terms of form. The form of a distribution includes three concepts: modality, skewness, and kurtosis. Here is what each one means.

## Modality

In Chapter 5, you learned that the mode is a measure of central tendency that identifies the most frequently occurring category of a variable. The modality of a distribution refers to how many modes a distribution has. If a distribution only has one value that occurs most frequently, then it is **unimodal**. If it has two modes, then its modality is **bimodal**. If the distribution has three or more modes, then this dimension of its form is called **multimodal**.

In graphic form, modality is identified by the number of peaks in a distribution. Unimodal distributions have one peak; bimodal, two; multimodal, three or more. Figure 7.1 illustrates what unimodal and bimodal distributions look like.

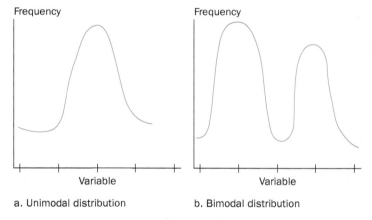

a. Unimodal distribution   b. Bimodal distribution

**FIGURE 7.1** Unimodal and Bimodal Distributions

At this time a technical point is worth noting. In a frequency distribution, the mode is interpreted strictly. Imagine a variable with seven attributes. If, for instance, one category of this variable has 23 cases and another category has 24 cases, then there is only one mode—and it is the name of the category that has 24 cases. In describing the modality of a distribution, the mode is interpreted more loosely. If you look at the example of the bimodal distribution in Figure 7.1, you will see that one peak is taller than the other. The height of the taller peak indicates it, in fact, has more cases. The frequency distribution of this variable, technically, has only one mode. However, since the concept of modality is used to give an approximate idea of the shape of a distribution, this distribution is called bimodal. In short, the modality component of a distribution's form employs a loose understanding of mode. If you examine a frequency distribution and see that more than one category stands out, then include it as part of the modality description.

## Skewness

Skewness is the second component of a distribution's form. The easiest way to understand skewness is in contrast to its opposite. The opposite of skewness is **symmetry**. Symmetry is present when the form of an object is balanced, when the halves of the object are identical. If you fold a piece of rectangular paper precisely in half, then the halves are symmetrical. If you draw an imaginary line down the centre of someone's face, you can see whether their face is symmetrical. The same applies to the distribution of a variable in a data set. If the halves are identical around the centre, then the distribution's form is symmetrical.

Not all objects are symmetrical. If you did a poor job folding a paper in half, the halves may not be identical. A comparison of the eyes, ears, and other components of people's faces shows that most are not symmetrical. Many objects, in fact are asymmetrical. **Skewed** distributions are asymmetric. Figure 7.2 illustrates symmetry and skewness in graphic form.

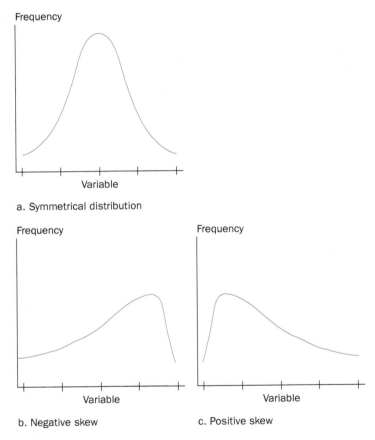

**FIGURE 7.2**  Symmetrical and Skewed Distributions

If you drew a vertical line down from the mode of the first distribution in Figure 7.2, you would produce two identical halves, which qualifies this distribution as symmetrical. If you performed the same procedure on the other two graphs, the halves would not be identical, which makes them skewed.

Notice that skewness refers to whether or not a vertical line can cut the distribution into identical halves. If this is possible, then the distribution is symmetrical; if not, then skewed. With this in mind, look at the bimodal distribution in Figure 7.3. Is it symmetrical or skewed?

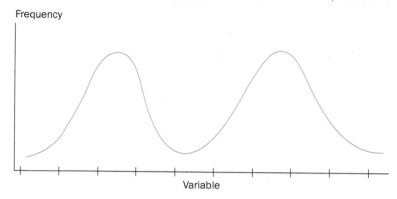

**FIGURE 7.3**  Bimodal Distribution

The distribution is Figure 7.3 is symmetrical. If you drew a vertical line at a point midway between the peaks, the halves of the distribution would be identical.

Skewed distributions are further categorized as either positively or negatively skewed. **Positively skewed** distributions are ones in which the right tail of the distribution is longer than the tail on the left side. These distributions are called positive because higher values of the variable are plotted on the right side, where more of the distribution is found. **Negatively skewed** distributions have the opposite feature—the left tail being longer than the right. The examples in Figure 7.2 provide illustrations of each.

## Kurtosis

The third feature of a distribution's form is its kurtosis. Look at the two distributions in Figure 7.4. They are the same in terms of their modality (unimodal) and skewness (both are symmetrical). But the form (shape) of these distributions is clearly not the same. Their difference is captured by **kurtosis**, which refers to how peaked a distribution is.

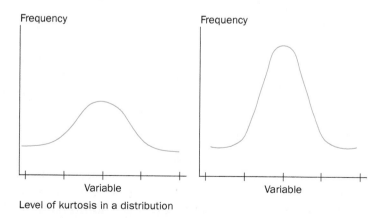

Level of kurtosis in a distribution

**FIGURE 7.4**  Different Kurtosis

As with skewness, there are precise statistics for calculating the amount of kurtosis in a distribution. Conceptually, however, the basic ideas are captured in the following terms. Distributions that are very peaked are called **leptokurtic**; those that are flat are labelled **platykurtic**, and those in between are **mesokurtic**. Examples of each type are illustrated in Figure 7.5.

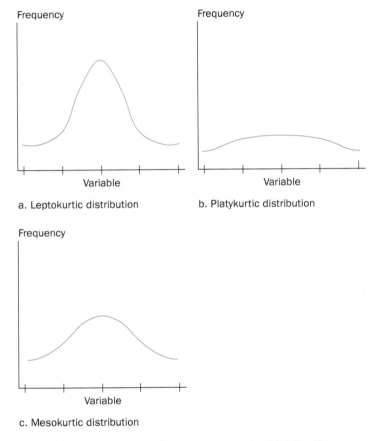

**FIGURE 7.5**  Leptokurtic, Platykurtic, and Mesokurtic Distributions

In summary, the form of any distribution is described by its modality, skewness, and kurtosis. If a researcher reports that the distribution of scores for a particular variable is bimodal, positively skewed, and mesokurtic, you have a ready approximation of the shape of the distribution. You could also draw the form of a distribution that is reportedly unimodal, symmetrical, and platykurtic. (If you have trouble visualizing how a distribution looks based on these descriptions, you should consider reviewing the material again.)

These descriptions of a distribution's form are summaries of the information contained in a frequency distribution (see Chapter 4). In other words, you identify modality, skewness, and kurtosis by examining how the frequency counts of a specific data set are distributed across the values of a variable. Since information in a frequency distribution is the same information used to calculate the summary statistics about central tendency (Chapter 5) and dispersion (Chapter 6), the ideas of central tendency and dispersion are also captured in descriptions of form. For instance, describing a distribution as either unimodal or bimodal tells you something about where its typical scores (centres) are. Likewise, as distributions become more platykurtic, their scores are more widely spread out and therefore more dispersed.

You have seen that the scores on a variable in a data set can be summarized in various ways. Frequency distributions summarize this information in the form of tables. Measures of central tendency and dispersion use calculations that result in summaries of this information in a single number. Summaries can also be provided in the form of pictures, which is what charts and graphs do.

# Charts and Graphs

Charts and graphs are visual displays of how the scores on a variable are distributed. Their popularity is based on the principle that "a picture is worth a thousand words." For most people, pictures are more familiar than tables or statistics. Because of this familiarity, most people find it easier to understand data when it is presented in a visual display.

One basic rule for determining which tools are appropriate data summaries concerns levels of measurement. Different tools are designed for variables at different measurement levels. In this regard, Chapter 4 introduced the distinction between categorical and continuous variables. This distinction essentially classifies the four levels of measurement (nominal, ordinal, interval, ratio) into two groups. The categorical grouping includes nominal and ordinal variables, since these are qualitative in nature. The continuous grouping includes interval and ratio variables, which are quantitative in nature.

The appropriate selection of charts and graphs relies on this categorical–continuous distinction. Specifically, if the variable under consideration is categorical, then the most common visual displays are pie and bar charts. For continuous variables, histograms and polygons are common graphing tools. Here is an introduction to each of these tools for illustrating univariate distributions.

## Pie Charts and Bar Charts

Qualitative (nominal and ordinal) variables are categorical in the sense that the values of the variable are discrete. In other words, each attribute of a categorical variable is distinct from every other. Pie and bar charts take advantage of this property, since they allow the frequency or percentages of a frequency table to be mapped as separate spaces.

Pie charts begin with a circle that is cut into pieces, as an apple pie would be. The pie represents the entire distribution of scores on a variable, each wedge of the pie represents an attribute (category) of the variable, and the size of each wedge is proportional to the frequency of scores in that category as a percentage of all scores in that distribution (sample). Figure 7.6 takes the frequency table for the variable grades and converts it into a pie chart.

Like the one in Figure 7.6, pie charts typically map the percentage distribution column from a frequency table. If you look in articles and reports, you will see that pie charts are common and often use vivid colours in the display. A key to successful pie charts is avoiding a cluttered look. In practice, a "clean" look appears when the pie is cut into six or fewer wedges.

Figure 7.7 exemplifies a second way of visually illustrating information for a categorical variable. The bar chart is organized around two axes. The horizontal axis identifies the attributes of the variable. The vertical axis identifies either the frequency or percentage information from a frequency table. The height of each column (bar) illustrates the frequency or percentage for each attribute of the variable. Notice that the bars are separated from one another, which illustrates the discrete nature of the categorical variable's attributes. Whether the vertical axis

| Grades | f | % |
|--------|-----|-----|
| A | 5 | 10 |
| B | 12 | 24 |
| C | 23 | 46 |
| D | 7 | 14 |
| F | 3 | 6 |
| Total (N) | 50 | |

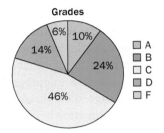

**FIGURE 7.6**   Frequency Table and Pie Chart

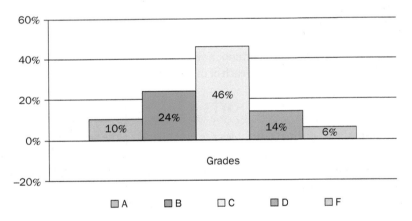

**FIGURE 7.7**   Bar Chart

includes frequency or percentage information, the picture provided in the bar chart displays the relative occurrence of each category in the data set.

Like pie charts, bar charts are readily understood; they have intuitive appeal. For displaying percentage information for qualitative variables with six or fewer attributes, either pie or bar charts will do. However, for displaying relative frequencies, or if the variable contains more than six categories, bar charts are preferable.

## Histograms and Polygons

These are two graphing tools commonly applied to continuous variables. In continuous (interval and ratio) variables, the attributes "run together"; they flow into one another in a steady progression. Applying the techniques of a bar graph to continuous variables results in a **histogram**, an example of which is provided in Figure 7.8.

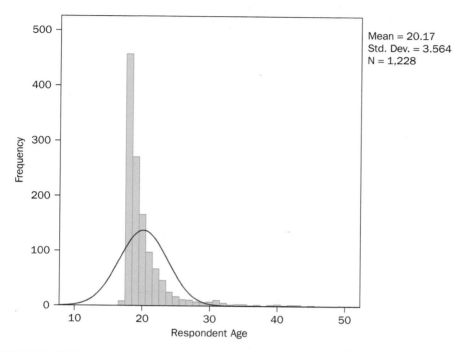

**FIGURE 7.8** Histogram

Notice that, like the bar graph, the histogram includes the categories of the variable on the horizontal axis and frequency or percentage information on the vertical one.[1] The principal difference is in how the bars (columns) are displayed. In bar graphs, the columns are separated; in histograms, the columns are butted against one another. This visual distinction is important since, for histograms, it illustrates the continuous nature of the categories of the variable.

A clearer illustration of the flowing nature of the attributes of a continuous variable is graphed in **frequency polygons**. Frequency polygons are sometimes called **line graphs**, which more clearly describes their presentation. The frequency polygon is an extension of the histogram. Each column in a histogram has a flat top, which illustrates the range of the category. If you identified the midpoint (centre) at the top of each column in a histogram, and connected these midpoints with a line, the result is a frequency polygon. Figure 7.9 provides an illustration.

Histograms and frequency polygons provide similar information in different forms. In large measure, choosing one over the other is largely a matter of aesthetic preference. Frequency polygons are preferred, however, in situations where a trend is being graphed. If, for example, divorce rates were plotted at five-year intervals over the last half-century, then a line graph would be preferable, since it readily displays the flow of the trend over time.

---

[1] Even though histograms can report either frequencies and percentages, the usual display is frequencies (unlike bar and pie charts).

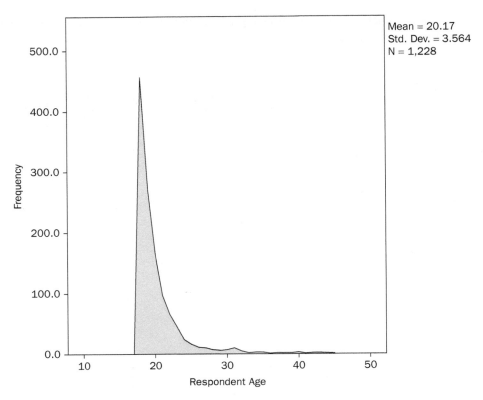

**FIGURE 7.9** Frequency Polygon (Line Graph)

This section introduced you to four tools for displaying frequency table information in a visual form. These "pictures" are intuitively appealing. The next section specifies the steps to follow in constructing and labelling each of these charts and graphs.

# STEP 2 Learning the Calculations

Computer technology has made hand-drawn graphs obsolete. When viewing computer-generated charts or graphs, you need to remember that they do *not* provide new information. Charts and graphs simply *display* the information from frequency distributions in pictorial form. While computers generate charts and graphs, the following are requirements for *presenting* these pictorial displays.

- *The chart or graph must have a clear title.* The title should reflect the content displayed in a clear way. Specifically, the title should report the kind of display (e.g. bar chart), the variable (e.g. ethnicity), the measure reported (e.g. percentages), the sample or population (e.g. Residents of Calgary), and the source of the data (e.g. Canadian Census). For example, a title of a bar chart might read: "Bar Chart of Ethnic Group Percentages in Calgary from 2011 Canadian Census."
- *The source of the data must be identified.* The source needs to clearly reflect where the data presented in the chart or graph came from. For example: "Source: 2011 Census of Canada." The source information must be sufficiently complete so that an interested reader could obtain the evidence and replicate the results.

- *Below the chart or graph, information about the sample size must be included.* It is important for readers to know how many cases are included in the image. Sample size information is typically reported as "$N =$ ____," where the space represents the number of cases.
- *With the exception of pie charts, the horizontal axis must display the variable and its attributes.* Although it is possible for the bars or other information to be displayed horizontally, it is conventional practice to display them vertically.
- *With the exception of pie charts, the vertical axis must identify the information being displayed (e.g. percentages) and its scale (e.g. in 10 per cent increments).* Again, this is conventional practice.

Figure 7.10 provides an example of a properly labelled bar graph.

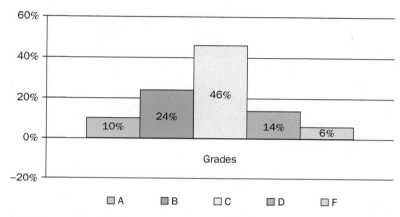

**FIGURE 7.10** Bar Chart of First Test Social Statistics Grades in Percentages

Source: Professor's Grade Book, $N = 67$

## STEP 3 Using Computer Software

As noted, charts and graphs are now almost always constructed using computer software. Here is how you can construct them using SPSS.

# Charts and Histograms

To produce a pie chart, bar chart, or histogram in SPSS, use the following steps.

1. Run a frequency distribution (Analyze → Descriptives → Frequencies), and select the variable(s) you want.
2. Select the Charts menu. In this section, you can select any of the three types of charts listed: bar chart, pie chart, and histogram, depending on the level of measurement (nominal, ordinal, interval, or ratio) of your variable.
3. If you select Histogram, an additional check box becomes highlighted. Check "Show normal curve on histogram" to include a normal curve in your output, in addition to the histogram.

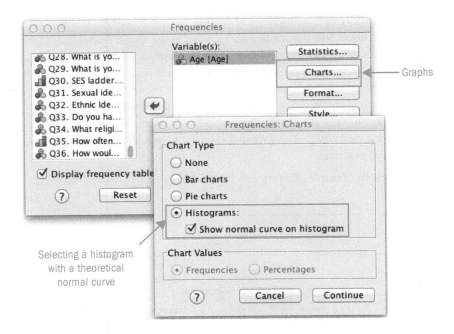

**FIGURE 7.11** SPSS Charting

Source: Reprint Courtesy of International Business Machines Corporation, © International Business Machines Corporation.

Figure 7.11 provides an SPSS image of this procedure.

The normal curve that SPSS generates is a *theoretical normal curve*, which can be used to compare and assess whether or not the shape of your histogram distribution approximates a normal one. As we can see in Figure 7.12, the age of the respondent variable is positively skewed, which we would expect since the sample is made up of primarily first-year university students.

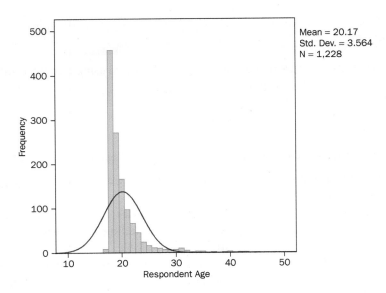

**FIGURE 7.12** Histogram of Age

# Polygons

SPSS generates polygons using the following steps, which are also illustrated in Figure 7.13.

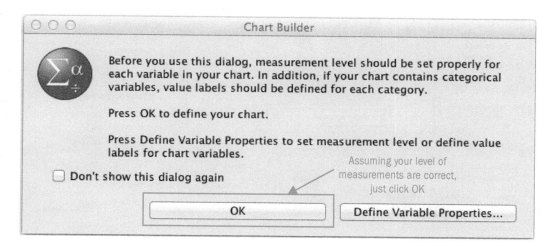

**FIGURE 7.13** SPSS Chart Builder

Source: Reprint Courtesy of International Business Machines Corporation, © International Business Machines Corporation.

1. Under Graphs in SPSS Menu, click Graphs → Chart Builder.
2. If this is the first time using the Chart Builder in SPSS, a warning box like the one in Figure 7.14 will probably open. Recall that polygons, like histograms, are for continuous variables (i.e. interval or ratio). In SPSS, this means that appropriate variables need to be set to the Scale option. To do this, or to double-check your work, go to the Variable View tab in the Data Editor window (refer to Chapter 3 if you are having trouble). Under the Measure column, you will see three options: Nominal, Ordinal, and Scale. All continuous variables should be set to *Scale*.
3. Select Histogram in the Gallery tab at the bottom left-hand side of the dialogue box.
4. Choose the Frequency Polygon graph, which is the third graph to the right. If you place your mouse over the graph, it should say Frequency Polygon. Click over the Frequency Polygon and then drag it with your mouse to the Chart preview area, directly above.
5. Select your variable, which is located in the upper left-hand corner of the Chart Builder window. Drag your desired variable to the X-axis area, which is also located in Chart Preview.
6. Select OK.

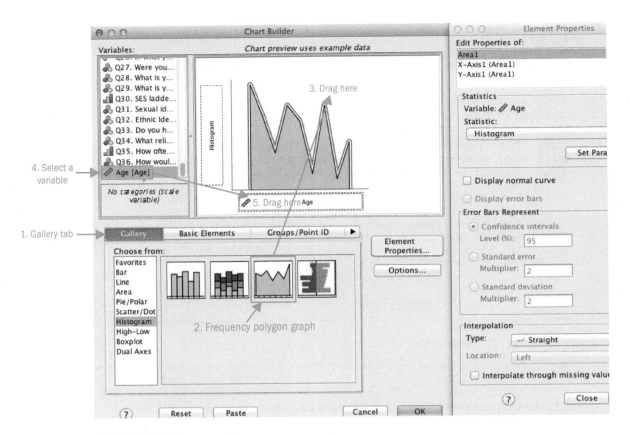

**FIGURE 7.14**  SPSS Polygons

Source: Reprint Courtesy of International Business Machines Corporation, © International Business Machines Corporation.

## STEP 4 Practice

You have been introduced to a variety of charting and graphing techniques. Computing technology has taken over the construction of charts and graphs, so there are no hand calculations to practise for this chapter. Instead, you should apply the appropriate SPSS procedures to the following questions. As always, check your answers, using the Answer Key in the back of the text. If your answer is incorrect, consult the Solutions section on the book's website. The Solutions provide a complete step-by-step analysis of how the answers are derived. As always, where appropriate, make sure you provide clear, written interpretations of your findings.

# SPSS Practice Questions

Use the Frequencies procedure to check the shape of the distributions for the following variables: Q16A, Q14C, Q22, and *Age*. Make sure to request a histogram with the normal curve shown for each variable.

1. Inspect the distributions. Notice that two of the variables approximate the normal distribution quite closely (symmetrical curve with little apparent skew or kurtosis); the other two variables, less so. What do you think accounts for the greater skew and kurtosis (deviation from symmetry or normality) of the latter two variables?

2. Use the Graphs procedure to make a pie chart for nQ1A and Q28.

3. Use the Graphs procedure to make a bar chart for Q35 and Q32.

4. Use the Graphs procedure to make a histogram for Q22 and Q17.

5. Create a pie chart of *How much in the past month have you felt calm* (Q2B). Which category has the highest frequency? Which has the lowest?

6. Create a bar graph of *My emotions seem to have a life of their own* (Q6B). Which category has the highest frequency? Which has the lowest?

7. Create a pie chart for *Satisfaction rating: Nutrition* (Q1C). Which category has the highest frequency?

8. Create a bar graph for *Feelings that people are untrustworthy* (Q20B). Which response has the highest response rate?

9. Create a histogram for *Felt that other people acted as if they thought you were dishonest* (Q23E). Describe the shape of the graph in terms of skew.

10. Create a bar graph of *How much in the past month have you felt happy* (Q2A). Which category has the highest frequency? Which has the lowest?

11. Create a pie chart of *I can usually relax when I want* (Q6E). Which category has the highest frequency? Which has the lowest?

12. Create histograms for both *Coping with stress by planful problem solving* (Q14B) and *Personal growth* (Q16C). Which is more skewed and how?

## STEP 5 Interpreting the Results

Since charts and graphs are simply visual displays of the information found in frequency tables, they do not require any additional interpretive understanding. Instead, this section reviews two issues related to the use of charts and graphs.

## Seeing Form

The form of a distribution refers to its shape, that is, how the actual scores are distributed across the categories of the variable. The beginning of the chapter noted that the form of a variable derives from its combination of modality, skewness, and kurtosis. One useful way of interpreting graphic displays is to see them in terms of these three characteristics of form.

Modality can be readily seen in pie and bar charts, histograms, and polygons. Simply look for exceptionally large pieces of the pie, or tall bars, or high lines to determine modality in these visual representations.

Skewness requires examining a distribution around its centre. This is clearly possible for interval and ratio variables, and can be done for ordinal variables. Since the categories of a nominal variable are arbitrary, skewness doesn't apply. So, for charts or graphs of variables at the three higher measurement levels, you can obtain a general sense of whether and how the distribution is skewed simply by looking at its degree of symmetry around the visual centre.

Kurtosis is another feature of form that only applies to graphs of variables at ordinal or higher levels. Although there are specific statistics that compute the precise degree of kurtosis, visual displays can provide a useful approximation. Simply compare the "peakedness" of the display to the leptokurtic, mesokurtic, and platykurtic models and make an interpretation.

In short, for all visual displays, you can discuss modality, and for most you can say something about all three features of the distribution's form.

## Being Careful

One attractive feature of charts and graphs is their intuitive appeal. Consumers find comfort in their familiarity. Being familiar with them, readers don't have to work as hard to "get the picture."

Such familiarity, however, contains a downside. When things are familiar we tend to examine them less carefully and, in doing so, risk being deceived. We are at greater risk of being fooled (of making incorrect interpretations) when we don't examine things carefully.

Figure 7.15 provides an example of two bar charts displaying the number of persons in a sample who love or hate tacos. Both charts come from the same frequency distribution and provide the same information—namely, that the sample contains twice as many people who love tacos as hate them. In every respect, both charts have followed the rules of presentation correctly (the labels are accurate, the bars are of equal width, etc.). However, for most observers, the graph on the right provides a more exaggerated appreciation of how many more taco lovers there are than haters.

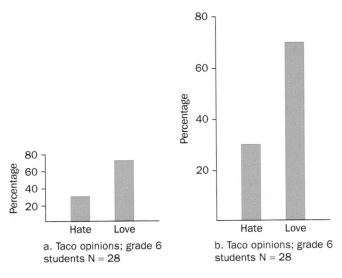

a. Taco opinions; grade 6 students N = 28

b. Taco opinions; grade 6 students N = 28

**FIGURE 7.15**  Bar Graph Examples

Or look at the two line graphs in Figure 7.16. Each of these graphs shows the hypothetical trend in suicides among soldiers for each year of a war. Both graphs have been properly constructed and display exactly the same information. However, most observers would not find the suicide trend in each graph equally disturbing. Most view the trend as more pronounced in the visual displays on the right.

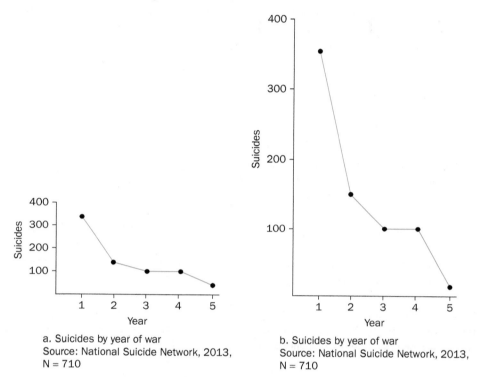

a. Suicides by year of war
Source: National Suicide Network, 2013,
N = 710

b. Suicides by year of war
Source: National Suicide Network, 2013,
N = 710

**FIGURE 7.16** Line Graph Examples

The point of these illustrations is that pictures (charts and graphs) can be deceiving. In both of these figures the exaggeration was produced by compressing or elongating the vertical or horizontal scale. In Figure 7.15 the bar chart on the left is less impressive because the vertical scale has been compressed and shows little difference between the bars. If you examine how the distances within the horizontal and vertical axes between the two graphs differ in Figure 7.16 you can see how the impact of the picture provided by the same data is produced.

In practice, readers do not receive alternative versions of a chart or graph; they receive one version. Their impressions are based on the version presented. What these examples illustrate is that impressions can be altered depending on how the chart or graph is scaled. There is no technical reason for choosing between the alternatives in Figures 7.15 and 7.16, so consumers have to be careful and presenters have to be fair.

The key to reducing deception is to examine the horizontal and vertical axes of any chart or graph and ask yourself: How would my impression change if the same information was scaled differently? A chart or graph does not create any new information; it only displays the information already present in a frequency table. So, when in doubt, always compare your interpretation of the visual display to the original source.

# Chapter Summary

This chapter took the information from recent chapters and discussed how to transform it into a visual display. Charts and graphs are simply ways of portraying frequency table information in a more accessible form. The chapter introduced you to two charting techniques for qualitative variables (pie and bar charts) and two graphing techniques for quantitative variables (histograms and frequency polygons). In addition, the chapter discussed the three central characteristics for describing the form of any univariate distribution (modality, skewness, kurtosis).

Having completed this chapter, you should appreciate that:

- Any univariate distribution of a data set is characterized by its form.
- Three features characterize the form of a univariate distribution—modality, skewness, and kurtosis.
- Modality refers to how many peaks a univariate distribution has.
- Skewness refers to how symmetrical a univariate distribution is.
- Kurtosis refers to how peaked a univariate distribution is.
- A variable's level of measurement determines the appropriateness of different charting and graphing techniques.
- For categorical variables, pie charts and bar charts are most appropriate.
- Histograms and frequency polygons are most appropriate for continuous variables.
- Because charts and graphs are intuitively appealing, it is important to examine their presentation for signs of deception.

Recent chapters have provided the basic tools used for describing, understanding, and displaying the basic features of univariate distributions. The next chapter concludes univariate analysis by showing how these tools can be brought together to provide new insights.

# 8

# The Normal Curve

## Overview

In recent chapters your path has taken you through the part of the statistical maze related to univariate distributions. This chapter takes you into the final part of the univariate distributions area. Univariate distributions focus on describing the features of a single variable in a data set. Collecting data from a sample of respondents results in a data set comprising many numbers, one for each variable from each person in the sample. Summarizing the findings for any specific variable relies on univariate analysis.

Recent chapters have introduced you to the ideas and statistical tools for summarizing single variables. These tools included measures of central tendency (to describe where typical scores are located), measures of dispersion (to describe how scattered the scores are around the variable's centre), and graphing techniques for displaying variables in pictorial form. This chapter brings these tools together in a powerful way by discussing the normal curve.

In this chapter you will learn about:

- the three features that characterize all normal distributions;
- how the features of normal curves allow comparison between variables; and
- how standard scores are calculated and interpreted.

Before beginning down this path, let's remember where the contents of this chapter are located in terms of the statistical selection checklist.

## Statistical Selection Checklist

1. Does your problem centre on a descriptive or an inferential issue?

   ☑ Descriptive

   ❑ Inferential

2. How many variables are being analyzed simultaneously?

   ☑ One

   ❑ Two

   ❑ Three or more

3. What is the level of measurement of each variable?

❏ Nominal

❏ Ordinal

☑ Interval

☑ Ratio

**Notice that the tools in the chapter are only relevant to continuous variables, which are those that are measured at the interval or ratio level of measurement.**

## STEP 1 Understanding the Tools

# Features of Normal Curves

You know that univariate distributions can differ in their modality (unimodal, bimodal, multimodal), their degree of lopsidedness (skewness), and their peakedness (kurtosis). These features can be combined in many ways, so variables can take a huge number of forms. One of these forms, however, is so common that it is labelled *normal*.

Before discussing the features of normal curves, it is important to recognize what type of variables they apply to. As you know, different statistical tools are tailored for variables at different levels of measurement. You saw this principle at work with respect to selecting appropriate measures of central tendency (Chapter 5), measures of dispersion (Chapter 6), and charting and graphing techniques (Chapter 7). Repeatedly, for instance, you saw that tools designed for interval variables are not applicable to nominal ones.

So, from the outset, it is important to appreciate that the ideas and tools related to the normal curve apply to quantitative, continuous variables: that is, variables that are at the interval or ratio level of measurement. Statistical tools related to the normal curve are powerful ones, and this power comes from exploiting a lot of information. Therefore, information-rich (quantitative) variables are required. So, in the following discussion, please keep in mind that we are using variables measured at either the interval or the ratio level of measurement.

*Normal variables* constitute a class; in other words, they share a set of common characteristics. Specifically, variables whose form is normal share the following three characteristics:

- All three measures of central tendency (mode, median, and mean) are the same (or, in practice, are approximately the same).
- The distribution is symmetrical; the halves are mirror images of one another, without skewness (again, in practice, the "more-or-less" provision applies).
- There is a fixed relationship between the mean and the standard deviation.

Each of these normal distribution features deserves some elaboration.

The first normal distribution feature concerns the first descriptive characteristic of univariate distributions—central tendency. In Chapter 5 you learned that there are three different ways of identifying the typical (central) scores in a distribution. The mode is the category of the variable that occurs most frequently; the median divides the rank-ordered distribution into two equal-sized groups; the mean identifies the value if all cases had an equal amount of the variable. Remember, the mode, median, and mean are very different ways of conceptualizing what is typical. Nonetheless, for normally distributed variables, these different measures of central tendency are the same (or, in practice, very similar). For example, if the mode of a normally distributed variable is 8, then its median and mean will also be 8. Figure 8.1 displays this characteristic in graphic form.

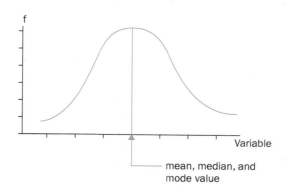

**FIGURE 8.1** Central Tendency Measures of a Normal Distribution

Among other things, this first feature tells us that normal variables are unimodal, because if they had more than a single mode the three central tendency measures could not be the same.

The second feature of normal curves concerns dispersion and notes that the distribution is symmetrical. Look at the example in Figure 8.1. If you folded the distribution on the centre (central tendency) line, the halves would be a matched set. In other words, normal distributions have no skew; they are not lopsided. Figure 8.2 displays two distributions that are *not* normal because they are skewed. The distribution on the left is negatively skewed; the one on the right, positively skewed. Notice how the asymmetry of these two distributions also creates a violation of the first feature of normal distributions.

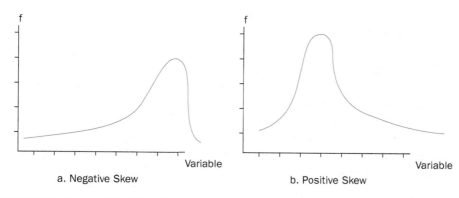

**FIGURE 8.2** Skewed Distributions

The negatively skewed distribution has a longer left tail, which indicates there are more extreme scores on the left side of the distribution. These scores, when included in the calculation of the mean, pull the mean to the left of the median. Similarly, in the positively skewed distribution, the extreme scores inflate the value of the mean relative to the median. From these images you can see that a skewed distribution generates violations in both the central tendency and symmetry characteristics of normal distributions.

Normal distributions, then, have a specific feature related to measures of central tendency (all measures are the same value) and a specific feature related to dispersion (the scatter is symmetrical). The third characteristic of normal distributions concerns the relationship between central tendency and dispersion.

Normal distributions apply to quantitative, continuous (interval and ratio) variables. As you learned in earlier chapters, for this type of variable, the mean is a common measure of central tendency; and the standard deviation, a common measure of dispersion. The third feature of normal distributions states that there is a fixed relationship between this central tendency and dispersion measure. In other words, for normal distributions, the mean and standard deviation are systematically connected. Understanding what this systematic connection means is best illustrated through an example.

Imagine collecting data on two variables for all members of your current class—*Weekly peanut butter consumption* (WPBC) (measured in spoonfuls) and *Last test grade* (LTG) (measured in percentage). If there are 100 students in your class, then we would have 200 scores: 100 WPBC scores and 100 LTG scores. The level of measurement for each of these variables is ratio.

We could summarize each of these 100-score distributions in terms of central tendency and dispersion, using the mean and standard deviation. Imagine computing these measures on each of these variables in the data set for your class, and obtaining the results shown in Table 8.1.

So far you are on familiar territory, since you could interpret each of these numbers using the tools provided in Chapters 5 and 6. The third feature of normal distributions takes you into new territory.

**TABLE 8.1**  Mean and Standard Deviation for Weekly Peanut Butter Consumption and Last Test Grade

|  | WPBC | LTG |
|---|---|---|
| Mean | 9 | 67 |
| Standard deviation | 1.7 | 8 |

This feature states that *every normal curve has a fixed relationship between the mean and standard deviation*. The word *every* in this statement is very important. It denotes that the relationship between this measure of central tendency and measure of dispersion is the same *no matter what variable is measured*. This relationship is the same for *Peanut butter consumption* (WPBC), *Test grades* (LTG), and every variable that is normally distributed. The content of the variables can be remarkably different (as it is for WPBC and LTG), but the mean–standard deviation relationship remains the same.

The fixed nature of the relationship between mean and standard deviation for every normally distributed variable is very specific. It means that if you *begin at the centre* of the distribution (identified by the mean) and *go out a fixed distance* (identified in standard deviations), you will find the *same percentage* of cases. This fixed connection holds true no matter what the specific mean and standard deviation values are. Some common benchmarks of this fixed relationship include the following:

- Mean ±1 standard deviation = 68 per cent of the cases
- Mean ±2 standard deviations = 95 per cent of the cases
- Mean ±3 standard deviations = 99 per cent of the cases

Figure 8.3 displays these fixed relationship benchmarks in graphic form.

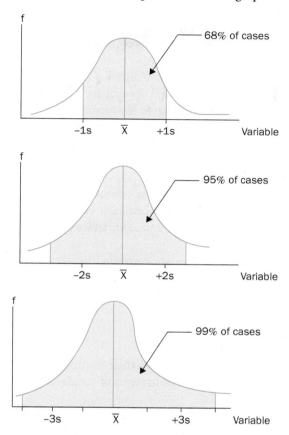

**FIGURE 8.3** Some Fixed Relationships for Normal Distributions

Applying these benchmarks to the peanut butter consumption and test grade data allows us to generate some interesting insights. For example, 68 per cent of students consumed between 7.3 and 10.7 spoonfuls of peanut butter in the previous week. This range was obtained by adding and subtracting the standard deviation for WPBC (1.7) from its mean (9). Similarly, we can calculate that 99 per cent of students obtained grades between 43 per cent and 91 per cent on the last test.

These three features of variables that are normally distributed allow us to perform a number of sophisticated techniques. We shall see these characteristics used extensively in later chapters that discuss inferential statistics. For now, let's turn our attention to one crucial descriptive statistical technique that utilizes normal distributions.

## Standard Scores

The fixed relationship between the mean and standard deviation of normally distributed variables permits powerful comparisons. Perhaps you know the colloquial expression "you cannot compare apples and oranges." This expression makes the valid point that when two things are different, comparison is problematic. In research, the things being compared are variables, and comparisons are problematic because variables are measured in different units. In our example, the variable *Weekly peanut butter consumption* is measured in spoonfuls, while *Last test results* is measured in percentages. Spoonfuls of peanut butter and test percentages are not the same thing.

It turns out, however, that you can compare different things. The trick is to *convert them to a common base*. For example, since apples and oranges are both fruits, they are comparable in terms of their fruitiness (setting aside, of course, the separate issue of how to actually measure fruitiness). Enter the power of the normal distribution. Remember that all normally distributed variables share a common characteristic; they all have a fixed relationship between their mean and standard deviation. Therefore, *all variables that are normally distributed are comparable, no matter what units the variable is measured in.*

It is thus possible to compare peanut butter consumption and test results—*or any other variables that are normally distributed.* The common base required for such comparisons comes from the fact that all normally distributed variables have the same, fixed relationship between mean and standard deviation.

Using this principle, when variables are converted to a common base, they move from being in their original units (e.g. spoonfuls of peanut butter, percentage grades) to a common unit. This common unit is *standard deviation units*, and the results are called *standard scores* or *z-scores*.

### z-Scores

Apples and oranges can be compared in terms of their fruitiness. Similarly, any score on a normally distributed variable can be compared in terms of standard deviation units. It is important to be clear about how this works.

Normally distributed quantitative variables have a central tendency captured by the mean and a level of dispersion captured by the standard deviation. In Chapters 5 and 6 you learned how these descriptive statistics were calculated and interpreted. In the *Weekly peanut butter consumption* (WPBC) example the mean of the distribution of students' scores was 9 spoonfuls per week and the standard deviation was 1.7. If a student in this class consumed 12 spoonfuls of peanut butter per week they would have been quite a bit above the average. In other words, they would be quite deviant. In fact their deviance can be calculated by computing a standard score.

Here is the formula for computing standard (*z*-) scores:

$$z = \frac{x - \overline{X}}{s}$$

As always, there is no need to be intimated by this formula. You already know its components. The numerator $(x - \overline{X})$ tells you to take the particular score (in this case, 12, the student's peanut butter intake) and subtract the mean *Weekly peanut butter consumption* (9) from it. Then divide this difference by the standard deviation, which is 1.7. The arithmetic result is 1.76. Any **standard (*z*-) score** tells you how many standard deviation units a particular score is away from the mean. In this particular case, the computed *z*-score of 1.76 tells us that the student who consumed 12 spoonfuls of peanut butter was 1.76 standard deviation units above average. Notice that this *z*-score in no longer in the original units (spoonfuls); it is converted to standard deviation units.

Any score from a normally distributed variable can be similarly converted. In the *Last test grade* (LTG) sample, a student who achieved 55 per cent is below the mean of 67 per cent. How much below? In percentage terms, this student is 12 per cent below the mean. In standard deviation terms, the student is 1.50 standard deviation units below average.

From these calculations, comparable situations have emerged. One student's peanut butter consumption is 1.76 standard deviation units above the mean, while another's test grade is 1.50 standard deviation units below the mean. These originally very different variables (one in spoonfuls, the other in percentages) are now comparable. Figure 8.4 graphs the relative distribution of these two variable scores on the normal curve.

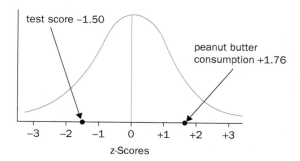

**FIGURE 8.4** Relative *z*-Scores

This graph illustrates two important points. First, it illustrates that variables measured in very different units are comparable as long as they are normally distributed. Second, the graph shows that one student's peanut butter consumption is more extraordinary (deviant) than another student's test grades.

### Using z-Score Tables

Standard (*z*-) scores make observations in different units comparable. There is another way to interpret standard scores that has intuitive appeal. This method uses a *z*-score table, found in Table 8.2.

The *z*-score table is set up with the *z*-scores (standard scores) listed in the left column, under *z*. The *z*-scores in this column are calculated to one decimal point. The row on the top of the table provides the second decimal place of the *z*-score. So, for example, a *z*-score of 1.76 is found at the intersection of the 1.7 row and the .06 column. (Check this out in Table 8.2, where you will find the number .4608).

The numbers *inside* the *z*-score table identify the *proportion* of cases that occur *between the z-score and the mean*. For example, the peanut butter consumption example had a standard score of +1.76, which has a table value of .4608. If you multiply this proportion by 100 (.4608 × 100), you will convert the proportion into a percentage. This tells you that 46.08 per cent of the persons in the sample consumed between 9 (the mean) and 12 spoonfuls of peanut butter. This conclusion is diagrammed in Figure 8.5.

In discussing the third characteristic of normal distributions, we noted some conventional benchmarks; for example, 68 per cent of all scores on a normal curve fall within a range of ±1 standard deviation from the mean. To convince yourself this is true, look up the *z*-score for 1.0 in Table 8.2 and read the number to the immediate right. It is .3413. This means that 34.13 per cent of all scores fall between the mean and one standard deviation above the mean. Since a

TABLE 8.2  Areas under the Normal Curve

| z | .00 | .01 | .02 | .03 | .04 | .05 | .06 | .07 | .08 | .09 |
|---|---|---|---|---|---|---|---|---|---|---|
| 0.0 | .0000 | .0040 | .0080 | .0120 | .0160 | .0190 | .0239 | .0279 | .0319 | .0359 |
| 0.1 | .0398 | .0438 | .0478 | .0517 | .0557 | .0596 | .0636 | .0675 | .0714 | .0753 |
| 0.2 | .0793 | .0832 | .0871 | .0910 | .0948 | .0987 | .1026 | .1064 | .1103 | .1141 |
| 0.3 | .1179 | .1217 | .1255 | .1293 | .1331 | .1368 | .1406 | .1443 | .1480 | .1517 |
| 0.4 | .1554 | .1591 | .1628 | .1664 | .1700 | .1736 | .1772 | .1808 | .1844 | .1879 |
| 0.5 | .1915 | .1950 | .1985 | .2019 | .2054 | .2088 | .2123 | .2157 | .2190 | .2224 |
| 0.6 | .2257 | .2291 | .2324 | .2357 | .2389 | .2422 | .24.54 | .2486 | .2517 | .2549 |
| 0.7 | .2580 | .2611 | .2642 | .2673 | .2704 | .2734 | .2764 | .2794 | .2823 | .2852 |
| 0.8 | .2881 | .2910 | .2939 | .2967 | .2995 | .3032 | .3051 | .3078 | .3106 | .3133 |
| 0.9 | .3159 | .3186 | .3212 | .3238 | .3264 | .3289 | .3315 | .3340 | .3365 | .3389 |
| 1.0 | .3413 | .3438 | .3461 | .3485 | .3508 | .3531 | .3554 | .3577 | .3599 | .3621 |
| 1.1 | .3643 | .3665 | .3686 | .3708 | .3729 | .3749 | .3770 | .3790 | .3810 | .3830 |
| 1.2 | .3849 | .3869 | .3888 | .3907 | .3925 | .3944 | .3962 | .3980 | .3997 | .4015 |
| 1.3 | .4032 | .4049 | .4066 | .4082 | .4099 | .4115 | .4131 | .4147 | .4162 | .4177 |
| 1.4 | .4192 | .4207 | .4222 | .4236 | .4251 | .4265 | .4279 | .4292 | .4306 | .4319 |
| 1.5 | .4332 | .4345 | .4357 | .4370 | .4382 | .4394 | .4406 | .4418 | .4429 | .4441 |
| 1.6 | .4452 | .4463 | .4474 | .4484 | .4495 | .4505 | .4515 | .4525 | .4535 | .4545 |
| 1.7 | .4554 | .4564 | .4573 | .4582 | .4591 | .4599 | .4608 | .4616 | .4625 | .4633 |
| 1.8 | .4641 | .4649 | .4656 | .4664 | .4571 | .4578 | .4686 | .4693 | .4699 | .4706 |
| 1.9 | .4713 | .4719 | .4726 | .4732 | .4738 | .4744 | .4750 | .4756 | .4761 | .4767 |
| 2.0 | .4772 | .4778 | .4783 | .4788 | .4793 | .4798 | .4803 | .4808 | .4812 | .4817 |
| 2.1 | .4821 | .4826 | .4830 | .4834 | .4838 | .4842 | .4846 | .4850 | .5854 | .4857 |
| 2.2 | .4861 | .4864 | .4868 | .4871 | .4875 | .4878 | .4881 | .4884 | .4887 | .4890 |
| 2.3 | .4893 | .4896 | .4898 | .4901 | .4904 | .4906 | .4909 | .4911 | .4913 | .4916 |
| 2.4 | .4918 | .4920 | .4922 | .4925 | .4927 | .4929 | .4931 | .4932 | .4934 | .4936 |
| 2.5 | .4938 | .4940 | .4941 | .4943 | .4945 | .4946 | .4948 | .4949 | .4951 | .4952 |
| 2.6 | .4953 | .4955 | .4956 | .4957 | .4959 | .4960 | .4961 | .4962 | .4963 | .4964 |
| 2.7 | .4965 | .4966 | .4967 | .4968 | .4969 | .4970 | .4971 | .4972 | .4973 | .4974 |
| 2.8 | .4974 | .4975 | .4976 | .4977 | .4977 | .4978 | .4979 | .4979 | .4980 | .4981 |
| 2.9 | .4981 | .4982 | .4982 | .4983 | .4984 | .4984 | .4985 | .4985 | .4986 | .4986 |
| 3.0 | .4987 | .4987 | .4987 | .4988 | .4988 | .4989 | .4989 | .4989 | .4990 | .4990 |

normal curve is symmetrical, then 34.13 per cent of all scores also fall between the mean and one standard deviation *below* the mean. Together then, 68.26 per cent of all scores in a normal distribution fall within plus or minus one standard deviation from the mean. (Okay, the 68 per cent benchmark isn't perfect, but it's pretty close!) Following the same logic, you should convince yourself that the 95 per cent and 99 per cent benchmarks are also quite accurate.

The standard score table can be used in various ways. We have just seen how it can be used to interpret a particular *z*-score. It can also be used to compare two different *z*-scores. Our previous example computed one student's peanut butter consumption as $z = +1.76$ and another student's test grade as $z = -1.50$. The peanut butter consumption score had 46.08 per cent of

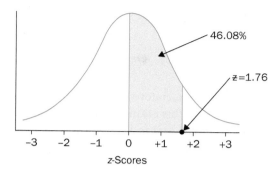

**FIGURE 8.5** *z*-Score Interpretation

the students between it and the mean (see Figure 8.5). Looking up the test grade *z*-score in the table reveals that 43.32 per cent of scores fall between it and the mean. Comparing these two percentages makes it clear just how much more deviant the student's peanut butter consumption was than her colleague's test grade.

Using the features of the standard score table allows for all kinds of comparisons. Imagine an IQ test of a large sample had a mean of 100 and a standard deviation of 10. In the next section, you will learn how to compute the answers to questions such as these:

- What percentage of persons have IQ scores between 115 and 117?
- What percentage have IQ scores over 125?
- What percentage have IQ scores between 90 and 108?

Standard (*z*-) scores use the characteristics of the normal distribution to create different kinds of insights. First, you can determine the relative location of two scores on the same variable. Second, you can compare the relative location of scores on different variables, since *z*-scores convert the variables to a common base. Third, you can determine the percentage of a sample that falls above, below, and between different standard scores.

The *z*-scores are standardized on a normal distribution that has a mean of zero. So, if an attribute's *z*-score is zero, it falls exactly on the mean of the distribution. If the *z*-score is positive, it falls above the mean and, if negative, below the mean. With these basic ideas of normal distributions and their application to the creation of standard scores, we can now turn to the computational procedures.

## STEP 2 Learning the Calculations

# Calculating z-Scores

Imagine you are an obstetrician and one of your patients gives birth to a healthy baby girl weighing 3.717 kilograms. The patient asks you if her new daughter is of average weight. What do you tell your patient? As a specialist, you know the following facts:

- mean average weight at birth of infants born in Canada: 3.462 kilograms
- standard deviation of infant birth weights in Canada: 0.177 kilograms

Given this information, here is how you would prepare an answer to the request.

1. Recall the formula for the z-score:

$$z = \frac{x - \overline{X}}{s}$$

2. Replace the symbols in the numerator with the appropriate numbers. $x$ is the score for the specific case, which, in this example is the baby's weight (3.717 kilograms). $\overline{X}$ is the mean of the distribution, and in this example the mean birth weight is 3.462 kilograms. Perform the subtraction (3.717 − 3.462) to get +0.255 kilograms—which tells you the baby weighs slightly more than average.

3. The patient's question asks how much bigger than average the baby is. Is she really heavy compared to other babies or is she within the range of normal? To answer this question you divide the numerator by the denominator, which is the standard deviation of the distribution (in this case 0.177 kilograms). The result is 1.441.

4. Interpret the meaning of the 1.441 result. The result means that our patient's new daughter is 1.441 standard deviations above the mean.

5. Although accurate, the result needs translation into simpler terms. Look up the result (1.441) in the z-score table. (Note that sometimes z-score tables are labelled with the title "Normal curve areas" or "areas under the normal curve.") Using the first column on the left of the table (the heading says z), find 1.4. Next, locate the second decimal place using the columns on the right. We are looking for a column heading of .04 (to match the second 4 in 1.441). At the intersection of this row and column in the body of the table, we locate the number .4251. Multiply this number by 100 and determine that 42.51 per cent of the population has a birth weight between the mean (3.462) and this baby's weight (3.717 kilograms).

6. Now you can answer the patient's question. Since half (50 per cent) of all babies have birth weights below average (the mean), and this particular baby's weight is 42.51 per cent above average, then this baby weighs more than 92.51 per cent of all babies.

Learning that her daughter's weight is quite a bit above normal (the average), this inquisitive mother then wants to know if her daughter is longer than average. Given that the average newborn is 51.3 cm long (standard deviation 2.7 cm) and this particular baby measures 46.8 cm, you can follow the steps to answer the mother's question:

1. Attend to the formula:

$$z = \frac{x - \overline{X}}{s}$$

2. Calculate the numerator: 46.8 − 51.3 = −4.5

3. Divide the numerator by the standard deviation (denominator): −4.5 ÷ 2.7 = −1.67

4. Interpret the result: This baby's length is 1.67 standard deviations *below* the mean.

5. Look up the result in a z-score table, multiply by 100, interpret the result: 1.67 = 0.4525 × 100 = 45.25 per cent of babies are between this baby's length and the mean.

6. Draw a conclusion: This baby's z-score result was negative, which indicates that she is shorter than average. We know half of all babies are longer than average, and that

over 45 per cent of babies have lengths between this one's and the mean. So we can conclude that over 95 per cent (50% + 45%) of all babies are longer than this one.

Earlier, three questions were posed related to a distribution of IQ scores that had a mean of 100 and a standard deviation of 10. Here are the questions again:

- What percentage of persons have IQ scores between 115 and 117?
- What percentage have IQ scores over 125?
- What percentage have IQ scores between 90 and 108?

Here is how you apply the $z$-score computations to answer each of these questions.

### IQ Scores between 115 and 117

1. Calculate and interpret the $z$-scores for each of these cases separately.

$Z_{115} = (115 - 100) \div 10 = 1.5 \rightarrow$ 43.32 per cent of IQ scores fall between 100 and 115
$Z_{117} = (117 - 100) \div 10 = 1.7 \rightarrow$ 45.54 per cent of IQ scores fall between 100 and 117

2. Draw a conclusion: If 45.54 per cent fall between 117 and the mean and 43.32 per cent fall between 115 and the mean, then 2.22 per cent (45.54 − 43.32) fall between IQs of 115 and 177. Figure 8.6 diagrams this result.

### IQ Scores over 125

1. Calculate and interpret the $z$-score:

$Z_{125} = (125 - 100) \div 10 = 2.5 \rightarrow$ 49.38 per cent of IQ scores fall between 100 and 125

2. Draw a conclusion: Half (50 per cent) of IQ scores fall below the mean. An additional 49.38 per cent fall between the mean (100) and 125. In short, 99.38 per cent of all IQ scores are below 125. This leaves only 0.62 per cent of the population with IQs above 125. Figure 8.7 diagrams this result.

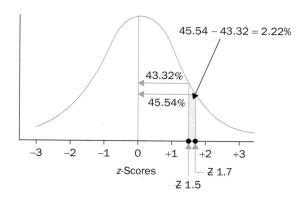

**FIGURE 8.6** Differences between Scores above the Mean

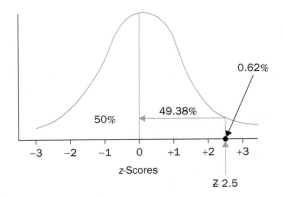

**FIGURE 8.7**  Results Greater Than a Particular Score

### *IQ Scores between 90 and 108*

1.  Calculate and interpret the *z*-scores for each of these cases separately.

    $Z_{90} = (90 - 100) \div 10 = -1.0 \rightarrow 34.13$ per cent of IQs fall between 90 and 100
    $Z_{108} = (108 - 100) \div 10 = 0.8 \rightarrow 28.81$ per cent of IQs fall between 100 and 108

2.  Draw a conclusion: If 34.14 per cent fall between 90 and 100 and an additional 28.81 per cent fall between 100 and 108, then 62.95 per cent of IQ scores occur between 90 and 108. Figure 8.8 diagrams this result.

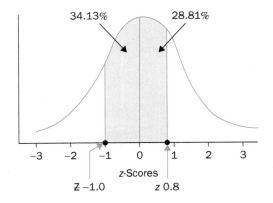

**FIGURE 8.8**  Differences between Scores Below and Above the Mean

## STEP 3 Using Computer Software

# Generating z-Scores

You now know how to calculate z-scores by hand. The nice thing about IBM SPSS Statistics software ("SPSS") is that the software converts all cases for variables you select to z-score values. This makes comparison within variables and between individual cases across variables quite simple. This is the good news! The bad news is that SPSS only creates z-score values. It does not provide z-score table results (as in Table 8.2, Areas under the Normal Curve) or make any of the various additional comparisons and calculations you learned in Step 2.

Recall that z-scores apply only to continuous variables (i.e. interval or ratio) that approximate a normal curve. As such, it is important to make sure your selected variables qualify on both accounts by completing the following steps.

1. As with all variables, first make sure that all applicable values are set to *missing*. See Chapter 3 for instructions on how to select desired values to *missing*.
2. Run a frequency distribution (Analyze → Descriptive Statistics → Frequencies) of your chosen variable.

   a. In the left-hand box, scroll to and highlight the variables (or variables) you wish to use. Click the arrow button to move the variables to the right-hand side. (Alternatively, you can double-click the variable and it will automatically move to the right-hand side).

   b. In the Charts option (located on the right-hand side), click the Histogram radio button, and then select "Show normal curve on histogram." Click Continue.
   Note: If you wish to reduce the amount of output generated in the Viewer window, turn off "Display frequency tables" in the Frequencies window. This tells SPSS that you only want the histogram. (Alternatively, if you select any measures of central tendency or dispersion, that table would appear in the Output/Viewer window—only the actual frequency table is suppressed.)

   c. Select OK.

As you can see in Figure 8.9, the variable Q16A. *Psychological well-being—Autonomy* approximates a normal distribution. Remember, it is almost impossible to achieve a perfectly normal distribution, especially in social science research. To this end, you should use your own judgment in deciding whether the variable resembles a normal distribution. Finally we see that the variable ranges from 1 to 24 (look along the *X*-axis), which means it is a continuous variable. We now are confident that we can proceed to generate z-scores for this variable.

To have SPSS create standard scores (z-scores) for any variable, use the following procedure. Make sure you are in the Data Editor window, in order to follow the procedures below.

## Procedure

1. Select Analyze from the menu bar.
2. Select Descriptive Statistics → Descriptives.

   a.  In the left-hand box, scroll to and highlight the variable you want to analyze (for example, Q16A).

   b.  Click the arrow button to move the variable from the left-hand column to the right-hand side.

   c.  Repeat steps a and b to select any additional variables you want to create z-scores for.

3.  Check the box "Save standardized values as variables" (see Figure 8.10).

4.  Click OK to create z-scores for the variable(s) you selected.

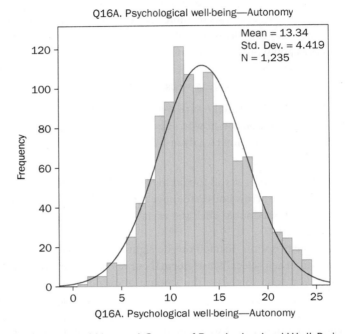

**FIGURE 8.9** Histogram and Normal Curve of Psychological Well-Being—Autonomy

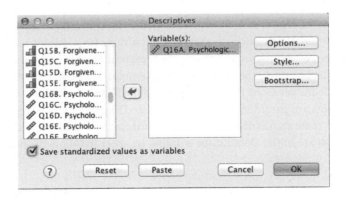

**FIGURE 8.10** Save Standard Values

Source: Reprint Courtesy of International Business Machines Corporation, © International Business Machines Corporation.

Note: By following this procedure, SPSS creates a *new variable*. Each new z-score variable will be added in a column at the end of your data set. The new variable will have the same variable name as the original variable, except that it will have a *Z* in front of it. For example, the variable Q16A becomes ZQ16A. To confirm whether the new standardized variable was in fact created, in the Data View of the Data Editor Window, use the horizontal arrow located at the bottom of SPSS and scroll to the end of the data set. In our example, ZQ16A should appear as the last variable in the data set.

# Assessing Normality Using Calculations

Chapter 7 discussed skew and kurtosis as features of non-symmetrical univariate distributions. Perfect normal distributions have no skew (i.e. are perfectly symmetrical) and no kurtosis. But real data are rarely perfect, so the practical question becomes, When does skew or kurtosis become a problem?

This section provides the procedures for generating skew and kurtosis statistics and general rules for interpreting the results.

## Procedure

1. Select Descriptive Statistics → Descriptives.
   a. In the left-hand box, scroll to and highlight Q16A, which is the variable we want to analyze.
   b. Click the arrow button to move the Q16A variable from the left-hand column to the right-hand side.
   c. In the upper-right side select the Options box.
   d. Under the Distribution title check *kurtosis* and *skewness* (see Figure 8.11).
2. Click Continue → OK.

## Interpretation

### Skew Score

Skewed distributions occur when one tail of a univariate distribution is longer than the other tail. Such imbalance indicates asymmetry of a univariate distribution around its mean. In a positive skew, the mean is being pulled toward the right or positive side of the curve. In negative skew, the mean is drawn to the left or negative side of the curve.

As a general rule, if the skew statistic is higher than ± 2, the skew of a distribution is *non-normal*. If the value of skew is ± 2 or lower, the amount of skew in the distribution is not considered a problem for data analysis. In the output shown in Figure 8.12, the skew value for variable Q16A (*Psychological well-being—Autonomy*) is .181. Since this value is well under 2,

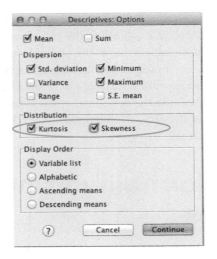

**FIGURE 8.11** Skew and Kurtosis

Source: Reprint Courtesy of International Business Machines Corporation, © International Business Machines Corporation.

**Descriptive statistics**

| | N | Minimum | Maximum | Mean | Std. Deviation | Skewness | | Kurtosis | |
|---|---|---|---|---|---|---|---|---|---|
| | Statistic | Statistic | Statistic | Statistic | Statistic | Statistic | Std. Error | Statistic | Std. Error |
| Q16A Q16A. Psychological well-being Autonomy | 1235 | 1 | 24 | 13.34 | 4.419 | .181 | .070 | −.309 | .139 |
| Valid N (listwise) | 1235 | | | | | | | | |

**FIGURE 8.12** Skew and Kurtosis Output

the distribution of this variable is near or approximately normal. If, for example, the value of skew were −3.71, we would conclude that the distribution is negatively skewed and, therefore, asymmetrical.

### Kurtosis Score

Kurtosis measures the heaviness or peakedness of a univariate distribution. In SPSS, a normal distribution has a kurtosis score of zero. Exceptionally non-normal distributions may have very high positive or negative values. As with skewness, kurtosis values higher than ±2 are considered to be problematic for analysis. Positive kurtosis ( > 2) indicates that the distribution is peaked, with too many respondent scores concentrated around the mean. Negative kurtosis scores ( < −2) indicate that the distribution is flat, with respondent scores spread more widely away from the mean.

In our example for Q16A, the kurtosis score is −.309 (see Figure 8.12). This kurtosis value is well within the ±2 band, so the distribution can be considered normal on this account.

You should note the interpretation guidelines for skew and kurtosis are not absolute. As with many matters in research, reasonable judgment needs to be applied. Sample size is an important consideration since, in larger samples, departures from normality matter less. In any case, you should always interpret normality scores alongside the visual inspection of the distribution's histogram with an imposed normal curve.

## STEP 4 Practice

You have been introduced to the features of normal distributions as well as the calculation and interpretation of z-scores, either through hand calculations (if the sample size is small) or computer software (for large sample sizes). To solidify your understanding you need to practise applying the statistical and software procedures.

This section provides you with opportunities to practise what you have learned about standard scores. The first set of questions uses hand calculations. The second set uses the SPSS procedures. For each set of questions:

1. Follow the procedural steps and complete the appropriate calculation (Set 1) or software application (Set 2).
2. Check your answers, using the Answer Key in the back of the text.
3. If your answer is incorrect, consult the Solutions section on the course website. The Solutions provide a complete step-by-step analysis of how the answers are derived.

After you have completed the next section (Interpreting the Results), return to your calculations or output and *provide complete, written interpretations of each of the statistics you have generated.*

# Set 1: Hand-Calculation Practice Questions

1. In Practice Question 5, Chapter 6, our mortgage shopper Sanjit had collected the five-year mortgage rates from the top 10 lenders in his city. The mean rate was 3.59 per cent with a standard deviation of 0.41 per cent.
   a. What would the z-score be for a rate of 4.15? For a rate of 2.8?
   b. What percentage of scores fall between 4.15 and the mean? Between 2.8 and the mean?
   c. What percentage of rates would be less than 3.0 per cent? Less than 4.0 per cent?
   d. What percentage of rates would be above 4.5 per cent? Above 2.75 per cent?
   e. What percentage of rates fall between 4.5 and 2.8?
   f. What percentage of rates fall between 3.0 and 3.5?
   g. What percentage of rates fall between 3.8 and 4.2?

2. City X is a popular coastal destination with a pleasant mild climate. The average price of detached houses in city X is $715,000, with a standard deviation of $369,000. On the other hand, city Y is land-locked in the northern heart of the continent, with a harsh continental climate. The average price of detached houses in city Y is $359,000, with a standard deviation of $228,000. The two markets for detached housing are normally distributed.
   a. What is the z-score of a $1 million house on each of these distributions?
   b. What percentage of houses cost over $1 million in city X? In city Y?
   c. What percentage of houses on market X falls between $820,000 and $1,500,000?

    d.   What percentage of houses on market Y falls between $100,000 and $250,000?

    e.   What percentage of houses on market Y falls between $250,000 and $500,000?

    f.   What percentage of houses is cheaper than $300,000 in each of these markets?

    g.   What percentage of houses falls between the mean and $1,200,000 in market X?

    h.   What percentage of the housing lies between ±1 standard deviation from the mean in each of those real-estate markets?

    i.   What is the price of a house that lies two standard deviations above the mean on each of those distributions?

# Set 2: SPSS Practice Questions

1. Using the Descriptives procedure, create a variable called ZQ22 that converts each subject's raw *Satisfaction with life* (Q22) score into z-scores. Then use the Frequencies procedure to compare the raw and standardized *Satisfaction with life* variables in terms of mean, standard deviation, variance, skewness, and kurtosis.

2. Using the Descriptives procedure, create a variable called ZQ18 that converts each subject's raw *Strength of religious faith* (Q18) score into z-scores. Then use the Frequencies procedure to compare the raw and standardized *Strength of religious faith* variables in terms of mean, standard deviation, variance, skewness, and kurtosis.

3. Using the Descriptives procedure, create a variable called *Zage* that converts each subject's raw age into a z-score. Then compare the raw and standardized age variable in terms of mean, standard deviation, variance, skewness, and kurtosis.

4. Using the Descriptives procedure, create a variable that converts each subject's raw *Self-esteem* score into a z-score. Then compare the raw and standardized *Strength of religious faith* variables in terms of mean, standard deviation, variance, skewness, and kurtosis.

5. Create a variable that converts each respondent's *Positive relations with others* score (Q16D) into a z-score. Then compare the raw and standardized *Positive relations with others* variables in terms of mean, standard deviation, variance, skewness, and kurtosis.

## STEP 5 Interpreting the Results

This chapter contains several important ideas but only a limited number of statistical tools that require interpretation. The first of these is z-scores. A z-score is a standard score; it converts the value of a variable from its original unit into standard deviation units. Standard scores are arrayed on a normal distribution that has a mean of zero, with positive z-scores falling above the mean and negative ones below it.

    Any specific z-score is interpreted in terms of how many standard deviation units the score is above or below the mean. For example, a case with a z-score of 2.36 is interpreted as follows: This case falls 2.36 standard deviation units above the mean of its distribution. A z-score interpretation expresses how discrepant (deviant) a particular observation is relative to the mean.

Interpretations of $z$-scores have limited intuitive appeal because they express results in terms of standard deviations. A more appealing interpretation expresses a $z$-score in terms of the proportion (percentage) of cases that fall between that score and the mean. This translation occurs by looking up the specific $z$-scores in the normal curve area table. Interpreting this table's results is easiest when the number (e.g. 0.2356) is multiplied by 100, which turns the result into a percentage. The percentage is then interpreted as follows: 23.6 per cent of the cases fall between this score and the mean value. Depending on how the normal curve area table is used, the resulting percentage can be expressed in terms of cases occurring above the value, below it, or between two scores.

# Chapter Summary

This chapter synthesized the tools related to central tendency, dispersion, and graphing covered in previous chapters. The synthesis occurred around the class of distributions called normal curves. The chapter discussed the three defining features of normal curves and how these features allow the creation of standard scores. Standard scores, in turn, are tools that allow the scores on variables to be compared in a variety of ways.

Having completed this chapter you should appreciate that:

* Normal distributions provide access to a set of powerful statistical tools.
* All normally distributed variables share three characteristics related to central tendency: measures, symmetry, and the relationship between mean and standard deviation.
* In normal distributions, the mode, median, and mean share the same value.
* Normal distributions are symmetrical.
* All normally distributed variables have a fixed relationship between the mean and standard deviation.
* Any score on a normally distributed variable can be converted into a standard score.
* Standard scores ($z$-scores) express a score's distance from the mean in standard deviation units.
* A $z$-score can be translated into the percentage of respondents occurring between that score and the mean.

Part II provided the tools for describing the characteristics of a single variable. These tools included the conversion of a variable's scores into frequency tables (Chapter 4), the calculation and interpretation of a distribution's measures of central tendency (Chapter 5), the calculation and interpretation of dispersion measures (Chapter 6), visual presentations using charts and graphs (Chapter 7), and this chapter's discussion and application of normal distributions. These chapters conclude the ideas and statistical tools related to univariate descriptions. Part III introduces the concepts, principles, and techniques related to bivariate analysis.

# PART III

## Bivariate Analysis

# 9

# Understanding Relationships

## Overview

You are about to enter a new region in the statistical maze. Part II introduced you to the ideas, tools, and interpretations of univariate, descriptive statistics. The tools and techniques you learned were applicable to variables at different levels of measurement. Part III shifts attention away from univariate analysis to bivariate procedures. Before making this shift, it is worth reviewing where we are on the map of statistical choices.

The toolkit of statistics contains a wide variety of tools. Different statistical tools are designed for different purposes. If you want to embed a nail in a wall, you select a hammer, not a wrench. Similarly, if a variable under analysis is nominal, you do not select analysis tools designed for interval variables. In short, one important criterion for statistical selection is level of measurement.

Another criterion is whether the issue at hand is descriptive or inferential. Descriptive statistics aim to characterize features of variables as they occur in data already collected (the sample). Inferential statistics aim to generalize from descriptive findings to all cases in the population of interest.

The final statistical selection criterion concerns the number of variables under consideration. Part II featured tools for examining single variables. These are univariate tools. We now shift our attention to bivariate statistical procedures, which examine two variables simultaneously. Specifically, Part III covers bivariate, descriptive statistical tools for various levels of measurement. In terms of our statistical selection checklist, here is where we are going next.

## Statistical Selection Checklist

1. Does your problem centre on a descriptive or an inferential issue?

   ☑ Descriptive

   ❏ Inferential

2. How many variables are being analyzed simultaneously?

   ❏ One

   ☑ Two

   ❏ Three or more

3. What is the level of measurement of each variable?

☑ Nominal

☑ Ordinal

☑ Interval

☑ Ratio

This chapter is a short, conceptual introduction to the notion of bivariate analysis. Because the chapter focuses on concepts rather than techniques, it does not use our standard five-step template. This chapter only includes a few conceptual questions at the end of the discussion. Later chapters in Part III introduce specific analysis tools and are presented using the five-step model. Understanding these tools, however, rests on a solid understanding of the core concept of bivariate analysis—the concept of relationship. This chapter discusses:

- what a relationship is; and
- the two features that characterize all relationships.

So let's begin your understanding of relationships by connecting it to your understanding of variables.

# Variables and Relationships

Univariate analysis focuses on single variables. Variables, of course, are properties of objects that can change. If we collected data on gender, social class, and income from 500 adults in your city, we could produce a variety of univariate statistical analyses. For instance, with respect to central tendency, for the sample we could compute the mean income, the median social class, or the modal gender. Alternatively, if we were interested in dispersion, we could compute the index of qualitative variation with respect to social class or the standard deviation of income. Through univariate descriptive analysis you learn about characteristics of a single variable in the sample.

Bivariate analysis is not about single variables. Bivariate analysis is concerned with describing the relationship between two variables. Bivariate analysis is not about the variables themselves; it is about the nature of the connection between the variables. The difference between focusing on single variables (univariate analysis) and focusing on the relationship between variables (bivariate analysis) can be sharpened by way of analogy.

All of the authors of this textbook have partners. The authors and their partners constitute four couples. Say you picked one couple (one pair) for analysis. Imagine you sent one person in the partnership to a psychologist for a complete psychological work-up. After conducting many hours of testing, the psychologist could provide you with a wide-ranging characterization of the person under study. You would know how much intelligence he or she has, how much introversion, how much self-esteem, etc. In other words, the file of results would tell you a great deal about that single individual. Now imagine having the same detailed analysis conducted on the other partner in the pairing. The results would provide you with a parallel, extensive description of that individual's character. These analyses of the separate individuals are the equivalent of conducting univariate tests. The focus is on the objects as separate entities.

Now imagine the two large files of information about the partners are sitting on your desk. You can study these files and learn much about the characters of the separate individuals. However, no matter how much you study these files, you will not learn about the *nature of the relationship* between the couple. The reason is simple: Relationships are something different than individuals.

Figure 9.1 diagrams the relationship between siblings. These brothers are separate, distinctive individuals. Their distinctiveness is captured by the many variables on which their scores differ: hair colour, age, height, etc. Their relationship is not something that is a part of them; it is something that exists *between them*.

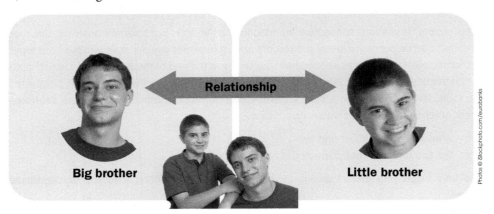

**FIGURE 9.1**  Components and Relationship

Figure 9.1 illustrates that a relationship is something different from the components. A **relationship** expresses a kind of connection between components. In the case of the couple who have had a complete psychological examination, their relationship could be of many possible types—including love, hate, indifference, abuse, etc. The same holds true of the relationship between the brothers. The important point to recognize is that you can't know what type of relationship connects the components from having only information about the parts. Understanding relationships requires a separate kind of analysis.

Among humans, analysis of a relationship (if it is a troublesome one) might be conducted by a marriage counsellor or social worker. Doing statistical analysis of relationships involves a different set of competencies. It requires bivariate analysis skills.

Bivariate analysis is a form of descriptive statistics that examines two variables simultaneously. In doing so, it characterizes the nature of the relationship between the variables. The situation is the same as that portrayed in Figure 9.2, except that the components are variables.

**FIGURE 9.2**  Variables and Relationships

Variables are properties of objects that can change. In Chapter 1 you learned that, in research terms, relationships express the connection between an independent and a dependent variable. Independent variables are those whose change initiates a difference or change in the dependent variable. Dependent variables are outcomes. The anticipated relationship between the independent and dependent variable is a hypothesis. For the sample of 500 adults in your city discussed earlier, a working hypothesis might be "the higher a person's social class, the higher their income." This hypothesis includes an independent variable (*Social class*), which is the initiating variable. It also includes a dependent variable (*Income*), which is the outcome of social class differences. Importantly, the hypothesis expresses the kind of connection between the variables (the more [*Social class*], the more [*Income*]). Figure 9.3 provides a diagram of this anticipated relationship.

**FIGURE 9.3**  Hypothesized Relationship between Social Class and Income

## Identifying Relationships

A hypothesis expresses an *anticipated* relationship. Based on some inductive or deductive reasoning, the researcher formulates what she expects is the connection between two variables. As you learned in Chapter 1, researchers collect data to test whether an anticipated hypothesis is aligned with reality. The task of statistical analysis is to examine the data and determine what (if any) patterns exist, and conclude whether the evident patterns are congruent with the hypothesis.

A central problem in bivariate analysis, then, is identifying whether or not a relationship exists between the variables under consideration. In this task, the key question becomes: How do you recognize the existence of a relationship? Returning to a human analogy helps answer this question.

Imagine you have a friend who has not had a date for an extended period. This person is currently not involved in a relationship. At this moment on campus, however, there are hundreds, perhaps thousands, of individuals who are potential partners. These people are moving around but their lives currently make no difference to your friend's. Now imagine that this coming weekend your friend reports: "I found a new partner. I am finally in a relationship with someone on campus!" What has changed?

Here is exactly what has changed. Your friend's life and the life of someone who was recently a stranger are now connected. Previously, as the former stranger went about living, it made no difference to your friend's life. Now that a relationship has been established, the lives of these former strangers now make a difference to one another.

In any circumstance, the sign of a relationship is that a change in one thing makes a systematic difference (change) in another. In the human case, relationships are between people. In research situations, relationships are between variables.

To be clear, let's apply this definition of *relationship* to a connection between independent and dependent variables. A relationship exists if a change in the independent variable produces a systematic change in the dependent variable. There are several important points embedded in this idea. First, both variables must change; if they didn't, then they wouldn't be variables and a relationship could not exist. If either variable quits changing (i.e. turns into a constant), then the relationship is terminated. (In human terms, this is what occurs if your friend's new partner is tragically killed in a car accident.) But changing isn't enough to establish a relationship; the change must be systematic. In other words, when one component in a relationship changes, then an orderly, patterned change occurs in the other component. Before your friend found a new partner, both persons were going about their lives; they were changing. What moved them from strangers to partners was the establishment of a systematic change in each other's lives. A pattern linking their lives now exists that wasn't previously present. When one goes to dinner or a movie then it is likely that the other is doing likewise. Variables, like people, are always changing. The key to identifying relationships between variables (and people) is to identify systematic change that links them. Bivariate analysis tools are aimed at identifying the presence of such linkages and characterizing them.

## Characterizing Relationships

When people's lives are systematically connected, they are in a relationship. The same is true of variables. If your mother reports she is in a new relationship with someone she met at the grocery store, what you know is that her life and that of the mystery person are systematically linked. Similarly, if you hear that the variables *Social class* and *Income* are related, then you know there is a pattern to their change. However, knowing that a relationship exists is only part of the story. We are interested in the character of the relationship. What sort of relationship does your mother have with the individual at the grocery store? How are *Social class* and *Income* linked?

Relationships of all kinds, whether between people or variables, are characterized by two features, form and strength. Understanding any relationship requires reporting on both these features.

Form reports on the structure of the relationship between components. Form tells you *what kind* of connection exists. For instance, if two people are reportedly in a loving relationship, we learn about the form of their connection. Knowledge of this kind of pattern helps us understand how these people will systematically change. If one becomes ill, we have a good idea of how the other will react. Alternatively, if a couple is in an abusive relationship, then the pattern and our expectations of how they will act toward one another will change accordingly.

Typically, three alternatives are used to describe the form of a relationship between independent and dependent variables. These are direct, inverse, and curvilinear forms. Direct relationships are also called positive relationships. In **direct relationships**, the variables move in the same direction. In other words, if one variable increases the other increases; if one decreases, the other decreases. These contrast with **inverse relationships**, where the variables change in opposite directions. If one increases, the other decreases. Inverse relationships are sometimes called indirect or negative relationships. Both direct and inverse relationships are of the linear type. In **linear relationships**, the same form holds across all attributes of the variable. You can see this clearly in the graphs of direct and inverse relationships in Figure 9.4 (*IV* means "independent variable" and *DV* means "dependent variable").

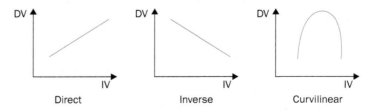

**FIGURE 9.4**   Relationships Forms

Notice that in the direct and inverse graphs the form of the relationship is the same (consistent), no matter what part of the line you examine.

This is not the case for the illustration of the curvilinear relationship in Figure 9.4. In curvilinear relationships the character of the connection between the variables changes, depending on what values of the variables you are examining. Notice that the lower values of the variables in the curvilinear graph have a direct (positive) connection, while the higher values have an inverse (negative) connection. It is more challenging to describe the form of curvilinear relationships than linear (positive and negative) ones. In linear relationships, one kind of connection characterizes the form. Curvilinear relationships require more elaborate descriptions of form.

Most of us would be very pleased if someone we cared about told us, "I love you." This characterization of form, of connection, makes us feel supported. But one reason we feel so supported is we make an assumption about the second element of any relationship, strength. Strength identifies how much difference one component makes to another. When someone says "I love you" we assume that the strength is *completely*. Sometimes this assumption is unwarranted, as those whose hearts have been broken will confirm. Loving relationships vary in strength. Sometimes the love is complete, sometimes it is moderate, sometimes it is little. The same is true of all human relationships. Sometimes hate is consuming, at other times there is just a tinge of it. Sometimes changing an independent variable makes a small difference to change in a dependent variable; in other cases it makes a huge difference.

In summary, the bivariate analysis tools focus on examining the nature of the relationship between two variables. These tools discern whether or not a connection exists between the variables, whether changing the independent variable results in a systematic change in the dependent variable. Because variables at different levels of measurement have different properties, various techniques exist for identifying connections. Once a relationship is identified, bivariate interpretations are required to characterize the relationship. Describing a relationship requires discussing both form (the kind of connection) and strength (how much difference the connection makes). The remaining chapters in Part III (chapters 10 through 14) discuss different bivariate statistical procedures related to variables at different levels of measurement.

# Practice Questions

*Identify* the probable form of the following relationships. Provide a short *justification* of the anticipated form of each relationship.

- study time and test mark
- vehicle horsepower and gas mileage
- interest rate and loan payments
- daily calories consumed and body mass
- daily calories burned and body mass
- level of education and income
- alcohol consumption and physical coordination
- age and physical strength

# Chapter Summary

This chapter introduced the central concepts behind any bivariate analysis. Bivariate analysis focuses on whether two variables are connected and, if so, on how to characterize this connection. Succeeding chapters discuss how to apply these ideas to variables at lower levels of measurement (Chapter 10) and at higher levels of measurement (Chapter 11). In addition Chapters 12 through 14 discuss computing and interpreting statistics that summarize different kinds of bivariate relationships.

Having completed this chapter you should appreciate that:

- Bivariate analysis is not about separate variables, but about the relationship between variables.
- Relationships exist when a systematic connection exists between two components.
- Relationships terminate when one or both components cease to change, or when the systematic link between their change is broken.
- All relationships are characterized by form and strength.
- Form describes the kind of connection that exists between the components.
- Strength captures how much difference the components make to one another.
- If you know the form of a relationship, but not its strength, you have an incomplete understanding. (e.g. "I love you" requires the answer to the additional question, "How much?")
- If you know the strength of a relationship, but not its form, you will be puzzled. (e.g. "You make a lot of difference to me" requires the answer to the additional question "What kind of difference?")

With these ideas in place, we can now move on to applying them to specific situations. The next chapter shows you how relationships are characterized in tables.

# 10

# Bivariate Tables

## Overview

The last chapter emphasized that bivariate analysis focuses on the character of the connection between the independent and dependent variables. Bivariate analysis is about understanding the *relationship* between the variables under consideration. A hypothesis expresses the expected (predicted) relationship between the variables. The job of bivariate statistics is to determine whether the expected pattern of a relationship is actually present in the data. *Association* is a synonym for the notion of *relationship*. When people associate with others they come together and, in doing so, establish a pattern in their interaction. In quantitative research, variables are the components that join together to form relationships. In bivariate analysis, the task is to determine what pattern of association emerges when the independent and dependent variables interact.

A variety of different statistical tools exist for determining the kind (form) and amount (strength) of association between two variables. As always, the appropriateness of the tools depends on the level of measurement of the variables. This chapter explains the use and interpretation of bivariate tools that are appropriate when the variables are at the nominal or ordinal levels of measurement. In terms of the statistical selection checklist, here is where the tools in this chapter are located:

## Statistical Selection Checklist

1.  Does your problem centre on a descriptive or an inferential issue?

    ☑  Descriptive

    ❏  Inferential

2.  How many variables are being analyzed simultaneously?

    ❏  One

    ☑  Two

    ❏  Three or more

3.  What is the level of measurement of each variable?

    ☑  Nominal

    ☑  Ordinal

    ❏  Interval

    ❏  Ratio

In this chapter you will learn how to:

- set up bivariate tables and identify their components;
- determine and interpret the form of the relationship expressed in a table; and
- determine and interpret the strength of the relationship existing in table evidence.

But before getting to these technical considerations, let's remind ourselves of the levels of measurement of the variables used for creating tables.

## STEP 1  Understanding the Tools

# Features of Categorical Variables

Nominal and ordinal variables are qualitative; the categories of the variables do not have quantitative properties. For nominal variables, the attributes are simply names. For example, the variable *Religion* might have values designed as (1) *Jewish*; (2) *Christian*; and (3) *Muslim*. The numbers associated with these categories merely designate the name. The variable could just as validly be constructed as (1) *Christian*; (2) *Muslim*; and (3) *Jewish*. Ordinal variables are nominal variables with the additional property that the values can be rank-ordered. Take the example of a five-point Likert scale (ranging from 1, *strongly disagree* to 5, *strongly agree*) responding to the statement: "Men are lazier than women." If one respondent scores a 2 on this scale and another scores a 5 we know that one person agrees with the statement more than the other. But that is all we know. We have no idea quantitatively how much more one respondent agrees with the statement than the other.

One feature of qualitative variables is that they typically contain a limited number of discrete response categories (values). This feature means that the categories have width, in the sense that the values are broadly defined along some dimension of difference. Among students who receive a B in a course, there are differences in their performance. The same holds true for nominal variable categories like *Christian* or *female*. In representational terms, the categories of qualitative variables are more like a line than a single point. For bivariate analysis interested in the intersection of the independent and dependent variables, the categorical width of qualitative variables is displayed in tables. Table 10.1 provides an illustration.

**TABLE 10.1**  Bivariate Table

| Income | Sex | | |
|---|---|---|---|
| | Male | Female | Total |
| Low | 35 | 75 | 110 |
| Medium | 45 | 75 | 120 |
| High | 70 | 50 | 120 |
| Total | 150 | 200 | 350 |

Notice in Table 10.1 that the values for the nominal variable *Sex* have width, which suggests that not all males (or females) are the same. But, clearly, the male and female columns are distinct. The same is true for the categories of the dependent variable, *Income*. Not all *high income* persons earn the same amount of money.

## Setting Up Bivariate Tables

Bivariate tables are constructed using standard conventions, including the following. First, the independent variable is listed on the top of the table, with the values of the independent variable identifying the columns. In Table 10.1, the independent variable is *Sex* and its categories are *male* and *female*. In bivariate tables, the dependent variable is listed on the left-hand side of the table, with its values as rows of the table. In our example, *Income*, with its values *low*, *medium*, and *high*, is the dependent variable.

In short, the top row and left column of a bivariate table identify the variables. The bottom row and right column include the marginals. Marginals (sometimes called marginal distributions) get their name from their location; they are on the margins of the table. The marginals are frequency distributions. The marginal on the bottom is a frequency distribution of the independent variable. The marginal on the right side is a frequency distribution of the dependent variable. Frequency distributions, you will recall from Chapter 4, are univariate distributions. Therefore, the marginals from Table 10.1 could be displayed as:

| Sex | f | Income | f |
|--------|-----|--------|-----|
| Male | 150 | Low | 110 |
| Female | 200 | Medium | 120 |
| | | High | 120 |

The number in the lower right corner of the bivariate table is the sample size (*N*). In the example, the sample contains 350 cases. Each of the marginal distributions, of course, adds up to this total.

The marginals (univariate distributions) are on the *outside* of the bivariate table. The cells *inside* the table display the coming together (intersection) of the univariate distributions. This makes sense because the independent and dependent variables need to come together (join) to form an association or relationship.

The numbers on the inside cells of the table are joint frequencies. Joint frequencies are also called conditional frequencies. **Joint (conditional) frequencies** are a count (frequency) of the *intersection* of a specific attribute of the independent variable and a specific value of the dependent variable. In Table 10.1, for example, the 35 in the upper-left cell is a joint frequency that indicates how many males (attribute of the independent variable) had low income (category of the dependent variable). In other words, this sample of 350 persons contains 35 males with low incomes. The other joint frequencies are similarly interpreted. There are 75 females with low income in the sample; 50 females with high income, etc. All joint frequencies express values that are required to meet two conditions—a specific value of the independent variable and a particular value of the dependent variable: in the case of Table 10.1, a specific *Sex* and a particular amount of *Income*.

Joint (conditional) frequencies are aggregated to form joint (conditional) distributions. You are already familiar with univariate frequency distributions. **Conditional distributions** are frequency distributions of the dependent variable *for a specific value of the independent variable*. The columns *inside* a bivariate table are conditional distributions. For example, the column with *low*, *medium*, and *high* values of 35, 45, and 70 is a frequency distribution of *Income* (DV) for those in the sample who meet the independent variable condition *male*. Likewise, the column with the values 75, 75, and 50 is a conditional distribution of *Income* for *female*.

This is how bivariate tables are conventionally displayed. They begin with univariate distributions of both variables in the marginals, and then display the bivariate intersection in the conditional distributions in the body of the table. We now turn to the basic tools for analyzing bivariate tables.

## Analyzing Bivariate Tables

Conditional distributions form the body of the bivariate table and represent the intersection (association) of the independent and dependent variables. Because of this intersection between the variables, bivariate tables are also called crosstabs. **Crosstabs** is an abbreviation for *cross-tabulations*, and indicates that a bivariate table tabulates the crossing (coming together) of the independent and dependent variables. The analysis of bivariate tables focuses on this crossing, which occurs in the conditional distributions. In a bivariate table, the more intensely crossed the variables are—the more they are intertwined—the greater the relationship between the variables.

In Chapter 9 you learned that a relationship between variables is characterized by a systematic connection. The more systematically the dependent variable changes as the independent variable does, the more evident a relationship is. In turn, describing any bivariate relationship requires identifying and characterizing both the form and strength of the connection. Form refers to the structure of the connection and answers the question, "What *kind* of relationship is it?" Strength addresses how much impact one variable has on the other. Determining form and strength in bivariate tables uses separate analytical tools. Before discussing these tools, some fundamentals need to be clarified.

### Table Orientation

Table 10.2 presents a bivariate table between two ordinal variables, *Social class* and *Income*. The table uses a conventional format with the independent variable and its attributes on the top, and dependent variable and its values on the left side.

**TABLE 10.2**  Social Class and Income

| Income | Social Class | | | Total |
|---|---|---|---|---|
| | Low | Medium | High | |
| Low | 200 | 50 | 5 | 255 |
| Medium | 75 | 100 | 25 | 200 |
| High | 25 | 50 | 70 | 145 |
| Total | 300 | 200 | 100 | 600 |

Chapter 9 stated that a relationship exists when a change in the independent variable is associated with a systematic change in the dependent variable. Let's be clear on what this means when you are looking at an actual table. First, we need to observe change in the independent variable. In Table 10.2 the independent variable is *Social class* and its values are *low*, *medium*, and *high*. Changing the independent variable means moving from *low* to *medium* or *high social class* (or the reverse). In Table 10.2 this literally means *moving across the table from left to right*; that's how we look at the table to see the independent variable change. In a parallel way, the dependent variable in Table 10.2 is *Income*, with its three values *low*, *medium*, and *high*. This variable is on the left side of the table and to observe it change we need to *move up or down*.

The "left to right" movement and the "up–down" movement let you observe change in either the independent or dependent variables. These are univariate changes. Bivariate change requires looking at these univariate changes simultaneously. How is that accomplished?

Let's begin by looking at the first conditional distribution of the independent variable, the column labelled Social Class: Low. This column displays differences in *Income* (the dependent variable) for a fixed value (*low*) of the independent variable (*Social class*). In this column, the independent variable is fixed while the dependent variable changes across its three categories (rows). Now look at the middle column under Social Class. Again, in this column, the independent variable is fixed (at *medium*) while the dependent variable changes. The same holds true for the final column (*high*) of the independent variable; the dependent variable changes while the independent variable is constant. This is what you observe when you look at the columns (conditional distributions of the independent variable) separately.

However, when you *compare the conditional distributions* (i.e. compare the first column, to the second and to the third) you are observing how *both* the independent variable changes (moving left to right) *and* the dependent variable changes (up–down). In short, looking for a bivariate relationship involves *comparing the conditional distributions of the independent variable*: comparing the table's columns.

Before discussing the techniques for establishing the form and strength of the relationship in the bivariate table, one other orientation point is necessary. This point relates to comparing the conditional distributions. Look at the conditional distributions in Table 10.2. Notice that the data set includes 300 individuals whose *Social class* is *low*, 200 who are *medium*, and 100 who are *high*. In other words, the distributions in these columns are of different bases; they are composed of different numbers of individuals. For this reason, you cannot directly compare the columns (conditional distributions) of a bivariate table; the *conditional distributions must be standardized before analysis*.

Standardizing the conditional distributions of the independent variable involves making them comparable. The easiest way to establish comparability is to *convert the columns to percentages*. Converting the columns in Table 10.2 to percentages results in Table 10.3.

**TABLE 10.3** Social Class and Income—Standardized

| Income | Social Class | | |
|---|---|---|---|
| | Low | Medium | High |
| Low | 66.7% | 25.0% | 5% |
| Medium | 25.0% | 50.0% | 25% |
| High | 8.3% | 25.0% | 70% |

The columns in Table 10.3 are now comparable because all the numbers are on the same base (percentages).

Comparing the standardized conditional distributions to one another provides evidence of the existence of a relationship between the independent and dependent variables. Setting up the bivariate table to perform the analysis is summarized in this refrain:

Percentage down, compare across

Following this rule will set you on the right path. To move further down the path, you need to apply the tools described below for determining both the form and the strength of the connection.

### *Establishing Form*

Determining the form of a bivariate relationship begins with a standardized (percentaged) table. Table 10.4 provides the example we will use for illustration. Notice that this bivariate table between *Sex* and *Income* is already standardized (percentaged).

**TABLE 10.4** Sex and Income—Standardized

| Income | Sex | |
|--------|------|--------|
|        | Male | Female |
| Low    | 11%  | 32%    |
| Medium | 29%  | 40%    |
| High   | 60%  | 28%    |

To start, look at the first column and imagine it is an apartment block with as many storeys as there are rows. So, in the example, you would look at the left-hand column and imagine it is an apartment block with three floors. Next, for that apartment block, let the percentages represent the number of people on each floor. In the example table, you would imagine there are 60 people on the bottom floor, 29 people on the middle floor, and 11 people on the top floor.

Next, imagine you leave the apartment block and return five minutes later. Now the distribution of the people in the apartment block is represented by the distribution of people (i.e. percentages) *in the next column*. In the example, when you return to the apartment block there are 32 people on the top floor, 40 people on the middle floor, and 28 people on the bottom floor.

Now you ask yourself this critical question: While you were away from the apartment block, which way did the people move? Your answer comes from making an *overall assessment from comparing the two columns*. The possible answers to this question are *up*, *down*, or *no change*. If you apply this technique to the example, the answer is *up*. Why? Because, in the right-hand column, there are more people on the upper floors than there are in the left-hand column.

In this case the *up* (or *down* or *no change*, in other cases) identifies the *form* of the relationship between the variables. Once you have determined what the appropriate form is, you need to draw an arrow illustrating the form below the two columns of the table. For our example, place the *up* arrow (i.e. your conclusion regarding the form) moving at a 45-degree angle from the lower left to the upper right, with the tip of the arrow at the upper right, as illustrated in Table 10.5.

**TABLE 10.5** Sex and Income—Form Analysis

| Income | Sex | |
|--------|------|--------|
|        | Male | Female |
| Low    | 11%  | 32%    |
| Medium | 29%  | 40%    |
| High   | 60%  | 28%    |

Next you are faced with the problem of interpretation—translating what this arrow identifying form means. To do this, you must remember that a statement about form has to state how changing the independent variable is connected to a systematic change in the dependent variable. To accomplish this, proceed as follows.

Imagine that the lower point of the arrow is point A and the tip is point B. In terms of our analogy, the change (movement) of the people in the apartment block was *up* in the sense that they walked up from point A to point B. This movement from points A to B can be decomposed into a horizontal component and a vertical component. In other words, you could move from point A to B by going from point A to point C (the horizontal component) and from point C to point B (the vertical component). This is illustrated in the following diagram.

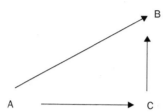

With this understanding, we can finally make an interpretation about form, using the following principles. The *horizontal* component represents the change in the *independent variable*; the *vertical* component represents the change in the *dependent variable*. Applying this to the example table, the horizontal arrow (A → C) tells us that *Sex* (the independent variable) changes from *male* to *female*. The vertical arrow (C → B) tells us that *Income* (the dependent variable) changes from *high* to *low*. To be clear, you can place the horizontal arrow above the independent variable in the table and the vertical arrow beside the dependent variable, as illustrated in Table 10.6.

**TABLE 10.6** Sex and Income—Form Specification

| Income | Male | Female |
|--------|------|--------|
| Low | 11% | 32% |
| Medium | 29% | 40% |
| High | 60% | 28% |

The arrows in Table 10.6 represent the direction of change in each of the variables. Putting these independent and dependent variable changes together creates a statement about the form of the relationship in the table. It is as follows: "As sex shifts from male to female, income changes from high to low." This is a statement about form because it tells us how changing the independent variable is related to a systematic change in the dependent variable.

One final note about determining form in a bivariate table: *You follow the same steps and logic no matter how large the table is.* If the table has more rows than the example table, you simply imagine that the apartment block has as many floors as there are rows in the table. If the table has more columns than this example, then you simply repeat the "while you were away, which way did people move?" process *between every pair of adjacent columns.* For example, if a table has four columns, you would compare column 1 with column 2, column 2 with column 3, and column 3 with column 4 to get the information required to determine the overall form. Remember that your final statement about form has to indicate how changing the entire independent variable is related to change in the dependent variable.

To solidify the identification of form, Table 10.7 illustrates a standardized table that contains three values of the independent variable and two attributes of the dependent variable. The appropriate arrows indicating how change occurs between the adjacent columns are provided, as is the summary trend decomposed into horizontal and vertical components. These components are included beside the independent and dependent variables.

**TABLE 10.7**  Standardized Table, Form Analysis

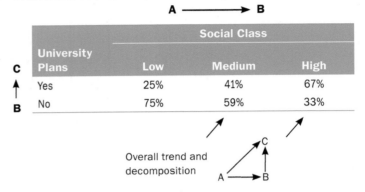

Here are some points about the analysis of Table 10.7. The trend arrow between the first and second columns is *up* and the trend arrow between the second and third columns is also *up*. This results in the overall trend between the variables as *up*, which is indicated by the A → C line in the diagram. The horizontal and vertical components of this diagram were transferred to locations beside the variables. Therefore, the form in Table 10.7 is the following: As *Social class* increases (i.e. changes from *low* to *high*), *Plans to attend university* also increases (i.e. changes from *no* to *yes*).

### Determining Strength

Once the form of the relationship in a bivariate table (i.e. what kind of relationship it is?) is determined and interpreted, the next task is to determine the *strength* of the relationship. In general, strength means answering the question, "How much difference does changing the independent variable make to the dependent variable?" Like form, the technique for determining strength begins with the standardized (i.e. percentaged) bivariate table.

Determining strength requires figuring out how much the dependent variable differs between categories of the independent variable. *To do this you calculate the difference between the adjacent columns of the table.* In arithmetic, differences are calculated through subtraction.

To illustrate the process, let's do the calculations for Table 10.6. Beginning at the top of the table, the difference between 11 and 32 is 21; the difference between 29 and 40 is 11; and the difference between 60 and 28 is 32. Note that you perform this operation without worrying about whether the differences are positive or negative.

Let's pause for a moment and think about what each of these calculations tell us. The first difference (21), tells you how much difference there is between males and females who have low income. The second difference (11) informs you of the amount of sex differences for medium income. Finally, the third difference (32) reports the male–female differences in high income. These are the pieces required for determining strength.

Your goal, however, is to put these pieces together to answer how much difference the independent variable (*Sex*) makes to the entire dependent variable (*Income*) (i.e. across all its categories). To obtain this result, you *calculate the mean of all the specific differences*. In this case 21 plus 11 plus 32 equals 64, then divided by 3 equals 21.3 per cent. This 21.3 per cent tells you the average difference *Sex* makes to *Income*; it is a measure of the strength of the relationship between the two variables.

This calculation of strength provides you with a number. As always, your task is to interpret what this number means and state your interpretation in a simple sentence. What does 21.3 per cent (or any other) difference mean? Let's begin with the extremes. If the calculated difference was zero, the interpretation would mean that changing the independent variable makes no difference to the dependent variable. In this case, no relationship exists. If person *X* makes no difference to person *Y*, these individuals are unrelated; they are strangers. At the other extreme, if the calculated difference is 100 per cent, then the independent variable makes all the difference to the dependent variable. The variables are completely connected, like conjoined twins.

But how are differences between these extremes to be interpreted? Table 10.8 is useful as a guide.

**TABLE 10.8** Guidelines for Interpreting Percentage Differences

| Amount of Difference | Interpretation |
| --- | --- |
| Under 5% | no relationship |
| 5–19% | small relationship |
| 20–34% | modest relationship |
| 35–55% | moderate relationship |
| 56–75% | strong relationship |
| Over 75% | very strong relationship |

When we apply these guidelines to our example, the 21.3 per cent difference would be interpreted as "Sex makes a modest difference to level of income."

One final note about calculating and interpreting strength in bivariate tables. If the table has more rows than the example table (Table 10.6), this only means you have more differences between the columns to use in computing the average difference. If the table has more columns than our example, follow these steps:

1. Compute the average (mean) difference between columns 1 and 2.
2. Follow the same procedure to determine the difference between columns 2 and 3.
3. Repeat this procedure until you have the average differences between all of the adjacent columns.
4. Take the mean (average) of these column differences (i.e. the mean of the mean differences) to get the overall strength in the table.
5. Use the guidelines table to state your conclusion.

Table 10.9 provides an illustration. This table contains a complete set of calculations. First, the mean differences between the adjacent columns were calculated. Specifically, the differences between *low* and *medium* years of school were 17 per cent for both levels of gross income, resulting in an average difference of 17 per cent between these columns. The same procedure comparing the *medium* and *high* school columns resulted in an average difference of 25 per cent. These mean differences were averaged to get the overall measure of strength: *(17 + 25) ÷ 2 = 21 per cent.* The interpretation of the calculated strength (21 per cent) is "Increasing years of schooling makes a modest difference to gross income."

**TABLE 10.9**  Schooling and Income, Standardized

| Gross Income | Years of School | | | | |
|---|---|---|---|---|---|
| | Low | | Medium | | High |
| Low | 67% | 17% | 50% | 25% | 25% |
| High | 33% | 17% | 50% | 25% | 75% |

$$\frac{17 + 17}{2} = 17\%$$

$$\frac{25 + 25}{2} = 25\%$$

$$\frac{17 + 25}{2} = \frac{42}{2} = 21\%$$

This covers all the ideas related to calculating and interpreting the form and strength of bivariate relationships when they are in table form. The following steps take these ideas and show you how these calculations are performed by hand and computer.

## STEP 2 Learning the Calculations

# Setting Up and Standardizing a Table

Here are the general rules for creating a table.

1.  The independent variable is always on the top, spanning all the columns. Each column in the table represents an attribute of the independent variable. So, if the independent variable has four values, then the table will have four columns.
2.  The dependent variable is always on the left side—beside the rows. Each value of the dependent variable appears as a separate row. For example, if the dependent variable has five values, then there should be five rows.
3.  Add labels for the values of the independent variable (over the columns) and dependent variable (beside the rows).
4.  Enter the conditional frequencies into the table. Add columns and rows to create totals.
5.  Standardize each column by creating percentages. Remember "percentage down"! Each percentaged column should add up to 100 per cent.

Let's work through an example. Say we are interested in figuring out if men attend religious institutions more frequently than women. We have the following data:

Among the men in our sample, 48 attend religious services weekly, 27 attend monthly, 96 attend less often, and 107 do not attend. Among women, 75 attend religious services weekly, 82 attend monthly, 111 attend less often, and 97 do not attend.

Where do we start?

First, we need to decide which variable is the independent variable and which is the dependent variable. This decision allows us to place the variables in the correct location on the table. A moment's thought guides our decision. It doesn't make sense that religious attendance would affect sex. On the contrary, it makes more sense that there are sex differences in religious attendance. This means that sex is likely the independent variable and religious attendance is likely the dependent variable.

We begin with a blank table that looks like Table 10.10a.

**TABLE 10.10a** Blank Table

| | | | |
|---|---|---|---|
| | | | |
| | | | |
| | | | |
| | | | |

It has four rows and two columns (plus two extra columns and two extra rows, used for labelling and totals). We label the columns and the rows as shown in Table 10.10b.

**TABLE 10.10b** Column and Row Labels

| Frequency of Religious Attendance | Sex | | |
|---|---|---|---|
| | Female | Male | Total |
| Weekly | | | |
| Monthly | | | |
| Less often | | | |
| Never | | | |
| Total | | | |

Next we add the frequency for each value of the variable using the information provided, as in Table 10.10c. Compute totals for all columns and rows.

But the table, as currently presented, doesn't give us much insight. Are men more likely than women to attend religious services? The next step toward answering this question involves standardizing the columns by calculating percentages. We reviewed this arithmetic in Chapter 4 (please refer to that chapter if you need a refresher).

**TABLE 10.10c** Frequency for Each Value of the Variable

| Frequency of Religious Attendance | Sex | | |
|---|---|---|---|
| | Female | Male | Total |
| Weekly | 75 | 48 | 123 |
| Monthly | 82 | 27 | 109 |
| Less often | 111 | 96 | 207 |
| Never | 97 | 107 | 204 |
| Total | 365 | 278 | 643 |

Table 10.10d is the standardized (percentage) table that results.

**TABLE 10.10d** Religious Attendance by Sex

| Frequency of Religious Attendance | Sex | |
|---|---|---|
| | Female | Male |
| Weekly | 21% | 17% |
| Monthly | 22% | 10% |
| Less often | 30% | 35% |
| Never | 27% | 38% |
| Total | 100% | 100% |

Finally, there are other guidelines that should be followed in creating a good table. First, all tables must have titles. The title should reflect the content of the table and should not be confusing to the reader. We added one (Religious Attendance by Sex) to the table above. Second, if there are missing cases, these should be included in the table prior to the total (we discussed this in Chapter 4). Finally, all the attributes (values) of each variable should be included in the table, even if a cell's value is zero.

The table is now standardized and ready for analysis to determine the form and strength. Here are the steps for determining each of these characteristics.

## Determining Form

Form refers to what kind of relationship (if any) exists between the variables in the table. It answers the question, "If you change the independent variable a specific way, what sort of change do you expect to see in the dependent variable?" Here are the steps for determining form.

1. Determine the direction of change between the adjacent columns in the table. To do this, apply the "movement within an apartment" analogy discussed earlier. The answer will be one of *up*, *down*, or *no change*.

2. If the independent variable has more than two values (i.e. the table has more than two columns), determine the overall trend by adding the effects of step 1 together. The result will be something like "up and up = overall up" or "down and down = overall down," etc.

3. Decompose the overall trend (step 2) into horizontal and vertical components.
4. Place the horizontal arrow from step 3 over the independent variable in the table.
5. Place the vertical arrow from step 3 beside the dependent variable in the table.
6. Write out a statement describing how the independent variable changes (by following the horizontal arrow) and the dependent variable changes (by following the vertical arrow).

Here are the six steps applied to the standardized (percentaged) Religious Attendance by Sex table.

1. *Down*, as evidenced by the arrow below the columns in the example below.
2. Not applicable, since the independent variable only has two values (*female, male*).
3. Horizontal effect: ⟶
   Vertical effect :   ↓

   These arrows result from the following decomposition:

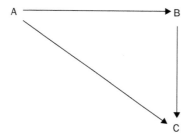

4 & 5. On Table 10.10e:

**TABLE 10.10e** Religious Attendance by Sex

| Frequency of Religious Attendance | Female | Male |
|---|---|---|
| Weekly | 21% | 17% |
| Monthly | 22% | 10% |
| Less often | 30% | 35% |
| Never | 27% | 38% |
| Total | 100% | 100% |

6. As *Sex* changes from *female* to *male*, the *Frequency of religious attendance* declines.

## Determining Strength

1. Calculate the differences between the columns of each row.
2. Calculate the mean of the differences between adjacent rows.
3. If the independent variable has more than two values (i.e. the table has more than two columns), perform steps 1 and 2 for every pair of adjacent columns.

4. If the dependent variable has more than two values (i.e. the table has more than two columns), calculate the mean of the step 2 differences.
5. Interpret the results using the interpretation table provided earlier.

Here are the steps applied to the standardized Religious Attendance by Sex table.

1. The differences are 4 per cent, 12 per cent, 5 per cent, 11 per cent.
2. Mean difference = (4 + 12 + 5 + 11) = 32 ÷ 4 = 8 per cent
3 & 4. Not applicable since the independent variable has only two values.
5. Sex makes a small difference to the level of religious attendance.

**TABLE 10.10f** Religious Attendance by Sex

| Frequency of Religious Attendance | Sex | | |
| | Female | Female–Male Difference | Male |
|---|---|---|---|
| Weekly | 21% | 4% | 17% |
| Monthly | 22% | 12% | 10% |
| Less often | 30% | 5% | 35% |
| Never | 27% | 11% | 38% |
| Total | 100% | | 100% |

## STEP 3 Using Computer Software

# Generating a Bivariate Table

Generating a bivariate table in IBM SPSS Statistics software ("SPSS") is a relatively simple task and can be done by following the procedures outlined below. Before proceeding further, however, it is important to reiterate that another name for bivariate tables is *crosstabulations*, or *crosstabs* for short. SPSS refers to bivariate tables as crosstabs.

## Procedure

In SPSS, crosstabs are identified as a descriptive statistic. As such, to produce a crosstab follow the steps below. Remember, however, you need to run two frequency distributions in order to determine if any values in your independent and dependent variables need to be set to *missing*. See Chapter 3 for instructions on how to select desired values to missing.

1. Under Analysis in the SPSS Menu, click Descriptive Statistics and then Crosstabs.
2. Move your independent variable from the left side to the right Column(s) box. In the example illustrated below, the independent variable is "Q27. Were you born in Canada?"
3. Select your dependent variable and move it to the Row(s) box on the right-hand side (e.g. "NQ30. Social Class").

4. Next, select the Cell(s) box (upper right corner).
   a. In the Percentages box, check Column. By selecting the column percentage, you will be able to analyze the output by following the rule: percentage down, compare across.
   b. Click the Continue box.
5. Select Continue and OK to execute the Crosstab command.

**nq30 NQ30.Social class * Q27 Q27. Were you born in Canada Crosstabulation**

| | | | Q27 Q27. Were you born in Canada | | DV Marginal |
| --- | --- | --- | --- | --- | --- |
| | | | 1 Yes | 2 No | Total |
| nq30 NQ30.Social class | 1 Lower | Count | 86 | 16 | 102 |
| *Compare Across* | | % within Q27 Q27. Were you born in Canada | 10.6% | 13.8% | 11.0% |
| | 2 Middle | Count | 567 | 77 | 644 |
| | | % within Q27 Q27. Were you born in Canada | 69.8% | 66.4% | 69.4% |
| | 3 Upper | Count | 159 | 23 | 182 |
| | | % within Q27 Q27. Were you born in Canada | 19.6% | 19.8% | 19.6% |
| Total | *Percentage Down to 100%* | Count | 812 | 116 | 928 |
| | | % within Q27 Q27. Were you born in Canada | 100.0% | 100.0% | 100.0% |

**FIGURE 10.1** Output for Q27 by Q30

From the output shown in Figure 10.1, you can see that students who were not born in Canada were slightly more likely to come from a lower social class background (13.8 per cent, compared to 10.6 per cent for Canadian-born participants), but there was no difference when it came to upper-class background, and thus no association is found between the independent and dependent variable (i.e. a difference under 5 per cent when comparing across).

## STEP 4 Practice

You now have the tools for analyzing bivariate tables either by hand or using computing software. To solidify your understanding you need to practise applying the statistical and software procedures.

This section provides you with opportunities to practise what you have learned about bivariate table analysis. The first set of questions uses hand calculations. The second set uses the SPSS procedures. For each set of questions:

1. Follow the procedural steps and complete the appropriate calculation (Set 1) or software application (Set 2).
2. Check your answers, using the Answer Key in the back of the text.
3. If your answer is incorrect, consult the Solutions section on the book's website. The Solutions provide a complete step-by-step analysis of how the answers are derived.

After you have completed the next section (Interpreting the Results), return to your calculations or output and *provide complete, written interpretations of each of the statistics you have generated.*

# Set 1: Hand-Calculation Practice Questions

1.  Use the "percentage down, compare across" method to analyze the data in Table 10.11. Describe the form and strength of the relationship between the variables.

**TABLE 10.11** Support for Universal Health-Care System

| Support for Universal Health Care | Nationality | |
|---|---|---|
| | American | Canadian |
| Low | 55 | 14 |
| Medium | 47 | 44 |
| High | 23 | 72 |
| Total | 125 | 130 |

2.  Table 10.12 reports hypothetical data for the survey question, "Do you think corporal punishment is an appropriate method of disciplining children?"
    a.  What is the sample size for this data set?
    b.  What are the univariate frequency distributions for each variable?
    c.  Describe the form and strength of this relationship.

**TABLE 10.12** Support for Corporal Punishment by Education Level

| Corporal Punishment | Highest Level of Education | | |
|---|---|---|---|
| | Less than High School | High School | Post-secondary |
| Never | 24 | 63 | 86 |
| Sometimes | 22 | 33 | 23 |
| Often | 12 | 11 | 6 |

3.  Describe the form and strength of the relationship in Table 10.13.

**TABLE 10.13** Perceived Quality of Life by Reported Level of Job Satisfaction

| Quality of Life | Job Satisfaction | | | |
|---|---|---|---|---|
| | Very Unsatisfied | Unsatisfied | Satisfied | Very Satisfied |
| Low | 48 | 43 | 21 | 10 |
| Moderate | 20 | 49 | 42 | 24 |
| High | 8 | 18 | 49 | 44 |

4.  *Social gradient in health* is a well-established term used in social sciences to describe the relationship between socio-economic status and health. One way of measuring social gradient in health is to examine the connection between a person's social class and their self-rated health. Use the hypothetical data in Table 10.14 to investigate the character (i.e. form and strength) of the relationship between the two variables.

**TABLE 10.14** Self-Rated Health and Social Class

| Self-Rated Health | Social Class | | |
|---|---|---|---|
| | Low | Medium | High |
| Poor | 28 | 18 | 7 |
| Fair | 54 | 32 | 15 |
| Good | 32 | 45 | 28 |
| Excellent | 13 | 21 | 21 |

5.  The hypothetical data from 100 countries shown in Table 10.15 displays *Infant mortality rates* by *Health-care spending per capita*. Are the two variables related? If so, what is the direction and strength of the relationship?

6.  Where in a bivariate table would you look for conditional distributions?

**TABLE 10.15** Infant Mortality Rates and Health-Care Spending per Capita

| Infant Mortality Rate | Health-Care Spending per Capita | | |
|---|---|---|---|
| | Low | Medium | High |
| Low | 2 | 10 | 16 |
| Medium | 13 | 18 | 8 |
| High | 26 | 6 | 1 |

# Set 2: SPSS Practice Questions

1.  Recode variables Q1D, Q1A, Q1B from the Student Health and Well-Being Survey into three-category (*dissatisfied*, *neutral*, *satisfied*) ordinal variables (nQ1D, nQ1A, nQ1B). Remember to set missing values to 9. (If necessary, see the SPSS section of Chapter 4 for review).

2.  Crosstabulate the recoded Satisfaction Rating: Mental Health (IV) with Q9 "In the past year, have you consumed soft drugs?" (DV). Remember to set missing values for Q9. What is the form of this relationship?

3.  Crosstabulate the recoded Satisfaction Rating: Level of Exercise (IV) with the recoded Satisfaction Rating: Overall Physical Health (DV).

4.  Assess the form and strength of the relationship in Question 3.

5.  Recode Q35 from a 5 category variable into a 4 category variable (nNQ35) by collapsing the *weekly* and *more than once a week* categories into one. Crosstabulate nNQ35 (IV) with nQ1D.

6.  Assess the form and strength of the relationship in Question 5.

7.  Crosstabulate Q6B—*My emotions seem to have a life of their own* (dependent variable) with Q6D—*I am able to control my level of anxiety* (independent variable). What is the form of this relationship?

8.  Crosstabulate Q6D—*I am able to control my level of anxiety* (dependent variable) with Q6E—*I can usually relax when I want* (independent variable). What is the form and strength of the relationship?

9.  Crosstabulate Q6E—*I can usually relax when I want* (dependent variable) with Q6B—*My emotions seem to have a life of their own* (independent variable). What is the form and strength of the relationship?

# STEP 5 Interpreting the Results

Bivariate tables are composed of qualitative independent and dependent variables: variables at either nominal or ordinal levels of measurement. Identifying the relationship between two variables requires calculating and interpreting both the form and strength of the relationship. Similarly, your interpretation of any association requires you to report on *both* the form and strength components.

After following the procedures for identifying form, you were able to decompose the effect into horizontal and vertical components. The horizontal effect was placed as an arrow over the independent variable, while the vertical effect was placed as an arrow beside the dependent variable. Expressing the form simply requires following these horizontal and vertical arrows and reporting the results. In general, form is expressed in this way:

As *X* [independent variable] changes from _____ [trend of the horizontal arrow], then *Y* [dependent variable] changes from _____ [trend of the vertical arrow].

For example, a statement of form might state:

As *Caloric consumption* increases from low to high, then *Weight* increases from light to heavy.

Regarding interpretations of form, here are some other useful observations:

- If the trend arrows between the adjacent columns are flat (i.e. neither up nor down), then this indicates that changing the independent variable makes no difference to the dependent variable. This is a sign of no association.
- If the independent variable includes three or more columns (values), it is not necessary that the form between the columns be consistent. In other words, the form trend is not always linear (i.e. *up, up* or *down, down*); it could easily be curvilinear (e.g. *up, down*). If it is, then simply report whatever trend is evident in the interpretation. For example, "Students' grades increase as they move from low to medium hours of part-time work, but decrease as they move from medium to high hours of part-time work."
- If both independent and dependent variables are ordinal, then it may be possible to describe the form using the *direct* and *inverse* terms introduced in Chapter 9. *Direct* means the attributes of the variables move in the same direction; *inverse* means they move in opposite directions. For example, there might be a direct relationship between amount studied and grades in a table. Remember, however, that your interpretations should make sense to an average person. Since most people are unfamiliar with the technical meaning of terms like *direct* and *inverse*, your form interpretation should still specify how the variables change in the relationship.
- If at least one of the variables in a table is nominal, then terms like *direct* and *inverse* will not apply—since the organization of categories of nominal variables is arbitrary.

Regarding strength, use the guidelines table to make your interpretations. Remember the calculated strength (average percentage differences) is not an interpretation. Interpretations of strength must report how much impact changing the independent variable has on the dependent variable. So write out your interpretations of strength between the independent and dependent variables in a simple English sentence.

# Chapter Summary

This chapter presented the tools for analyzing bivariate relationships when both the independent and dependent variables are qualitative (i.e. nominal or ordinal). The chapter discussed the procedures for obtaining and interpreting the form and strength of such associations. Having completed this chapter you should appreciate that:

- When associations are based on qualitative variables, the evidence is presented in table form.
- Bivariate tables present the intersection (crossing) of two univariate distributions.
- Tables are set up using standardized conventions.
- The analysis of a bivariate relationship includes calculating and interpreting both the form and strength of the association.
- A protocol exists for establishing the form of the connection in bivariate tables.
- A separate procedure exists for establishing the strength of the connection.

You now have the tools for examining bivariate relationships when the variables are qualitative. The next chapter presents the tools for examining bivariate relationships when the independent and dependent variables are quantitative.

# 11

# Scatterplot Analysis

## Overview

The previous chapter began your exploration of the region in the statistical maze dealing with the connection between variables. This chapter extends your understanding of statistical relationships. Bivariate analysis seeks to determine whether changing one (independent) variable is associated with a systematic (reliable) change in another (dependent) variable. If an association is detected, bivariate analysis seeks to identify what kind of connection exists between the variables and determine how much difference one makes to the other. The last chapter showed you how bivariate analysis is performed when both variables are qualitative (i.e. at the nominal or ordinal levels of measurement). These procedures centred on the analysis of data tables.

This chapter deals with bivariate analysis with quantitative variables, those that are at either the interval or the ratio level of measurement. In terms of the statistical selection checklist, here is the territory explored in this chapter:

## Statistical Selection Checklist

1. Does your problem centre on a descriptive or an inferential issue?

   ☑ Descriptive

   ☐ Inferential

2. How many variables are being analyzed simultaneously?

   ☐ One

   ☑ Two

   ☐ Three or more

3. What is the level of measurement of each variable?

   ☐ Nominal

   ☐ Ordinal

   ☑ Interval

   ☑ Ratio

In this chapter, you will learn about:

- how scatterplots are constructed;
- how tables are related to scatterplots; and
- how to identify and interpret the relationships in scatterplots.

Before we discuss scatterplots, a short reminder about levels of measurement is helpful.

Remember that variables measured at higher levels of measurement contain more information than variables measured at lower levels. Quantitative variables, therefore, contain more information than qualitative ones. Because qualitative variables contain less information, measurements using them are less precise. *Male* and *female*, for example, are loosely defined categories of the variable *Sex*. So is the ordinal difference between *disagreeing* and *strongly disagreeing* with the statement "I hate social statistics." Because quantitative variables contain more information, measurements using them are more precise. There is a specific difference between temperatures of 15 and 20 degrees Celsius. Likewise, there is a precise difference between having 1,000 or zero dollars in your bank account.

To take advantage of the information-rich measurements in quantitative variables, bivariate analysis employs a specific set of statistical techniques. These techniques are based on a type of graph called a scatterplot.

## STEP 1 Understanding the Tools

# Scatterplot Structure

Chapter 7 introduced you to charting and graphing techniques for single variables. These were univariate techniques for variables at different levels of measurement. Charts and graphs follow the axiom "a picture is worth a thousand words." In many instances, charts and graphs are preferred presentation techniques because these visual representations have intuitive appeal.

Scatterplots use the same principle. Scatterplots are bivariate graphs; instead of displaying changes in a single variable (which univariate charts and graphs do), scatterplots represent how two variables change simultaneously. In this sense, scatterplots are analogous to bivariate tables, which also display simultaneous change in two variables. Just as bivariate tables follow a set of conventions in their production, scatterplots have their own conventions. These are illustrated in Figure 11.1.

Several points about Figure 11.1 are worth highlighting. First, note that the independent variable and its attributes are arranged along the horizontal axis, and the dependent variable and its values along the vertical axis. This is very similar to what occurs in a bivariate table. The only difference is that, on a table, the independent variable and its scores are listed across the top, while the scatterplot has these listed on the bottom.

Another important difference relates to how the values of the variables are displayed. Bivariate tables display nominal or ordinal variables, whose categories have width. For this reason the attributes in bivariate tables are displayed as columns or rows. Scatterplots display quantitative (interval and ratio) variables, whose categories are incremental and precise. Therefore, these scores are displayed as points on the horizontal or vertical axis. These axes are lines which, in fact, are composed on an infinite number of points, indicating that the variables in a scatterplot are continuous.

The independent and dependent variables displayed in Figure 11.1 begin at zero and increase horizontally (for the independent variable) or vertically (for the dependent variable). It is possible, however, for continuous variables to have negative values (e.g. temperature, net worth) and scatterplots can accommodate this possibility by extending their axes. Full scatterplot displays are composed of four quadrants, as illustrated in Figure 11.2. In cases where the lowest scores of independent and dependent variables are zero, it is conventional to display just the upper-right quadrant, as in Figure 11.1.

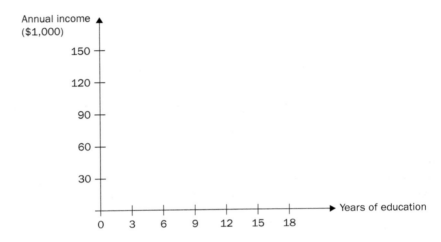

**FIGURE 11.1**  Conventional Scatterplot Configuration

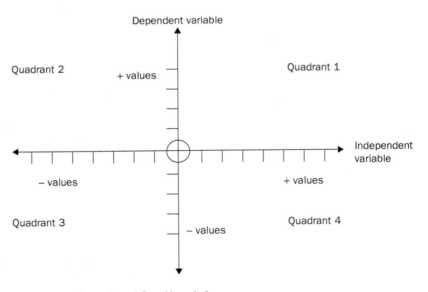

**FIGURE 11.2**  Four-Quadrant Scatterplot

# Scatterplot Content

With the structure of a scatterplot established, the next task is to enter its content. This is analogous to entering numbers into the cells of an empty bivariate table. In scatterplots, however, instead of entering digits that represent a number of cases, each case in the data set is entered as a specific point. This is accomplished through the use of ordered pairs.

To understand the use of ordered pairs, you need to remember that scatterplots are bivariate tools; they focus on the relationship between two variables. Therefore, the information on a scatterplot has to include information about *both* the independent and the dependent variables. Ordered pairs accomplish this task.

Ordered pairs are representations of the independent and dependent variable scores for every case in the data set. They are called pairs because, for each case, the independent variable score is paired (linked) to the dependent variable score. They are called ordered because each pair is expressed, in parentheses, as the independent variable score followed by the dependent variable score. The general form of ordered pairs is: (independent variable score, dependent variable score). So, for the case of a person who studied 15 hours for an exam (IV) and achieved a test grade of 76 per cent (DV), that case's ordered pair is (15, 76). Ordered pairs can be created for every case in a data set. So, for example, a sample of six students might have ordered pairs linking their study time and grades as follows: (15, 76) (10, 55) (5, 33) (20, 90) (13, 70) (9, 44).

Once ordered pairs are created, they can be graphed on a scatterplot, with the intersection of each pair of independent and dependent variable scores shown by a dot on the graph. Figure 11.3 plots the 6 ordered pairs for the study time–grades connection.

Notice that when ordered pairs from a data set are plotted, they are scattered over the quadrant. The range of horizontal scattering represents change on the independent variable, while vertical scattering represents dispersion on the dependent variable. Most importantly, the pattern of the dots on a scatterplot represents the relationship (intersection) of the independent and dependent variables.

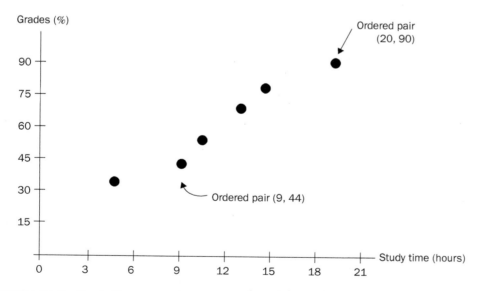

**FIGURE 11.3** Study Time and Grades Scatterplot

# Bivariate Tables and Scatterplots

Bivariate tables and scatterplots present the same thing in different forms. Both display the relationship between two variables. Bivariate tables display the association between categorical variables, while scatterplots display the connection between continuous variables. The difference in display is due to levels of measurement. Bivariate tables use lower levels of measurement, which contain less information and are therefore more rudimentary. Scatterplots employ higher levels of measurement, which contain more information and are therefore more sophisticated. Since higher levels of measurement have lower levels embedded in them, it should be possible to translate scatterplots into bivariate tables.

Figure 11.3 is a scatterplot relating the independent variable *Study time* to the dependent variable *Test grade*. Both are ratio variables: *Study time* measured in hours and *Test grade* measured in percentage. These variables could be reduced to ordinal levels by introducing some cutting points. For example, *Study time* could be organized into *low* and *high* categories, with *low* ranging from zero to 14 hours and *high* as above 14 hours. Similarly, *Test grade* could be organized into *low* and *high*, with *low* including scores under 56 per cent and *high* covering scores above that point. Figure 11.4 places these cutting points on the scatterplot.

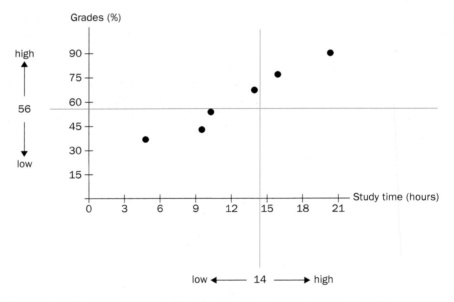

**FIGURE 11.4**  Study Time and Grades Scatterplot, Including Cutting
                 Points

The ordered pairs on the scatterplot can now be counted and translated into a bivariate table. In Figure 11.4, notice that there are three points in the area of *low study time–low test grade*. Similarly, there is one point in the *low study time–high test grade* area; zero points in the *high study time–low test grade* area; and two points in the *high study time–high test grade* area. These counts can be used to create a bivariate table of the same data, as illustrated in Table 11.1.

**TABLE 11.1** Bivariate Table of Study Time and Test Grade

| Test Grade | Study Time | |
|---|---|---|
| | Low | High |
| High | 1 | 2 |
| Low | 3 | 0 |

This exercise illustrates that both bivariate tables and scatterplots are simply different ways of presenting the same thing—the association (connection) between two variables. The higher level of measurement information contained in scatterplots can always be translated into lower-level bivariate tables. This translation, however, comes at the cost of a loss of information. For example, in Table 11.1 you know there are three cases in the *low–low* cell, but you no longer know precisely where these are located—which you do know in the scatterplot. That is the reason why you cannot create accurate scatterplots from bivariate tables. Qualitative variables do not provide enough information to plot the precise location of cases.

# Interpreting Scatterplots

A scatterplot is a graph of the relationship between two continuous variables. Each dot on the graph represents a specific case in the data set. The collection of dots (the scatter) represents the pattern of the relationship between the two variables. Interpreting the relationship between the variables in a scatterplot requires making sense of the pattern of dots. So, the question becomes: How do you look at the scatterplot to understand its pattern?

In the last chapter you learned that interpreting a bivariate table required speaking to two considerations, form and strength. Scatterplots use variables at higher levels of measurement than those of a bivariate table; they contain more information. Therefore, interpreting a scatterplot requires speaking to three considerations: form, extent, and precision.

## Interpreting Form in a Scatterplot

The idea of form in scatterplots mirrors the interpretation in bivariate tables. Form refers to the structure or nature of the connection between the variables. The form tells you how the dependent variable changes when you change the independent variable in a particular way. In scatterplots, researchers try to identify one of four possible forms: direct, inverse, curvilinear, and formless. The identification of a direct (positive) relationship means that the independent and dependent variables change in the same direction. In direct relationships, both variables increase or they both decrease. Inverse (negative) relationships portray the variables as moving in opposite directions; if one increases, the other decreases. In both direct and inverse relationships, the same form occurs across all values of the variables. The relationships are linear. This is not the case in curvilinear relationships where, as the independent variable moves in one direction, the dependent variable moves in differing directions—sometimes increasing, sometimes decreasing. Formless relationships indicate that changing the independent variable has no systematic connection to change in the dependent variable.

In scatterplots, form refers to the *underlying direction of change* in the data. The ordered pairs (dots) on a scatterplot display the data, so determining form is a matter of examining the scatter on the graph and identifying the underlying direction of change between the variables. Note that this is not a matter of connecting the dots; it is a matter of identifying a pattern in the dots. Determining the pattern is assisted by familiarity with the following general templates:

- Direct relationships: The dots are patterned from the lower left to the upper right of the graph.
- Inverse relationships: The dots are scattered from the upper left to the lower right.
- Curvilinear relationships: Several patterns are possible, including U-shaped curves, inverted U-shapes, and others.
- Formless relationships: The dots on the scatterplot are random, indicating no underlying patterns (form) to the relationship between the variables.

Figure 11.5 illustrates each of the form templates.

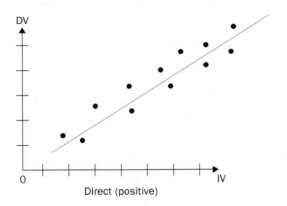
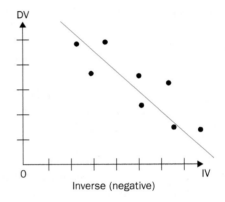

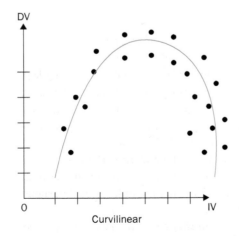
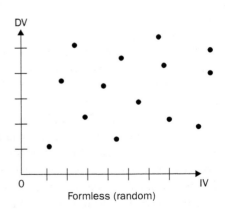

**FIGURE 11.5**  Scatterplot Forms

In Figure 11.5 lines have been imposed on the ordered pairs to help you see the underlying structure of the relationship. These lines are called regression lines. **Regression lines** identify the underlying structure of the relationship between variables displayed in a scatterplot. Regression lines are also called best-fit lines. This synonym is used because the regression line comes closest to all the data points and, in that sense, it best fits (approximates) the actual data. Chapter 14 provides you with the technical tools for determining exactly where the regression line should be drawn for any particular data set. For current purposes, it is sufficient that you appreciate the general form, which you can draw freehand on any scatterplot.

A useful way to think about regression lines is in terms of signal and noise. Often the dots on a scatterplot are messy; in the sense that they are scattered about, they are *noisy*. A best-fit regression line tries to discern the underlying *signal* (message) in the noise. This parallels what you do when you try to interpret the message of someone who is speaking to you in a noisy room, or when you try to make sense of some piece of abstract art or the bizarre behaviour of some individual. In all cases you are trying to determine the signal within the noise—trying to find out what the underlying message is.

## Interpreting Extent in a Scatterplot

The second concept used to interpret the relationship between the continuous variables in a scatterplot is extent. **Extent** refers to how much impact (difference) changing the independent variable has on the dependent variable. Let's say one research study reports a direct relationship between attending class and grades, and another study reports a direct relationship between studying and grades. In terms of form, both independent variables (attending class and studying) have the same kind (form) of relationship to the dependent variable (grades). Both connections are direct. However, these connections might be very different in terms of extent. The first study might report that every hour of attending class increases your grades by 2 per cent, while the second study may find that every hour of studying increases grades by 6 per cent. In this situation, studying makes three times as much difference to grades as attending class; its impact is much greater.

To determine the extent of a relationship in a scatterplot, look at the slope of the regression line. The steeper the slope the greater the impact (extent) of one variable on another variable. Chapter 14 introduces you to the technical calculation of slope, but the following formula provides the general idea.

$$\text{Slope} = \frac{\text{Change in the dependent variable}}{\text{Change in the independent variable}}$$

Applying this formula to the earlier illustration of how attending class and studying affect grades makes the point. In the first study, grades (the dependent variable) changed 2 per cent for every hour of class attendance. The slope of this relationship is *2 ÷ 1 = 2*. In the second case, grades changed 6 per cent for every hour of studying, yielding a slope of *6 ÷ 1 = 6*. Figure 11.6 plots these slopes as regression lines.

It is clear in Figure 11.6 that the notions of form and extent are different. While form tells you the kind of connection between the variables, extent tells you about impact. In general, extent is only relevant for linear relationships (i.e. direct and inverse ones) because, in these forms, the slope is constant. The slope of a curvilinear relationship (see Figure 11.5) differs depending on what part of the connection you are examining. Therefore, extent cannot be captured in curvilinear relationships with a general term.

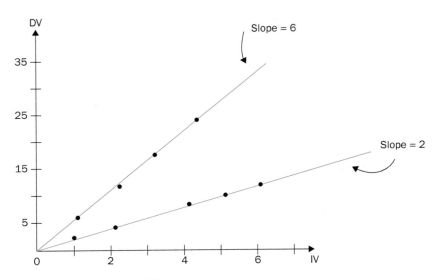

**FIGURE 11.6**  Comparison of Slopes

## Interpreting Precision in a Scatterplot

Both the form and extent of a relationship displayed in a scatterplot are determined by examining and interpreting the regression (best-fit) line. The regression line, however, is a theoretical construct. The actual data are the ordered pairs (dots) on the scatterplot; the regression line is something drawn (imposed) on the graph after the fact to help us discern the signal in the pattern. It is common that a regression line touches few (if any) of the actual data points on a scatterplot.

The regression line is drawn so it comes closest to the actual data points. It is a best approximation. The concepts of form and extent are derived from looking at and interpreting the regression line. They are not determined by looking directly at the actual data points. Since the regression line is the best fit, it is reasonable to ask, How good a fit is it? Precision addresses this question. **Precision** refers to how well the regression line approximates the actual data and, by extension, how appropriate the form and extent descriptions are.

An analogy is helpful for understanding the concept of precision. If you buy a new suit at a typical store, the clerk will try to find the suit on the rack that best fits you. (This best fit is equivalent to the best-fit [regression] line on a scatterplot.) Note, however, that just knowing the selected suit (or regression line) is the best fit *does not tell you how good a fit it is*. If you are an exceptionally tall or large person, for example, the best-fitting suit in a conventional store might be a terrible fit. Alternatively, if you are fortunate, an off-the-rack suit might fit you perfectly.[1]

The precision of a scatterplot tells us how well the best-fit line approximates the location of the actual data points. The closer the data points are to the regression line (i.e. the more tightly they are located around the line), the greater the precision of the relationship. In the case of perfect precision, all the data points fall exactly on the line. From this extreme, the precision gets progressively weaker until the points approach a random pattern. Figure 11.7 illustrates these possibilities.

---

[1]Another analogy might be helpful. If your mother asks you how you did on the last test and you answered "I did my best," your mother does not know how well you actually performed. Your "best" on that day could have been an excellent performance or a very poor one. Similarly, attending only to the best-fit regression line does not tell you how well the line approximates the actual data, which are the dots on the scatterplot.

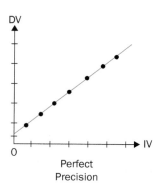
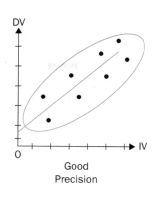
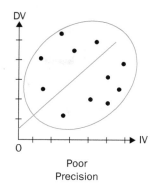

**FIGURE 11.7** Differing Precision

Again, Chapter 14 will provide you with the technical tools for calculating and interpreting precision. For present purposes, it is sufficient to understand the general idea of precision. As illustrated in Figure 11.7, it is helpful to draw a bubble around the actual data points of a scatterplot that contains a regression line. The shape of this bubble gives you a sense of the precision of the relationship. As noted, if the precision is perfect and all data points fall exactly on the line, then there will be no bubble. Perfectly precise relationships are exceedingly rare, however. In reality, the graph will almost always contain some scatter. If the bubble is cigar-shaped, then the precision is quite good. As the bubble expands and becomes more ball-shaped, the precision becomes poorer.

This covers the basic concepts and principles you need to know in order to understand and interpret scatterplots. The next sections provide you with details of how to apply these tools, first through hand calculations and then by using computer software. These are followed by opportunities for you to practise.

## STEP 2 Learning the Calculations

# Creating a Scatterplot

Here are the steps for taking data from a data set and organizing it into a scatterplot. Note: It is much easier to perform these steps if you do so on graph paper, which provides a standardized grid.

1. Draw and label the axes of the scatterplot graph. The name of the independent variable goes below the horizontal ($x$) axis, and the name of the dependent variable beside the vertical ($y$) axis. The marks on the horizontal and vertical axes represent the categories (scores) of the variables. Remember that, since scatterplots graph interval or ratio variables, the categories must be equidistant from one another. These units must be identified. Finally, include an appropriate title.

2. Create ordered pairs connecting the independent and dependent variables for all cases. Ordered pairs identify, in parentheses, the score on the independent variable, followed by the score on the dependent variable, for all cases in the data set.

3.    Place the ordered pairs on the scatterplot. For each ordered pair, identify the point on the graph at which each pair of independent and dependent variable scores intersects. Place a dot on the graph at each of these points.

Here is an example that applies these steps to a sample of 10 cases, examining the relationship between *Years worked* (IV) and *Income* (DV). The independent variable is measured in *years*, while the dependent variable is measured in *thousands of dollars*.

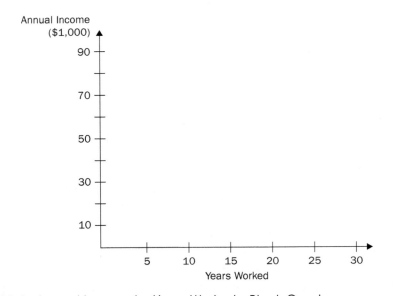

**FIGURE 11.8**  Annual Income by Years Worked—Blank Graph

1.    Draw and label the graph.
2.    Create ordered pairs.

From a data set, the following information has been extracted and placed into ordered pairs. Note that, for each pair, the first number is the independent variable score for that case, and the second number is the dependent variable score. Remember that the dependent variable is in *thousands of dollars* units, so 55, for example, means $55,000.

Participant 1 (11, 41)
Participant 2 (8, 55)
Participant 3 (5, 23)
Participant 4 (17, 66)
Participant 5 (12, 67)
Participant 6 (4, 30)
Participant 7 (1, 18)
Participant 8 (3, 24)
Participant 9 (27, 67)
Participant 10 (13, 38)

3.    Place the ordered pairs on the scatterplot.

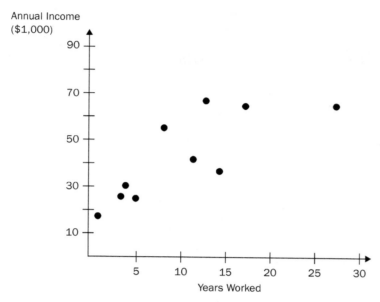

**FIGURE 11.9**    Annual Income by Years Worked—Scatterplot

# Analyzing a Scatterplot

Note: Some of the points in this section involve interpretation, which are also reviewed in Step 5: Interpreting the Results.

1.  Draw a regression line on the scatterplot. Look at the shape of the points on a scatter-plot and draw a *straight line* that you think comes closest to all the data points. If the pattern of dots is random, the line will be horizontal. Otherwise the line will move either from the lower left to the upper right, or from the upper left to the lower right.

    This hand-drawing technique is often called the "eyeball method," since you are drawing the regression line by sight. While the results are not extraordinarily accurate (since different people might sight the line slightly differently), it is often good enough for visual representation.

2.  Draw a bubble around all of the data points. Here you are drawing an envelope around the actual data points. The shape that results will vary. Sometimes it will be a circle, but it can be the shape of an elongated ellipse (cigar-shaped). It might be fat or thin.

3.  Interpret the form of the regression line. Pay attention to *only* the regression line drawn on the graph. Decide whether the line is positive (moves from the lower left to the upper right) or negative (moves from the upper left to the lower right). Write a simple sentence that interprets this form with respect to the variables under consideration. Note: If the line is flat, this signifies no relationship between the variables.

4.  Interpret the extent of the regression line. Again, pay attention to *only* the regression line. Make a judgment about *how steep* it is—regardless of whether it is positive or negative. Write a simple sentence that interprets the extent (impact) of the independent variable on the dependent variable. The steeper the line, the higher the impact.

5.  Interpret the precision of the regression line. Pay attention to *only* the circle (enve-lope) you drew around the data points. Write a simple sentence that interprets how well (precisely) the regression line approximates the actual data points. If the circle is large, it is a poor fit. As the shape of the envelope becomes more elongated and thin, the better the fit.

Here are the steps in scatterplot interpretation applied to our example.

1.& 2.  Draw the regression line and envelope. In Figure 11.10, a regression line and bubble that encircles the data points have been added to our scatterplot in Figure 11.9.

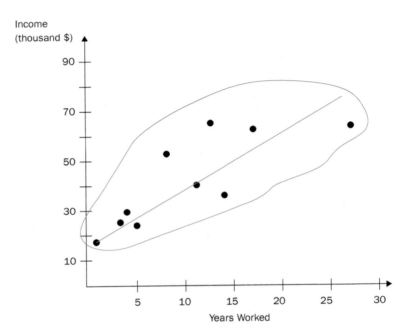

**FIGURE 11.10**  Annual Income by Years Worked—Regression
Line and Bubble

3.  Interpret the form. In this graph, the regression line flows from the lower left to the upper right, which indicates that the relationship between the variables is positive (direct). In direct relationships, the variables change together *in the same direction*. Therefore, the form interpretation is as follows:

As the number of *Years worked* increases, the level of *Income* also increases.

4.  Interpret the extent. In this graph, the regression line is not very steep. This modest slope suggests that changing the independent variable makes only a modest difference (impact) on the dependent variable. Therefore, the extent interpretation is as follows: Increasing the number of years worked makes only a modest difference to improvements in income.

5.  Interpret the precision. The envelope around the data points is quite tightly clustered around the regression line. The envelope has a thin cigar shape. This suggests that the regression line is a fairly good approximation of the actual data. Therefore, the interpretation of precision is as follows: The descriptions of the form and extent of the relationship between *Years worked* and *Income* is a good approximation of the actual case.

Notice that this construction and interpretation of a scatterplot is quite labour intensive, and this was the case when the sample included only a small number of cases (10). For realistic sample sizes involving hundreds or thousands of cases, this approach would be very time-consuming. Fortunately, as the next section shows, these steps can be applied to large samples with relative ease by employing computing software. It is important, however, to understand what the computer is doing with such ease, and the best way of appreciating that is through the hands-on steps just completed.

## STEP 3 Using Computer Software

# Generating a Scatterplot

The relationship between quantitative (i.e. interval and ratio) variables can be visually displayed in IBM SPSS Statistics software ("SPSS") through the creation of a scatterplot. Below are the instructions to generate a scatterplot using SPSS.

## Procedure

As with all statistical analysis, first you should run frequency distributions of your independent and dependent variables to ensure that all values not needed in the scatterplot are set to *missing* (e.g. *no response* values). See Chapter 3 for instructions on how to set desired values to *missing*.

The procedure for producing a scatterplot between an independent and a dependent variable is as follows.

1.  In the SPSS menu, select Graphs → Legacy Dialogs → Scatter/Dot.
2.  A new window will appear. Click once over the Simple Scatter box (it is the default selection) and click Define.
    a.  Move the independent variable Q16F (*Psychological well-being—Self-acceptance*) to the *X*-axis.
    b.  Move the dependent variable Q22 (*Rosenberg self-esteem index*) to the *Y*-axis.
    c.  Select OK.
    d.  In the Output window of SPSS, the scatterplot will be displayed (see Figure 11.2).
3.  Next, add a best-fit line (i.e. the regression line).
    a.  To do this, first double-click the scatterplot in the Output window to activate the pivot table (see Chapter 3).

b.  A new window will appear, which is titled the Chart Editor. If this procedure is done correctly, new menu selections should appear in the SPSS Menu bar (see Figure 11.11).

i.  In the new Menu bar, go to Elements → Fit Line at Total.

ii.  Check the Linear radio button, which is found in the Fit Method section.

iii.  Select Apply.

c.  To exit the pivot table, click anywhere in the Output window. The best-fit line should now appear in your scatterplot (see Figure 11.12).

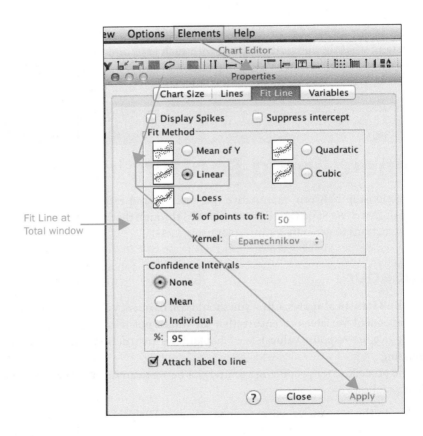

**FIGURE 11.11**  Creating Scatterplots

Source: Reprint Courtesy of International Business Machines Corporation, © International Business Machines Corporation.

## STEP 4 Practice

You now have the tools for constructing scatterplots either by hand or using computing software. To solidify your understanding you need to practise applying the statistical and software procedures.

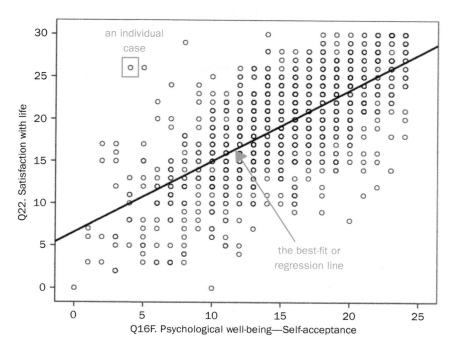

**FIGURE 11.12** Output for Scatterplot

Source: Reprint Courtesy of International Business Machines Corporation, © International Business Machines Corporation.

This section provides you opportunities to practise what you have learned about scatterplot analysis. The first set of questions uses hand calculations. The second set uses the SPSS procedures. For each set of questions:

1. Follow the procedural steps and complete the appropriate calculation (Set 1) or software application (Set 2).
2. Check your answers, using the Answer Key in the back of the text.
3. If your answer is incorrect, consult the Solutions section on the book's website. The Solutions provide a complete step-by-step analysis of how the answers are derived.

After you have completed the next section (Interpreting the Results), return to your calculations or output and *provide complete, written interpretations of each of the statistics you have generated.*

# Set 1: Hand-Calculation Practice Questions

1. Use the data below to complete this question.

   Years of Schooling

   | | |
   |---|---|
   | Bob | 8 |
   | John | 17 |
   | Mary | 20 |

| Ivan | 21 |
|---|---|
| Belinda | 12 |
| Rina | 14 |

Annual Income

| Bob | $35,000 |
|---|---|
| John | $50,000 |
| Mary | $55,000 |
| Ivan | $70,000 |
| Belinda | $40,000 |
| Rina | $45,000 |

a.　Construct ordered pairs for the data.

b.　Construct and label a scatterplot representing the relationship between these variables.

c.　Draw a freehand best-fit line and a precision bubble on the scatterplot.

d.　Interpret the relationship.

2.　Use the data below to complete this question.

Hours of Exercise per Week

| Tom | 8 |
|---|---|
| Kyle | 7 |
| Malik | 2 |
| Ian | 3 |
| Bjorn | 12 |
| Lu | 5 |
| Trong | 10 |
| Perry | 1 |

Body Mass Index (BMI)

| Tom | 24 |
|---|---|
| Kyle | 29 |
| Malik | 27 |
| Ian | 29 |
| Bjorn | 21 |
| Lu | 31 |
| Trong | 22 |
| Perry | 35 |

a.　Construct ordered pairs for the data.

b.　Construct and label a scatterplot representing the relationship between these variables.

c.　Draw a freehand best-fit line and a precision bubble on the scatterplot.

d.　Interpret the relationship.

3. Table 11.2 shows actual data from 2011 from 10 countries. It contains information about actual life expectancy at birth for both sexes and the total health expenditure per capita in each country.

   a. Construct a scatterplot representing the relationship between these two variables.
   b. Does the relationship appear linear? If so, draw a freehand best-fit line and a precision bubble on the scatterplot.
   c. Interpret the relationship. Comment on its form and extent.
   d. Do your freehand best-fit line and the precision bubble change if we eliminate the United States from the scatterplot? If so, how and why?

**TABLE 11.2** Country, Health Expenditure per Capita, and Life Expectancy

| Country | Total Health Expenditure per Capita (PPP int. $)** | Life Expectancy* |
|---|---|---|
| Australia | 3691.6 | 82 |
| Canada | 4520.0 | 82 |
| Chile | 1292.2 | 79 |
| Czech Republic | 1922.8 | 78 |
| Japan | 3174.3 | 83 |
| Mexico | 940.1 | 75 |
| Peru | 496.2 | 77 |
| Slovenia | 2518.9 | 80 |
| United Kingdom | 3321.7 | 80 |
| United States of America | 8607.9 | 79 |

*Life expectancy at birth for both sexes in 2011. Data retrieved from World Health Organization. http://apps.who.int/gho/data/node.main.688?lang=en

**Per capita total expenditure on health (PPP int. $) in 2011. Data retrieved from World Health Organization. http://apps.who.int/gho/data/node.main.78?lang=en

> **Tips for graphing:**
> Round the *Total health expenditure per capita* to the nearest hundred dollars (e.g. 3,691.6 = 3,700) and set the scale to 2 centimetres per each thousand dollars of health expenditure. Since the difference between the minimum (Peru at 500) and maximum (US at 8,600) scores equals 8,100, your X-axis would be about 16 centimetres long.

# Set 2: SPSS Practice Questions

Use the graphs procedure to construct a scatterplot with a best-fit line for each of the following bivariate relationships:

1. a. Q16F *Psychological well-being—Self-acceptance* (IV) and Q4 *Depression* (DV).
   b. Interpret the relationship.
2. a. Q4 *Depression* (IV) and Q14C *Avoidant coping* (DV).
   b. Interpret the relationship.
3. a. Q16B *Psychological well-being—Environmental mastery* (IV) and Q3B *PANAS positive affect* (DV).
   b. Interpret the relationship.
4. a. Q18 *Strength of religious faith* (IV) and Q16E *Psychological well-being—purpose in life* (DV).
   b. Interpret the relationship.

5.  Create a scatterplot with *Age* as the independent variable and *Satisfaction with life* (Q22) as the dependent variable. Describe the relationship.
6.  Create a scatterplot with *Purpose in life* (Q16E) as the dependent variable and *Coping with stress by seeking social support* (Q14A) as the independent variable. Describe the relationship.
7.  Create a scatterplot with *Age* as the independent variable and the *Self-esteem scale* variable (Q17) as the dependent variable. Describe the relationship.
8.  Create a scatterplot with *Strength of religious faith* (Q18) as the dependent variable and the *Depression index* (Q4) as the independent variable. Describe the relationship.

## STEP 5 Interpreting the Results

Scatterplots show the bivariate relationship between quantitative variables. Because these variables are at higher levels of measurement, they contain more information. This information allows more extensive interpretation of the connection between the variables. Whereas the interpretation of bivariate tables focuses on two considerations (form and strength), the interpretation of a scatterplot centres on three issues: form, extent, and precision.

This chapter introduced you to the ideas underlying each of these concepts. The technical calculation and interpretation of each is deferred to Chapter 14. So, for present purposes, your interpretations of scatterplots are limited to the following considerations.

# Interpreting Scatterplot Form

Here the choices are similar to those used in bivariate table analysis and utilize similar interpretations. You examine the regression line and decide whether it is direct, inverse, curvilinear, or formless. Remember, it is not enough to identify the kind of connection between the variables as one of these forms. You must interpret the selected form with respect to the variables under consideration. For direct relationships, you can follow the model:

As *X* [independent variable] increases, *Y* [dependent variable] also increases.

Example: As years of education increases, income also increases.

For inverse relationships:

As *X* [independent variable] increases, *Y* [dependent variable] decreases.

Example: As the number of children increases, marital happiness decreases.

For curvilinear relationships, you need to create a sentence that characterizes the changes. For example:

As *X* [independent variable] increases to a specific value, *Y* [dependent variable] increases, but after a specific value on *X* [independent variable] *Y* [dependent variable] decreases.

Example: As hours of studying increases to six hours per day, test performance increases, but after studying more than six hours daily, test performance decreases.

Finally, if the pattern is formless (i.e. no connection) then a statement like the following is appropriate:

> X [independent variable] and Y [dependent variable] have no connection to one another. As one changes, it makes no systematic difference to the other.

> Example: The amount of peanut butter consumed makes no systematic difference to the number of traffic tickets received.

# Interpreting Scatterplot Extent

This term applies to the slope of the regression line in cases where form is direct or inverse. If the form is identified as formless or curvilinear, extent does not have a meaningful interpretation. So look at the steepness of the direct or indirect regression line, and create a statement such as the following:

> Changing X [independent variable] makes [a lot of, a moderate, some, little—depending on steepness of slope] difference to Y [dependent variable].

> Example: Calorie consumption makes a lot of difference to body weight.

Notice that whether the form is direct or inverse is irrelevant to the interpretation of extent.

# Interpreting Scatterplot Precision

This term informs readers how accurately the form and extent interpretations capture the actual data pattern. At this point, simply look at the shape of the bubble drawn around the scatterplot and make an interpretation using the following statement:

> The descriptions of form and [if applicable] extent are a [very precise, adequate, poor—depending on shape of bubble] characterization of the relationship between X [independent variable] and Y [dependent variable].

> Example: The descriptions of form and extent are adequate characterizations of the relationship between income and happiness.

# Chapter Summary

This chapter presented the tools for analyzing bivariate relationships when both the independent and dependent variables are quantitative (i.e. interval or ratio). The chapter discussed the procedures for understanding and interpreting the form, extent, and precision of such associations. Having completed this chapter you should appreciate that:

- Bivariate quantitative analyses are conducted on scatterplots.
- Scatterplots are constructed using a set of standardized conventions.

- On a scatterplot, ordered pairs represent the data set evidence.
- Scatterplots are interpreted in terms of form, extent, and precision of the bivariate relationship.
- Form refers to the nature (structure) of the connection between the variables.
- Form is identified by examining the best-fit regression line.
- Form possibilities include direct, inverse, curvilinear, and formless relationships.
- Extent refers to the impact one variable has on another.
- Extent is identified by examining the slope of the regression line for direct and inverse relationships.
- Extent is not applicable for curvilinear or formless relationships.
- Precision refers to how well the regression line fits the actual data points (ordered pairs).
- Precision ranges from perfect (where all data points are on the regression line) to poor (where the data points are randomly distributed).
- Interpretations of form, extent, and precision require simple statements expressed in terms of the connection between the independent and dependent variables.

The previous chapter provided you with the tools for examining bivariate relationships between qualitative variables. This one provided the tools for doing similar work for quantitative variables. These basic tools become more sophisticated with the addition of statistical calculations. The next chapter introduces you to the features of a family of such statistical summaries for bivariate relationships.

# 12

# Proportional Reduction in Error Statistics

## Overview

This chapter takes a break from learning specific statistical procedures. On your trip through the statistical maze, you can think of this as a rest-stop, where you pause to get your bearings. Chapters 13 and 14 introduce an entire set of statistical tools and it is important to orient yourself to where you are headed.

The last two chapters introduced the basic tools for bivariate analysis. For qualitative variables, these tools centred on analyzing tables; for quantitative variables, scatterplot analysis tools were introduced. The goal in both cases was the same—to identify whether a relationship exists between the variables and, if so, to describe the nature of the connection.

But, as you probably learned from doing the practice questions, analyzing crosstabs and scatterplots is somewhat awkward. The awkwardness stems from arbitrariness. Consider the issue of interpreting strength in tables, which requires using a table of guidelines. These guidelines inform you that if a table difference was 55 per cent the relationship is *moderate*, but if the difference reaches 56 per cent, then the relationship is *strong*. Like all the interpretive guidelines for tables, this distinction is an arbitrary social construction. Similar vagueness is evident when you interpret extent or precision in scatterplots. How steep does a slope have to be in order for the impact to be *strong*? How tight do the dots have to cluster around the regression line to qualify as *very precise*? Again, there is a large element of arbitrariness in these decisions.

This chapter introduces a family of statistics that move the analysis of bivariate relationships, whether in tables or scatterplots, from vagueness to clarity. This set of statistical procedures is in the "proportional reduction in error" family. This chapter introduces the central concepts, principles, and logic that this family of statistics shares, including:

- marginal and relational predictions rules;
- ways of counting prediction errors; and
- generic calculation formulas.

Understanding these root ideas will ease your understanding of the specific statistics covered in upcoming chapters. This understanding begins with appreciating the link between association and prediction.

# Association and Prediction

Bivariate analysis focuses on understanding the character of a relationship between an independent variable and a dependent variable. The signature of a relationship appears when change in an independent variable is associated with a systematic change in a dependent variable. Relationships are associations. What is true in social life holds true for statistical connections. When your life makes no systematic difference to someone else's existence, you have no relationship; you and the other are strangers. The more systematic difference your life makes to another's, the stronger the association between the two of you.

The systematic difference that characterizes relationships has implications for prediction. Chapter 1 noted that prediction is one of the central purposes of science. Prediction is the flip side of explanation. Where explanation provides an understanding of events that occurred in the past, prediction centres on forecasting future events. If two events are systematically linked, then it should be possible to predict one from the other. Everyday examples help us appreciate this important point.

Imagine a person randomly chosen from the population of the nation of Kyrgyzstan. Since you probably don't even know where this nation is located, it is very likely that the randomly selected person is a complete stranger to you. You and that stranger have no relationship. This lack of connection has implications for prediction. Imagine you are attending your best friend's wedding this Saturday evening. Based on this information about where you will be on Saturday, can you predict what the Kyrgyzstan stranger will be doing? Of course, you have no idea. The reason you cannot make a reasonable prediction is that your experience and that of the other halfway round the world are not linked.

By contrast, imagine you are getting married and that you and your partner have a strong relationship. Because of this relationship, if someone asks where your spouse will be on Saturday evening, you can confidently predict their attendance at the wedding. The general principle is the following: *Relationships can be translated into predictions; the stronger the relationship, the better the predictions.* This principle is the one that governs the family of statistical procedures called proportional reduction in error.

# PRE Family Characteristics

Proportional reduction in errors statistics are identified by the acronym *PRE*. PRE statistics are a family, since there is a group of them. And, like other statistical procedures, selecting the appropriate PRE statistic depends on the level of measurement of the variables involved. Later chapters will introduce you to the details of calculating and selecting appropriate PRE statistics. Our current interest lies in understanding their foundation.

The idea of a family of PRE statistics indicates that, although each statistical procedure in the family is unique, they share a set of common characteristics. In fact, all PRE statistics share four common features, which are linked to the principle that stronger relationships result in better predictions. These features include a marginal prediction rule, a relational prediction rule, ways of counting errors, and a standardized calculation formula.

## Marginal Prediction Rule

The first characteristic of every PRE measure of association is that it utilizes a marginal prediction rule. To understand this point, you need to remember that the term *marginal distribution* is a synonym for *frequency distribution*. A marginal distribution is simply a frequency distribution of a particular variable.[1] In this case the marginal distribution of interest is that of the dependent variable.

A dependent variable is the outcome variable in a bivariate relationship. As a variable, it can take on different scores. The particular score it takes, however, depends on the value of the independent variable. Our interest is always in predicting the value of the dependent variable. For instance, we might be interested in predicting whether or not you will have a date this weekend, how much you like chocolate ice cream, or what score you will achieve on the next test.

One way we can predict a particular person's score on the dependent variable is by knowing about the distribution of that variable in a sample. The distribution of a variable in a sample occurs in the form of a frequency (marginal) distribution. If we knew what percentage of students typically have weekend dates, or what proportion of a representative sample likes or dislikes chocolate ice cream, or what the average test performance is, then we could make a prediction about any individual's score. In other words, we could use general information about the dependent variable (its marginal distribution) to make predictions about particular cases.

There are different ways of using marginal distribution information to predict dependent variable scores of specific cases, but the easiest way to appreciate the approach comes from measures of central tendency. In Chapter 5 you learned that a single variable (univariate distribution) can be summarized in terms of where its typical cases occur. This is what measures of central tendency do; they tell you what the average case of a variable looks like. For example, if you didn't know anything about a particular student but knew that the class average on a test was 67 per cent, then it is reasonable to predict that the student's test score was 67 per cent. This is what the **marginal prediction rule** does; it predicts the values for particular cases on the dependent variable from information about the dependent variable distribution (i.e. the distribution of all cases across the values of the dependent variable).

This procedure is called a marginal prediction rule because it predicts the dependent variable from information about its frequency distribution. The procedure is called a rule because the prediction procedure is followed for all cases in a sample. In other words, a marginal prediction rule is not simply used to predict the dependent variable value of one specific case; it is used to predict the dependent variable score of every case in a sample.

All PRE statistics use some marginal prediction rule. They don't use the same marginal prediction rule, but all have one. Note that the marginal prediction rule makes predictions about the dependent variable using only information about the dependent variable. Marginal predictions do not rely on any outside information. They basically predict that all cases will have a typical or average dependent variable score.

---

[1]Recall that marginal distributions get their name from their location on the sides of a bivariate table. See Chapter 10.

## Relational Prediction Rule

Chapter 1 emphasized that knowledge is a form of mental map. If you know about some topic, you will have a better understanding of how it operates. Take the case of a car engine. If your car doesn't start one morning, and you are completely ignorant of internal combustion engines, then you have no clue about what the trouble might be. What you need is someone who is knowledgeable in this matter—an auto mechanic. The mechanic's knowledge provides her or him with a mental map of how the system works, and this knowledge can be used to predict the problem and remedy it.

In short, knowledge informs us how things are connected. The mechanic knows how the fuel injection is related to ignition; a medical doctor knows how pill consumption is related to symptom relief; an economist knows how interest rates are related to unemployment. Such knowledge (understanding) improves the predictions these practitioners make. The doctor can predict that if you take the prescription, you will experience symptom relief. Similarly the economist can predict that if interest rates rise, unemployment will increase.

Knowledge uses relational prediction rules. A **relational prediction rule** is a way of predicting dependent variable scores using information about independent variable scores. This rule is called *relational prediction* because it uses knowledge about an independent–dependent variable relationship. It is called a rule because this approach is used to make predictions about dependent variable scores for every case in a data set.

All PRE statistics have some sort of relational prediction rule. Different specific PRE statistics use different relational prediction rules, but they all have some such rule. Whenever you give people advice, you are employing your own version of a relational prediction rule. The message "Don't drink and drive" is predicting that if you consume alcohol before driving (the independent variable) some undesirable dependent variable consequences will follow (accidents, fines, etc.). The idea that "more studying improves grades" includes the prediction that both variables will move in the same direction.

Relational prediction rules specify how two variables are connected. As such, relational prediction rules rely on understanding the form of the relationship. Chapters 10 and 11 emphasized that form specifies the character of the connection between two variables. This form is used to make predictions about dependent variable values based on information about an independent variable score.

Both marginal and relational prediction rules aim to predict the dependent variable score of each case in a data set. The difference in how they make their dependent variable predictions is very important. A marginal prediction rule only utilizes information about how the dependent variable scores are distributed. A relational prediction rule utilizes information about the nature of the connection between the independent and dependent variable.

Take the example of the long-married couple, *X*–*Y*. Imagine you want to predict where *Y* will be next Saturday evening. Based on information about the 52 Saturday nights of the past year, you understand that *Y* spent 42 of them watching television at home. Based on only this information, you would predict that *Y* will be at home watching TV next Saturday evening. This is equivalent to using a marginal prediction rule. You only have information about the dependent variable (i.e. *Y*'s general Saturday evening behaviour), and based on this information you make your prediction.

Now you are provided with some additional information: *Y*'s partner, *X*, is having a birthday celebration with his or her family in Montreal next Saturday. You are now in a position to change your prediction based on a relational prediction rule. You know that this couple has a long-standing relationship and that the birthday party is a family celebration. In all probability, then, $X$ and $Y$ will be attending the party together. So you will now predict that $Y$ will be in Montreal next Saturday. This prediction is based not on $Y$'s typical Saturday night conduct (the marginal prediction rule), but on knowing (1) that $X$ and $Y$ are strongly connected and (2) that $X$ is going to be in Montreal on Saturday night (the relational prediction rule).

This example illustrates how relational prediction rules perform: They use knowledge of the form of the connection and the specific independent variable value to make a prediction about the dependent variable. The root idea is that knowledge (i.e. understanding relationships) helps improve prediction.

## Ways of Counting Errors

In reality, almost no predictions are perfect. Mechanics often are mistaken about why cars don't start; sometimes physicians prescribe pills that don't bring relief; economists' predictions about future unemployment rates are commonly in error. Predictions are mistaken because reality is complicated and dynamic, while our understanding is limited. But knowledge doesn't have to be perfect to be useful. While every additional hour you study does not necessarily translate into better grades, studying does help academic performance. It is bad advice to completely disregard what physicians, economists, or any other experts say.

Deciding the relevance of anything requires answering the "compared to what?" question. Answering this question with "compared to perfection" is not reasonable, since all knowledge is imperfect. Would you ever have a date or buy a car or take a vacation if your standard was perfection? A more sensible approach is to compare something with something else *in the real world*. Is this date, car, or vacation better than others that are accessible to you?

PRE measures of association use this "compared to what?" principle in determining whether and how strongly an independent variable is related to a dependent one. The logic used involves counting errors. Both marginal and relational prediction rules involve predicting the expected dependent variable score for each case in a data set. Some of these predictions will be accurate, others will be in error. The logic of PRE measures of association requires counting these errors. An example helps illustrate the principles involved.

A marginal prediction rule predicts the dependent variable scores for each case in a data set by stating that each case will be the average. If the goal is to predict the last test score for each student in your class, a marginal prediction rule might use the mean which, say, is 63 per cent. So, person 1's test grade would be predicted to be 63 per cent; person 2's score would also be predicted to be 63 per cent; person 3's test score prediction would be the same; and so on for each student in your class.[2] For the students in the class whose last test grade was actually 63 per cent, the marginal prediction rule would be accurate. However, for students whose grade was something other than 63 per cent, the marginal prediction rule would have generated an error. For PRE purposes, errors using the marginal prediction rule are designated as $E_1$. $E_1$ stands for the number of errors using the first prediction rule, which is the marginal prediction rule.

---

[2] Note that it is called a marginal prediction "rule" because the same prediction procedure is used for every case.

PRE measures also make predictions of dependent variable scores using a relational prediction rule. A relational prediction rule predicts a dependent variable score for each case based on knowledge of each case's independent variable score. If we believed that studying increases grades, then, if person 1 reported they studied 100 hours for the test, we might predict their grade to be 96 per cent (i.e. higher than average). And if person 2 reported studying only 20 minutes for the test, their predicted grade might be 35 per cent (i.e. lower than average). Using the direct relationship form between studying and grades, the relational prediction rule would predict test scores for each student in the class. And, again, sometimes these predictions would be accurate and sometimes they would be in error. The number of errors using the relational prediction rule is designated as $E_2$. $E_2$ stands for the number of errors using the second (relational) prediction rule.

To summarize, every PRE measure of association has a way of predicting dependent variable scores for each case using only information about the dependent variable (marginal prediction rule). Each has a way of predicting dependent variable scores for each case based on information about the case's independent variable scores (relational prediction rule). And every PRE measure has ways of counting the number of errors made using the marginal prediction rule ($E_1$) and the relational prediction rule ($E_2$). The final characteristic shared by all PRE measures is a standardized calculation formula based on this information.

## Standard Calculation Formula for PRE

The calculation of every PRE measure of association can be reduced to the following formula:[3]

$$\text{PRE} = \frac{E_1 - E_2}{E_1}$$

Let us look at this formula and see what it says. First you determine how many errors you make using the marginal prediction rule ($E_1$). This tells you how many errors you make predicting peoples' scores on the dependent variable, *when you do not have any information about the independent variable*. Then, you determine how many errors you make using the relational prediction rule ($E_2$). This tells you how many errors you make predicting dependent variable scores *when you know the independent variable and its relationship to the dependent variable*. The *difference* between these errors (i.e. $E_1 - E_2$) tells you *how many fewer errors (mistakes) you make predicting the dependent variable when you know the independent variable, as opposed to not knowing the independent variable*.

Applying this logic to our example, if 30 errors were made predicting students' grades by guessing that each student's score is the mean (i.e. using the marginal prediction rule), and 15 errors were produced by predicting each student's score on the basis of how much he or she studied (i.e. using the relational prediction rule), then errors have been reduced by 15 mistakes ($E_1 - E_2 = 30 - 15$).

Let's be very clear what this means: This improvement in prediction (i.e. reduction in errors) occurs *because of the relationship of the independent variable to the dependent variable*. In other

---

[3]As you will see in later chapters, not every PRE measure uses this formula in its computations. Sometimes there are more efficient calculation procedures. But whatever specific formula is used, it could be translated into this standardized form.

words, if the independent variable was unrelated to the dependent variable (for instance, if instead of having information about how much each student studied we had information on how much they liked ice cream), there would then be no reduction in errors between $E_1$ and $E_2$. Knowing how much a person likes ice cream has no connection to test grades. This is a critical point: *The stronger the relationship the independent variable has to the dependent variable, the greater will be the reduction in prediction errors.* If, for instance, studying was perfectly related to grades, then knowing how much a student studied would completely eliminate all prediction errors; knowing how many hours students studied would allow us to correctly predict their grades 100 per cent of the time.

Next notice that the PRE formula divides $E_1$ minus $E_2$ by $E_1$. The numerator of the equation provides the *actual* reduction in errors. The magnitude of this result, however, is entirely a function of how many cases there are in the sample (i.e. how many predictions are being made). A study using 50 subjects would almost certainly get a smaller $E_1 - E_2$ result than a study containing 500 subjects. The PRE calculation formula makes a correction to adjust for this fact. Instead of being a reduction in error statistic, it is a *proportional* reduction in error statistic.

Proportions help make results comparable and, when multiplied by 100, become percentages. So, by using the PRE formula, we can determine the percentage of error reduction based on our understanding of how the independent and dependent variables are related. This calculation can be interpreted in terms of the extent to which the independent and dependent variables are related. More extensive relationships provide better predictions (i.e. a greater reduction in the error rate).

To summarize, all statistics included in the PRE family of statistics share the following four features:

- a marginal prediction rule;
- a relational prediction rule;
- a way of counting errors for both marginal and relational prediction rules; and
- the use of a common formula: PRE $= (E_1 - E_2) \div E_1$.

These four features are shared by all PRE measures of association. As you shall see in the next two chapters, there are differences in the application and calculation of different statistics in this family of measures. The *differences* in these statistics are derived from the fact that the *specific* marginal and prediction rules are different for each statistic, and the *specific* ways of counting errors are different. In short, the specific members of the family of PRE statistics are unique but they all share a common template.

As always, different statistical tools are developed for variables of different levels of measurement. Chapter 13 presents PRE measures used for qualitative variables.

# Chapter Summary

This chapter introduced the concepts and principles of the proportional reduction in error (PRE) family of statistics. The next chapters show you how specific PRE statistics are computed and interpreted. Understanding these statistics, however, requires an appreciation of the fundamentals discussed in this chapter. Having completed this chapter you should appreciate that:

- If two variables are associated, then this knowledge can improve the accuracy of dependent variable predictions.
- The PRE family of statistics share four common features.

- Marginal prediction rules predict the value of the dependent variable for every case utilizing only information about the dependent variable's distribution.
- Relational prediction rules predict the value of the dependent variable for every case utilizing knowledge of each case's independent variable score.
- Errors in marginal prediction rule predictions are designated as $E_1$ and prediction errors using the relational prediction rule as $E_2$.
- All PRE statistical calculations can be reduced to the $(E_1 - E_2) \div E_1$ formula.
- The stronger the relationship the independent variable has to the dependent variable, the greater will be the proportional reduction in prediction errors.

With these fundamental concepts and principles in place, you are ready to learn how to calculate and interpret specific PRE measures of association. The next chapter introduces the PRE measures that are appropriate for qualitative variables.

# 13

# Statistics for Categorical Relationships

## Overview

The last chapter oriented you to the proportional reduction in error (PRE) family of statistics. The statistics in this family are measures of association; that is, they quantify the relationship between an independent and a dependent variable. Like all statistics, PRE measures are summaries; they distill into an interpretable form the information about how two variables in a data set are related.

Bivariate relationships occur in the form of tables and scatterplots. Bivariate tables use categorical variables measured at the nominal and ordinal levels. Scatterplots use continuous variables measured at the interval and ratio levels. The discussion of tables (Chapter 10) and scatterplots (Chapter 11) provided the basic ideas for understanding bivariate connections, but also revealed that rough-and-ready interpretations contain considerable ambiguity. PRE measures reduce this ambiguity by providing a more reliable understanding of the nature of the independent–dependent variable connection.

As you know, selecting an appropriate statistical tool requires answering three questions: Does your problem centre on a descriptive or an inferential issue? How many variables are being analyzed simultaneously? What is the level of measurement of each variable? Selecting the appropriate statistic from the PRE family requires answering these questions. All PRE measures have the same answers to the first two questions, since (1) they deal with descriptive issues and (2) they are concerned with bivariate connections. Therefore, selecting among different PRE measures reduces to identifying the level of measurement of the variables involved in the relationship.

Different PRE measures exist for different possible pairings of independent and dependent variables. In this chapter you learn about PRE measures for nominal and ordinal variables. Specifically, you will learn about two PRE measures, lambda and gamma. In terms of the statistical selection checklist, the contents of this chapter are:

## Statistical Selection Checklist

1. Does your problem centre on a descriptive or an inferential issue?

   ☑ Descriptive

   ❏ Inferential

2. How many variables are being analyzed simultaneously?

    ☐  One

    ☑  Two

    ☐  Three or more

3. What is the level of measurement of each variable?

    ☑  Nominal—lambda

    ☑  Ordinal—gamma

    ☐  Interval

    ☐  Ratio

In this chapter you will learn about:

- the marginal and relational prediction rules for both lambda and gamma;
- the ways to calculate the values of lambda and gamma statistics; and
- how to interpret lambda and gamma in proportional reduction in error terms.

The journey begins with lambda.

## STEP 1   Understanding the Tools

# Lambda

The symbol representing lambda is $\lambda$. **Lambda** is an appropriate PRE measure when *both* the independent and dependent variable are at the *nominal* level of measurement. Since lambda is part of the PRE family of statistics, it shares the four features that characterize this family. The easiest way to understand lambda is to discuss its marginal and relational prediction rules, its method of counting errors, and its calculation. We will discuss these four features by applying them to Table 13.1.

**TABLE 13.1**  Sex and Religiosity

| Regular Religious Practice | Sex | | Total |
|---|---|---|---|
| | **Male** | **Female** | **Total** |
| Yes | 47 | 375 | 422 |
| No | 83 | 249 | 332 |
| Total | 130 | 624 | 754 |

Before discussing lambda's features, let's review the features of this table. First, *Sex* is the independent variable and includes the categories *male* and *female*. It is a nominal variable. The dependent variable, *Religiosity*, is also a nominal variable, covering the categories *yes* and *no*.[1]

---

[1]*Religiosity* refers to how actively engaged an individual is in his or her spiritual commitment. In this case, this property is measured by whether or not an individual is engaged in regular religious practice.

The sample on which this table is based includes 754 respondents. The independent variable includes 130 males and 624 females; the dependent variable includes 422 who regularly participate in religious practice and 332 who do not. The conditional distributions inside the table indicate how the marginal distributions (*Sex* and *Religiosity*) are related.

## Marginal Prediction Rule for Lambda

A marginal prediction rule is a method of predicting a dependent variable score for each case based only on information about the dependent variable's distribution. In Table 13.1, the distribution of the dependent variable is its marginal distribution, which states that 422 respondents regularly participate in religious practice and 332 do not. The following scenario operationalizes lambda's marginal rule.

Imagine you are in a room and the 754 respondents are in the hallway. All respondents are going to individually enter the room and, when they do, you have to predict whether they do or do not regularly engage in religious practice. Based on the marginal distribution of the dependent variable, you have to develop a rule that you use for predicting each case. The rule you develop is a marginal prediction rule.

The marginal prediction rule for lambda is *predict the mode of the dependent variable*. The mode of a distribution is the category of a variable that occurs most frequently.[2] In Table 13.1, the mode is *yes*, since this category has more cases than the *no* value. Applying this rule to the scenario, as each case entered the room, using this marginal prediction rule, you would predict that every one of 754 cases does regularly engage in religious practice.

## Relational Prediction Rule for Lambda

A relational prediction rule is a method of predicting the dependent variable score for each case based on knowledge of each case's independent variable score. The prediction scenario now changes. Each case still enters the room individually, but before you predict their dependent variable score, they tell you their independent variable score. In the case of Table 13.1, they tell you whether they are male or female.

The relational prediction rule for lambda is *predict the mode of each conditional distribution of the independent variable*. Recall that conditional distributions of the independent variable are the *columns inside the table*. Table 13.1 contains two conditional distributions, one for each category of *Sex*. In the prediction scenario, before you make a dependent variable prediction, the respondents are telling you which column (conditional distribution) they are in. The relational prediction rule for lambda tells you to use this information and predict the mode of that column (conditional distribution).

In Table 13.1, for the *male* conditional distribution, the mode is *no*; for the *female* conditional distribution, the mode is *yes*. The relational prediction rule for Table 13.1 becomes "If *male*, then *no*; if *female*, then *yes*." Notice how this relational prediction rule differs from the marginal prediction rule, which stated "*yes* for all cases." Now you are making a contingent prediction based on your understanding of the relationship between the independent and dependent variables. If the respondent is male, then your prediction is quite different than if the respondent is female.

---

[2]See Chapter 5 for a review of the mode.

## Method of Counting Errors for Lambda

Lambda's third characteristic is a way of counting errors in prediction. All PRE statistics have two error counts, $E_1$ and $E_2$. $E_1$ is the way of counting errors made when the marginal prediction rule is used. $E_2$ is the way of counting errors when the relational prediction rule is used.

For **lambda**, $E_1$ is *the number of cases that do not fall into the modal category of the dependent variable*. For lambda, $E_2$ is *the number of cases that do not fall into the modal categories of the independent variable's conditional distributions*. Applying these methods of counting errors to Table 13.1 illustrates their use.

The mode of the dependent variable (*Religiosity*) is *yes*, since it contains 422 cases. $E_1$ is the number of cases that do not fall into this category—in other words, the number of cases that fall into the *no* category. Therefore, $E_1$ is 332. $E_1$ tells you that if you use the marginal prediction rule (predict every case as *yes*), you will be wrong for 332 of the cases.

The relational prediction rule for Table 13.1 is "If *male*, then *no*; if *female*, then *yes*." This rule recognizes that the mode of the *male* conditional distribution is *no*, and the mode of the *female* conditional distribution is *yes*. $E_2$ is the number of cases that do not fall into the modes of these conditional distributions. In the case of males, the relational prediction rule predicts *no* for all 130 cases, and is in error for 47 of these cases. For females, the relational prediction rule predicts *yes* for all 624 cases, and is wrong 249 times. Using the relational prediction rule for Table 13.1 results in *47 + 249 = 296 errors*, which is $E_2$. $E_2$ tells you that, across all 754 cases in this sample, if you know a respondent's *Sex* before predicting their *Religiosity*, you will make 296 errors.

## Lambda Calculation Formula

Lambda uses the standard PRE calculation formula $(E_1 - E_2) \div E_1$. Applying the $E_1$ and $E_2$ errors just determined to the formula results in the following calculation:

$$\text{lambda} = \frac{332 - 296}{332} = \frac{36}{332} = 0.11$$

A couple of points about this calculation are worth emphasizing. First, the numerator (36) informs us how many fewer errors we make in predicting *Religiosity* if we know a respondent's *Sex*, compared to not having such information. In this case, we see that knowing *Sex* does improve predictions about *Religiosity*. It does so by generating 36 fewer prediction errors across all 754 cases. Making fewer errors is evidence of a relationship.[3] Second, the overall result (0.11) expresses this error reduction in *proportional* terms. Step 5, Interpreting the Results, reports on how this proportional result is interpreted.

# Gamma

**Gamma** is a PRE statistic used when both the independent and dependent variables in a table are at the ordinal level of measurement. The symbol for the gamma statistic is $\Upsilon$. Like all

---

[3]If your professor guarantees you would make 36 fewer errors on the next test by wearing a green hat, you would probably do it. However mysterious, you would conclude that wearing a green hat is related to getting an improved grade.

statistics in the PRE family, gamma uses marginal and relational prediction rules, ways of counting errors, and a standard calculation formula. Before turning to how gamma specifically uses each of these considerations, two central points about the logic of gamma need to be established.

## Pairs of Cases for Gamma

Chapter 4 introduced the idea of sample size, expressed by the symbol $N$. Whether in univariate analyses (Chapters 4 through 8) or bivariate analyses (since Chapter 9), you are accustomed to working with data sets containing a sample of a particular size. The sample size reports the number of cases the study is based on: the number of respondents from whom data are collected. Table 13.1, for instance, includes 754 respondents, which is its sample size ($N$). It is conventional to think of sample size in terms of specific cases, in the sense that Table 13.1 reports evidence from 754 respondents.

Understanding the logic of gamma requires you think of the sample size in a bivariate table in a different way. Instead of thinking of the sample size as discrete, individual cases, gamma examines the sample in terms of *pairs of cases*. The small sample size in Table 13.2 illustrates this point. Table 13.2 contains a sample of only three cases. Formally, the number 1 should appear in each of the cells instead of a name, but including names helps personalize the point about pairs of cases. Imagine the data in Table 13.2 comes from students in a grade 1 class, where their reading and arithmetic competence have been rated by the teacher.

**TABLE 13.2** Reading and Arithmetic Competence, Sample Size 3

| Arithmetic Competence | Reading Competence | | |
| --- | --- | --- | --- |
| | Low | Medium | High |
| Low | John | | |
| Medium | | Paula | Lester |
| High | | | |

In this table the independent variable is *Reading competence* and the dependent variable is *Arithmetic competence*. Both variables are ordinal, as evidenced by the fact that the categories are rank-ordered. The underlying question addressed by the evidence is whether reading and arithmetic competencies are associated.

As noted, the first step in understanding the logic of gamma is to appreciate that this PRE statistic examines pairs of cases. It does not look at John, Paula, and Lester as individuals; it focuses on pairs. So, for Table 13.2, gamma would focus on the following pairs: John–Paula; John–Lester; Paula–Lester. Notice that this sample of three cases generates three pairs, not six pairs. Reversing the names in a pair does not create a new pairing. A Paula–John is no different than a John–Paula pair.

As a general rule, the number of distinct pairs in a sample is determined by the following formula: $N(N - 1) \div 2$. Applying this formula to Table 13.2, where the sample size ($N$) is 3, creates $3 \times (3 - 1) \div 2 = 3 \times 2 \div 2 = 6 \div 2 = 3$ pairs. Table 13.1, whose sample size is 754 contains $754(753) \div 2 = 567,762 \div 2 = 283,881$ pairs.

## Comparing Pairs Ordering for Gamma

The second distinct feature of gamma builds upon its attention to pairs of cases. Once pairs are established, gamma *compares the ordering of each pair on the independent and dependent variables*. The small sample in Table 13.3 helps illustrate this point.

The first column of Table 13.3 lists the three pairs derived from the sample in Table 13.2. The names of each pair are abbreviated by their first letters. The second column in Table 13.3 lists the *ordering of each pair on the independent variable*. Examining this column informs us that John has less reading competence than Paula, Paula has less reading competence than Lester, and John has less reading competence than Lester.[4] The third column in Table 13.3 lists the *ordering of each pair on the dependent variable* and informs us that John has less arithmetic competence than Paula, Paula has more arithmetic competence than Lester, and John has less arithmetic competence than Lester.

**TABLE 13.3**  Comparing Ordering of Pairs

| Pairs | Pair Ordering on Reading | Pair Ordering on Arithmetic | Ordering on Independent and Dependent Variables |
|---|---|---|---|
| J–P | J < P | J < P | same |
| P–L | P < L | P > L | opposite |
| J–L | J < L | J < L | same |

## MATH TIPS

### More or Less Than

Symbols are useful shorthand; a single squiggle can convey the meaning of several words. Here are two useful symbols.

The symbol < means "less than," meaning that the score of the number on the left is lower than the score to the right. In the example above, John's score is lower (<) than Lester's.

The symbol > means "more than," meaning that the score of the number on the right is lower than the score on the left. In the example above, Paula's score is more than (>) Lester's score.

The fourth column in Table 13.3 illustrates the second distinctive feature in understanding the logic of gamma. This column *compares the ordering of each pair on the independent and dependent variables*. Notice that this column contains the terms *same* and *opposite*. The term *same* means that the specific pair has the same ordering on independent and dependent variables. So, for the J–P pair, J has less than P on the independent variable, and J also has less than

---

[4]The same point could be stated as: Lester has more reading competence than John, Paula has more reading competence than John, Lester has more reading competence than Paula. Either formulation is equally acceptable.

P on the dependent variable. This pair is ordered the same way on both variables. This contrasts with the P–L pair. On the independent variable P has less than L, but on the dependent variable P has more than L. This pair has an opposite ordering on the two variables.

To summarize, the logic of gamma requires a different perspective on two accounts. First, instead of looking at the sample in terms of individual respondents, gamma examines it in terms of pairs of cases. Second, gamma compares the ordering of each pair on the independent and dependent variables—and reports this ordering as either *same* or *opposite*. From these two basic points, we can turn to the four distinctive PRE features of gamma.

## Marginal Prediction Rule and $E_1$ for Gamma

All PRE measures have a marginal prediction rule, a way of predicting scores on the dependent variable without information about the independent variable. They also have a means of counting the number of errors made using this rule ($E_1$). To understand gamma's marginal prediction rule you must keep in mind that this statistic focuses on *pairs of cases* and the *ordering of pairs of cases*. So, for gamma, the marginal prediction rule question translates into a rule for predicting *the order of each pair* on the dependent variable.

To appreciate this point, let's return to the simple situation in Table 13.2, which contained three pairs. For this situation the marginal prediction rule question is, What rule can be used for predicting the *order* of each pair (i.e. J–P, P–L, J–L) on *Arithmetic competence*? Imagine these three pairs of grade 1 students in the hallway. The first pair walks into your office with paper bags covering their entire bodies. (The paper bags are required because you must make your prediction *without any information about other variables*. The paper bags keep such additional information [e.g. sex, appearance, social-class indicators, etc.] from you.) The marginal prediction rule question asks, What rule would you follow to predict which student in the pair has greater arithmetic competence? Remember, you are required to use the rule you develop for every pair of cases.

The rule developed under these conditions is the marginal prediction rule. It is the rule used to predict the ordering of pairs on the dependent variable for all cases in the data set. *For gamma, the marginal prediction rule is "guess."* Now, this may seem like an odd rule, but what else can you do? You have no idea which one of the pair has greater arithmetic competence, since you are ignorant of any information about them. So all you can do is guess.

Marginal prediction rules are imperfect; they always result in errors. The number of prediction errors made using the marginal prediction rule is identified as $E_1$. If you follow gamma's marginal prediction rule ("guess"), how many errors will occur? Since guessing the ordering of any pair has a 50–50 chance of being correct, $E_1$ is *half the number of pairs*. In the Table 13.2 example which contains three pairs, $E_1$ is 1.5 errors (i.e. 3 ÷ 2). For Table 13.1, which contains 283,881 pairs, $E_1$ is 141,940 errors.

In summary, gamma's marginal prediction rule is to guess the ordering of each pair on the dependent variable. This rule generates errors for half the pairs for which predictions are made. The next task is to determine gamma's relational prediction rule and error count.

## Relational Prediction Rule and $E_2$ for Gamma

A relational prediction rule is a method for predicting the dependent variable using knowledge of the independent variable and its connection to the dependent variable. Since gamma focuses on pairs of cases and their ordering on the variables, the relational prediction rule asks: What rule predicts the order of pairs on the dependent variable when you know their ordering on the independent variable? This relational prediction rule becomes clearer when applied to the simple example in Table 13.2. For example, if you know that Paula has less reading competence than Lester (P < L), what rule predicts this pairs' ordering on arithmetic competence? Remember that the relational prediction rule is used across *all pairs* of cases. Whatever rule is used on the P–L pairing must also be used on the J–P pairing and the J–L pairing.

In fact, there are only two choices for gamma's relational prediction rule. The practical task is to select one of these choices. The first alternative is "always predict *same* order," while the second alternative is "always predict *opposite* order." The first alternative means that whatever ordering a pair has on the independent variable, predict it will have the same ordering on the dependent variable. If P < L on the independent variable, predict P < L on the dependent variable; if J < P on the independent variable, predict J < P on the dependent variable; etc.

The second alternative relational prediction rule states that whatever ordering a pair has on the independent variable, predict the *opposite* ordering on the dependent variable. If P < L on the independent variable, predict P > L on the dependent variable. Repeat this same prediction process for every pair of cases in the data set.

Every pair of cases in a data set will have either the same or the opposite ordering on the independent and dependent variables. This fact is illustrated in the final column of Table 13.3, which compares the ordering of each pair on the two variables. For every pair, the result is either *same* or *opposite* ordering. The relational prediction rule, however, requires you to select *one* of these alternative rules and apply it to *all* the cases. The question becomes: How do you know which relational prediction rule to select? The answer is: Select the alternative which occurs more frequently in the data set. This is illustrated in Table 13.3.

The final column in Table 13.3 reports how each pair is ordered on the independent and dependent variables. Each pair has either the same or the opposite ordering. Since the relational prediction rule has to apply to all cases, select the alternative which occurs most frequently in this comparison column. In the final column of Table 13.3, *same* occurs twice while *opposite* occurs only once. Therefore, for Table 13.2, the relational prediction rule should be "always predict same order." In other words, for every pair in the data set, whatever ordering they have on the independent variable, you should predict they have the same ordering on the dependent variable.

Like marginal prediction rules, relational prediction rules are imperfect; applying them results in prediction errors, which are identified as $E_2$. For gamma, $E_2$ is the number of pairs for which the relational prediction rule yields incorrect predictions. If the prediction rule is "always predict same order," $E_2$ will be the number of pairs which have the opposite order. Similarly, if the prediction rule is "always predict opposite order," $E_2$ will be the number of pairs which have the same ordering. Table 13.3 indicated that the best prediction rule is "always predict same order." The last column of this table indicates that this prediction rule would yield correct results for two of the pairs, and would generate an error for one pair.[5] So, for Table 13.2, $E_2$ is 1.

---

[5]Specifically, for the P–L pair, P < L on reading competence but P > L on arithmetic competence. Using the "predict same order" relational prediction rule for this pair results in a prediction error, since the pair has the opposite ordering on the independent and dependent variables.

## Gamma Calculation Formula

Gamma is a member of the PRE family of measures of association. As part of this family, its calculation utilizes the $(E_1 - E_2) \div E_1$ formula. Applying the $E_1$ and $E_2$ errors related to Table 13.2 produces a gamma result of 0.33. $E_1$ is 1.5 and $E_2$ is 1, so:

$$(E_1 - E_2) \div E_1 = (1.5 - 1) \div 1.5 = 0.5 \div 1.5 = 0.33$$

## A First Complication: Tied Pairs

The basic ideas for gamma are now in place. You know the marginal prediction rule and the procedure for obtaining $E_1$. You also know how to determine the correct relational prediction rule and how to use it to obtain $E_2$. The $E_1$ and $E_2$ results can then be used in the standard computational formula to generate the gamma result. You have seen how all these steps apply in a simple example that included three cases.

Table 13.4 makes the example slightly more complicated by adding one more student to the sample. This student is Bono. Adding Bono to the mix results in Table 13.4.

**TABLE 13.4** Reading and Arithmetic Competence, Sample Size 4

| Arithmetic Competence | Reading Competence | | |
|---|---|---|---|
| | Low | Medium | High |
| Low | John | | |
| Medium | | Bono | Lester |
| High | | Paula | |

Let's review the gamma procedure on this new table. Using the $N(N - 1) \div 2$ formula, it is evident that the table contains six pairs. The pairs and their ordering on the independent and dependent variables are listed in Table 13.5.

**TABLE 13.5** Comparing Ordering of Pairs

| Pairs | Pair Ordering on Reading | Pair Ordering on Arithmetic | Ordering on Independent and Dependent Variables |
|---|---|---|---|
| J–P | J < P | J < P | same |
| P–L | P < L | P > L | opposite |
| J–L | J < L | J < L | same |
| J–B | J < B | J < B | same |
| P–B | P = B | B < P | ? |
| L–B | L > B | L = B | ? |

Comparing this table to the previous one (Table 13.3) reveals some important differences. First, when the P–B pair is ordered on the independent variable, and when the B–L pair is ordered on the dependent variable, we see that the equal sign (=) is used. The equal sign designates that the pair is tied on the independent variable. **Tied pairs** are those that have the same score on either the independent or dependent variable, or both.

Tied pairs create a problem for gamma. The problem is evident in the fourth column of Table 13.5, where the ordering of each pair on the independent and dependent variables is compared. The only legitimate alternatives for this column are *same* or *opposite*. When pairs are tied they cannot be ordered in this way—which is why the fourth column contains the question marks. Ordering of the pairs with question marks is not possible because they are neither the same nor the opposite on both variables.

Gamma handles the problem of tied pairs in a simple way; it ignores them. *The calculation of gamma is restricted to untied pairs.* For calculation purposes, Table 13.4 uses four pairs, not the six identified in Table 13.5. The P–B and B–L pairs are not included in the analysis. Consequently, gamma for Table 13.5 is $(E_1 - E_2) \div E_1 = (2 - 1) \div 2 = 0.5$.

## A Second Complication: Practical Calculations

The examples in Tables 13.3 and 13.5 used unrealistically small sample sizes. Real statistical studies don't contain just three or four cases; they typically contain hundreds, often thousands, of cases. Since gamma deals with pairs of cases, a large number of pairs becomes a serious management problem. A sample size of 500, for instance, generates 124,750 pairs of cases. The task of creating a comparison table that lists the pairs (like Table 13.5) becomes practically impossible. So another method is required.

The practical solution is linked to the fact that gamma can be calculated from two pieces of information—the number of same-ordered pairs, and the number of opposite-ordered pairs. In fact, generating these two pieces of information (contained in the final column of Table 13.5) is the purpose of creating comparison tables. What is really required, then, is an alternative way of generating the number of same- and opposite-ordered pairs in a table. Examining Table 13.6 clarifies how this is accomplished.

**TABLE 13.6**  Reading and Arithmetic Competence, Realistic Sample Size

| Arithmetic Competence | Reading Competence | | | |
| | Low | Medium | High | Total |
| --- | --- | --- | --- | --- |
| Low | 50 | 28 | 12 | 90 |
| Medium | 35 | 42 | 20 | 97 |
| High | 16 | 19 | 47 | 82 |
| Total | 101 | 89 | 79 | 269 |

The rules that generate the number of same- and opposite-ordered pairs in a bivariate table of two ordinal variables require a particular table set-up. Specifically, the table must be organized so that, from the upper-left corner, *the variables change in the same direction.* This is highlighted by the coloured arrows imposed on Table 13.6. Notice that, from the upper-left corner, the independent variable (*Reading competence*) increases and the dependent variable (*Arithmetic competence*) increases. Here the variables are moving in the same direction (both increasing), which is what's required.[6]

––––––––––

[6]It would also be acceptable for the table to be arranged so both ordinal variables decrease from the upper-left corner. What is not acceptable is if one variable increases from that point and the other decreases.

Now let's look at the cells *in the table*, which are highlighted in colour on the table. It is important to focus only on cells in the table (i.e. conditional distributions) and exclude the marginal distributions from consideration. With this cells-only focus, here are the rules for generating the number of pairs with same and opposite ordering.

- Same ordering: Multiply each cell frequency by the sum of the cell frequencies down *and* to the *right*. Sum the products.
- Opposite ordering: Multiply each cell frequency by the sum of the cell frequencies down *and* to the *left*. Sum the products.

Applying these rules to Table 13.6 helps demonstrate their operation and logic. Let's begin by calculating the number of same-ordered pairs in the table.

Following the same-ordered rule requires the following computations:

1. Begin with 50 and multiply it by the sum of 42 + 20 + 19 + 47.
2. Move to the next cell frequency (28) and multiply it by the sum of 20 + 47.
3. Move to the next cell frequency (12) and multiply it by nothing!—because there are no frequencies down *and* to the right of this number *within* the table.
4. Move to the next cell frequency (35) and multiply it by the sum of 19 + 47.
5. Move to the next cell frequency (42) and multiply it by 47.
6. Multiply the remaining cell frequencies (20, 16, 19, 47) by nothing because there are no frequencies down *and* to the right of these frequencies *within* the table.

Each of the computations results in the following totals:

1. $50 \times 128 = 6{,}400$
2. $28 \times 67 = 1{,}876$
3. $12 \times 0 = 0$
4. $35 \times 66 = 2{,}310$
5. $42 \times 47 = 1{,}974$
6. Nothing

The last part of the rule states to sum these products. Doing so generates *12,560 same-ordered pairs* in the table.

To convince yourself this is correct, let's examine the pairs generated in the first computation:

$$= 50 \times (42 + 20 + 19 + 47)$$

Think about what the 50 represents. It represents 50 students in the study who had both *low reading competence* and *low arithmetic competence*. Although we don't have access to the information, these 50 individuals have names. The 50 in the upper-left cell could be substituted with actual names. John, from Table 13.4, might be one of these students. Let's compare these 50 students to the 42 students whose *Reading competence* and *Arithmetic competence* were both *medium*. If you compare the 50 and the 42 students in terms of *Reading competence*, you see that the 50 (who score *low*) had less competence than the 42 on the independent variable. If you compare the 50 and the 42 in terms of *Arithmetic competence*, you see that the 50 have less competence than the 42 on the dependent variable. In other words, comparing these 50 students with the 42 students, we observe the *same ordering* on the independent variable and the dependent variable. On both variables the 50 have less competence than the 42. If you perform the same analysis for every other set of cells following the rule, you will see that they are all correct; they all have the same ordering on both variables.

Generating the number of opposite-ordered pairs in a table uses a similar logic but a slightly different rule. Opposite-ordered pairs are generated by multiplying each cell frequency by the sum of the cell frequencies down *and* to the *left*, and then summing the products. Here is this rule applied to Table 13.6; following the opposite-ordered rule requires the following computations:

1. Begin with 12 in the upper-right cell and multiply it by the sum of 42 + 35 + 19 + 16.
2. Move to the next cell frequency (28) and multiply it by the sum of 35 + 16.
3. Move to the next cell frequency (50) and multiply it by zero.
4. Move to the next cell frequency (20) and multiply it by the sum of 19 + 16.
5. Move to the next cell frequency (42) and multiply it by 16.
6. Multiply the remaining cell frequencies (35, 16, 19, 47) by nothing because there are no frequencies down *and* to the left of these frequencies *within* the table.

Each of the computations results in the following totals:

1. $12 \times 112 = 1{,}344$
2. $28 \times 51 = 1{,}428$
3. $50 \times 0 = 0$
4. $20 \times 35 = 700$
5. $42 \times 16 = 672$
6. Nothing

The last part of the rule states to sum these products. Doing so generates *4,114 opposite-ordered pairs* in the table.

In short, following specific rules generates the number of pairs in a table that have the same ordering on both variables, as well as the number of pairs that have opposite ordering. The final task is using this information to calculate gamma.

## Gamma Calculation

As part of the PRE statistical family, gamma uses marginal and relational prediction rules, ways of counting errors, and a conventional formula. The marginal prediction rule for gamma is to guess the ordering of each pair on the dependent variable, and $E_1$ is half the number of pairs for which predictions are made. Gamma's relational prediction rule is "use the greater of same or opposite pairs ordering" and $E_2$ is the smaller of the number of either same- or opposite-ordered pairs. Using $N_s$ to stand for the number of same-ordered pairs and $N_o$ to stand for the number of opposite-ordered pairs, these points can be expressed as follows:

$$E_1 = \frac{N_s + N_o}{2}$$

$$E_2 = \text{Smaller of } N_s \text{ or } N_o$$

Applying these ideas to the calculations on Table 13.6, where $N_s$ is 12,560 and $N_o$ is 4,144, produces the following estimates of errors:

$$E_1 = (N_s + N_o) \div 2 = (12{,}560 + 4{,}144) \div 2 = 16{,}704 \div 2 = 8{,}352 \text{ errors}$$

$E_2$ = Smaller of $N_s$ or $N_o$ = 4,144 errors (since it is smaller than 12,560)

These $E_1$ and $E_2$ numbers can now be used to calculate gamma for Table 13.6, using the standard PRE formula:

$$\text{gamma} = (E_1 - E_2) \div E_1 = (8{,}352 - 4{,}144) \div 8{,}352 = 4{,}208 \div 8{,}352 = 0.50$$

Step 5: Interpreting the Results shows you precisely how to interpret this result. For now, you can tell that Table 13.6 shows some relationship between students' reading competence and their arithmetic competence. This is evident in the numerator, which shows that a reduction of 4,208 errors in predicting *Arithmetic competence* occurs when you know *Reading competence*. This error reduction occurs because the variables are related. If they were not associated then the errors would remain at 8,352.

You now have a thorough understanding of the logic gamma uses. The next section provides you with a systematic approach to calculating this PRE statistic. Then you will see how to have IBM SPSS Statistics software ("SPSS") generate the result. This is followed, as always, by practice questions that solidify your understanding. The final section, Step 5, explains interpretation.

## STEP 2 Learning the Calculations

# Calculating Lambda

Here are the steps required to calculate lambda.

1.  Two preliminary points:
    a.  Double-check to make sure the table is set up in the conventional form—independent variable over the columns, dependent variable along the left side.
    b.  Remember that the computational formula for lambda is $(E_1 - E_2) \div E_1$.
2.  Calculate $E_1$, the number of errors made using the marginal prediction rule. $E_1$ is the number of cases that are *not* in the modal category of the marginal (frequency) distribution of the dependent variable.
3.  Calculate $E_2$, the number of errors made using the relational prediction rule. $E_2$ is the number of cases that are *not* in the modal categories of each conditional distribution (column) of the independent variable.
4.  Compute lambda by entering $E_1$ and $E_2$ into the computational formula.

Let's apply these steps for calculating lambda using an example that has more categories in the independent and dependent variables. It is a bit more complicated than the 2 × 2 tables illustrated earlier, but the steps are still the same.

In this fictitious example (Table 13.7), we are examining the influence of religious affiliation on political party support.

1.  The independent variable (*Religious denomination*) is on the top and the dependent variable (*Political affiliation*) is on the left side, so the table is set up correctly. The formula for lambda is $(E_1 - E_2) \div E_1$.

**TABLE 13.7** Political Party Affiliation by Religious Denomination

| Political Affiliation | Religious Denomination | | | | |
|---|---|---|---|---|---|
| | Christian | Muslim | Other | None | Total |
| Conservative | 10 | 9 | 5 | 14 | 38 |
| Liberal | 14 | 12 | 10 | 6 | 42 |
| NDP | 11 | 4 | 25 | 10 | 50 |
| Total | 35 | 25 | 40 | 30 | 130 |

2. The marginal distribution of *Political affiliation* (the dependent variable) includes 38 (Conservatives), 42 (Liberals), and 50 (NDP). The modal category is NDP, which has 50 cases. So $E_1$ is $38 + 42 = 80$. In other words, if we predicted *Political affiliation* for the 130 people in the sample without any knowledge of *Religious denomination*, we would be in error 80 times.

3. The independent variable has four categories (*Christian, Muslim, other, none*) and $E_2$ involves the conditional distributions of each of these values. For *Christian*, the modal category is 14, so its contribution to $E_2$ is $10 + 11 = 21$. For *Muslim*, the modal category is 12, so its contribution is 13 $(9 + 4)$. For *other*, the modal category is 25, so its $E_2$ contribution is 15. And for *none* the $E_2$ contribution is 16. Adding these $E_2$ contributions generates an $E_2$ of 65 $(21 + 13 + 15 + 16)$. This means we make a total number of 65 errors when we use *Religious denomination* to predict *Political affiliation*.

4. Placing the $E_1$ and $E_2$ calculations in the formula results in the following:

$$(E_1 - E_2) \div E_1$$

$$\text{lambda} = \frac{80 - 65}{80}$$

$$= 0.19$$

If we multiply this result by 100 we get 19 per cent. When we look this up our strength chart, Table 10.8, we see that there is a small association between *Political affiliation* and *Religious denomination*.

# Calculating Gamma

There are actually two ways to calculate gamma. The earlier discussion introduced one way, which relied on the standard PRE formula. Here are the steps for using the first alternative.

## Gamma: Computing Strength and Adding Form

1. Three preliminary points:
   a. Double-check to make sure the table is set up in the conventional form—independent variable over the columns, dependent variable along the left side.
   b. Make sure that, from the upper-left corner, the independent and dependent variables move in the *same direction* (i.e. either both increase or both decrease).
   c. Remember the formula: gamma $= (E_1 - E_2) \div E_1$.
2. Calculate the number of same-ordered pairs $(N_s)$ in the table.
3. Calculate the number of opposite-ordered pairs $(N_o)$ in the table.

4. Compute $E_1$, using the following arithmetic: $N_s + N_o \div 2$.
5. Identify $E_2$ as the *smaller of* $N_s$ or $N_o$.
6. Solve the gamma formula: $(E_1 - E_2) \div E_1$. This result provides you information about the strength of the relationship.
7. Identify the form of the relationship. If $N_s$ is larger than $N_o$, then a direct relationship is indicated. If $N_o$ is larger than $N_s$, then an inverse relationship is in place. (This final step is elaborated in Step 5: Interpreting the Result).

Note: In this approach the form of the gamma relationship had to be determined *after* the result was calculated. There is an alternative way of computing gamma in which the strength and form of the relationship are both included in the statistical calculation. The alternative calculation method is explained below.

## Gamma: Computing Strength and Form with the Alternative Calculation Method

To solidify your understanding, the following example sets out the steps and applies them to the following table of results investigating the effects of social-class membership on life satisfaction (Table 13.8).

**TABLE 13.8** Life Satisfaction by Social Class

| Satisfaction | Social Class | | |
| --- | --- | --- | --- |
| | Lower Class | Middle Class | Upper Class |
| Low | 2 | 4 | 7 |
| Medium | 5 | 9 | 4 |
| High | 8 | 3 | 1 |

1. Calculate $N_s$, the number of pairs having the *same ranking on the independent and dependent variables*. Do this by taking the frequency in each cell in the table and multiply it by the sum of all the cells appearing below and to the right of it. At the end, we sum all the results. Here is how you do this.
   a. Let's calculate the sum for the first cell in our table:

**TABLE 13.8A** Calculating Sum I

| Satisfaction | Social Class | | |
| --- | --- | --- | --- |
| | Lower Class | Middle Class | Upper Class |
| Low | 2 | 4 | 7 |
| Medium | 5 | 9 | 4 |
| High | 8 | 3 | 1 |

What we are doing here is calculating the rankings for the light blue cell by using the data in the dark blue cells. We take 2 (the value of the cell we are interested in—the light blue one) and multiply it by the numbers in the dark blue cells. We get the following answer:

$$2(9 + 3 + 4 + 1) = 34$$

b.  Now we move to the next cell of interest, the one with *lower class–medium satisfaction*. We have highlighted it in light blue for you. We highlighted the cells you are using to do the calculation in dark blue, just like we did in the previous step, a.

**TABLE 13.8B**  Calculating Sum II

| Satisfaction | Social Class | | |
| --- | --- | --- | --- |
| | Lower Class | Middle Class | Upper Class |
| Low | 2 | 4 | 7 |
| Medium | 5 | 9 | 4 |
| High | 8 | 3 | 1 |

Just as we did before, we take the value of the light blue cell and multiply it by the sum of the dark blue cells.

$$5(3 + 1) = 20$$

c.  Let's repeat this step for the first cell in column 2.

**TABLE 13.8C**  Calculating Sum III

| Satisfaction | Social Class | | |
| --- | --- | --- | --- |
| | Lower Class | Middle Class | Upper Class |
| Low | 2 | 4 | 7 |
| Medium | 5 | 9 | 4 |
| High | 8 | 3 | 1 |

Notice that there is only one reference column (all cells down and to the right). Let's repeat the calculation using the light blue and dark blue cells.

$$4(4 + 1) = 20$$

d.  If you have worked it out properly, you now realize that we have only one cell left to go—that is the second cell of column two. We have illustrated that here.

**TABLE 13.8D**  Calculating Sum IV

| Satisfaction | Social Class | | |
| --- | --- | --- | --- |
| | Lower Class | Middle Class | Upper Class |
| Low | 2 | 4 | 7 |
| Medium | 5 | 9 | 4 |
| High | 8 | 3 | 1 |

Repeating the formula:

$$9(1) = 9$$

e. We now add up the four values we have calculated from steps a through d:

$$34 + 20 + 20 + 9 = 83$$

2. Calculate $N_o$, the number of pairs having the *opposite ranking on the independent and dependent variables.* This is done by multiplying the frequency in each cell of the table by the sum of all cells appearing below and to the left of it, with all the products being summed. Note that we are now moving in the opposite direction to steps 1 a through d.

**TABLE 13.8E** Calculating Sum V

| Satisfaction | Social Class | | |
| --- | --- | --- | --- |
| | **Lower Class** | **Middle Class** | **Upper Class** |
| Low | 2 | 4 | 7 |
| Medium | 5 | 9 | 4 |
| High | 8 | 3 | 1 |

Notice that we start from the opposite end of the table—the cell in the top right-hand column. Notice as well that the pattern remains the same. We take the information from all the cells to the bottom and to the left of our cell of interest (we have marked this one in light blue). The information from the cells we use in the calculation are in dark blue.

a. Thus, the formula for the first calculation is:

$$7(9 + 3 + 5 + 8) = 175$$

b. We repeat this process with the second cell in the last column.

**TABLE 13.8F** Calculating Sum VI

| Satisfaction | Social Class | | |
| --- | --- | --- | --- |
| | **Lower Class** | **Middle Class** | **Upper Class** |
| Low | 2 | 4 | 7 |
| Medium | 5 | 9 | 4 |
| High | 8 | 3 | 1 |

Let's plug in the numbers to complete the calculation:

$$4(8 + 3) = 44$$

c. And let's do this once again for the first cell in the middle column (the one shaded in light grey).

**TABLE 13.8G** Calculating Sum VII

| Satisfaction | Social Class | | |
| | Lower Class | Middle Class | Upper Class |
|---|---|---|---|
| Low | 2 | 4 | 7 |
| Medium | 5 | 9 | 4 |
| High | 8 | 3 | 1 |

$$4(5 + 8) = 52$$

d. Finally, we can calculate the results for the last cell—the second cell in the second column (we've highlighted this cell in light grey).

**TABLE 13.8H** Calculating Sum VIII

| Satisfaction | Social Class | | |
| | Lower Class | Middle Class | Upper Class |
|---|---|---|---|
| Low | 2 | 4 | 7 |
| Medium | 5 | 9 | 4 |
| High | 8 | 3 | 1 |

$$9(8) = 72$$

e. Now we add the four figures together:

$$175 + 44 + 52 + 72 = 343$$

3. Now we use the alternative form of the gamma formula. It looks like this:

$$\text{gamma } \Upsilon = \frac{N_s - N_o}{N_s + N_o}$$

This translates into

$$\Upsilon = \frac{\text{same} - \text{opposite}}{\text{same} + \text{opposite}}$$

Using the numbers we calculated as part of steps 1 and 2, we plug in the following formula: 343 (opposite pairs) and 83 (same pairs)

$$\frac{83 - 343}{83 + 343}$$

$$= \frac{-260}{326}$$

$$= -0.798$$

In Interpreting the Results below, you will see that −0.798 represents a very strong association between *Social class* and *Life satisfaction*. But what about that negative sign? This gives us another piece of information—the direction of the relationship. Because it is negative, the relationship indicates that as *Social class* increases (goes up) then *Life satisfaction* moves in the other direction—it decreases (goes down).

## STEP 3 Using Computer Software

To calculate lambda and gamma, the first step is to generate a crosstab (bivariate table) in SPSS in order to get a visual estimation of the relationship between the independent and dependent variables (see Chapter 10 for instructions on how to do this). Then we request the measure of association (lambda or gamma, depending on the level of measurement of your selected variables), which summarizes that relationship in a single number.

# Generating Lambda

The following procedure illustrates the steps for generating lambda for the research question, Does participation in volunteering differ by religious service attendance?

## Procedure

1. Select Analyze from the SPSS menu bar.
2. Go to Descriptive Statistics → Crosstabs.
   a. Move the dependent variable Q19 (*Volunteering*) to the Row box.
   b. Move the independent variable NQ35 (*Religious attendance*) to the Column box.
   c. Select Cells → Column Percentages → Continue.
   d. Select Statistics (located above the Cells tab)→ Lambda.
3. Select Continue → OK.

This procedure parallels the steps used to generate percentaged bivariate tables found in Chapter 10. The only addition is the request for the lambda statistic to be calculated for the crosstab. Following the above instructions should produce a crosstab and a table of Directional Measures in the Output/Viewer window (see Figure 13.1). To interpret the crosstab table, follow the same instructions as in Chapter 10.

To interpret the lambda statistic, look under the Value column in the Directional Measures table. In this column, you will notice that there are various lambdas, depending on which row you use. This is because the three rows under the Value column describe different forms of lambda. The first row is a symmetrical lambda, while the second and third row report asymmetrical lambdas. In most cases, one of the asymmetrical lambdas will be most appropriate. *Asymmetrical* means that you can distinguish which of the variables in the bivariate table is the independent variable and which is the dependent (*symmetrical* means that you are unable to make this distinction). In this case, you want the lambda that has Q19 as the dependent variable.

Q19 Q19. In the past 6 months, have you volunteered or participated in non-paid service * NQ35
NQ35. Do you regularly attend religious services?  Crosstabulation

| | | | NQ35 NQ35. Do you regularly attend religious services? | | Total |
| | | | 1 Yes | 2 No | |
|---|---|---|---|---|---|
| Q19 Q19. In the past 6 months, have you volunteered or participated in non-paid service | 1 Yes | Count | 204 | 401 | 605 |
| | | % within NQ35 NQ35. Do you regularly attend religious services? | 66.2% | 43.9% | 49.5% |
| | 2 No | Count | 104 | 513 | 617 |
| | | % within NQ35 NQ35. Do you regularly attend religious services? | 33.8% | 56.1% | 50.5% |
| Total | | Count | 308 | 914 | 1222 |
| | | % within NQ35 NQ35. Do you regularly attend religious services? | 100.0% | 100.0% | 100.0% |

Directional Measures

| | | | Value | Asymp. Std. Error[a] | Approx. T[b] | Approx. Sig. |
|---|---|---|---|---|---|---|
| Nominal by Nominal | Lambda | Symmetric | .110 | .018 | 5.775 | .000 |
| | | Q19 Q19. In the past 6 months, have you volunteered or participated in non-paid service Dependent | .165 | .027 | 5.775 | .000 |
| | | NQ35 NQ35. Do you regularly attend religious services? Dependent | .000 | .000 | .[c] | .[c] |
| | Goodman and Kruskal tau | Q19 Q19. In the past 6 months, have you volunteered or participated in non-paid service Dependent | .038 | .011 | | .000[d] |
| | | NQ35 NQ35. Do you regularly attend religious services? Dependent | .038 | .011 | | .000[d] |

lambda statistic

a. Not assuming the null hypothesis
b. Using the asymptotic standard error assuming the null hypothesis.
c. Cannot be computed because the asymptotic standard error equals zero.
d. Based on chi-square approximation

**FIGURE 13.1**  Generating Crosstab and Lambda

In our example, we know the independent variable is NQ35 (*Religious attendance*) and the dependent variable is Q19 (*Volunteering*). As such, the appropriate lambda is 0.165, since this value uses Q19 as the dependent variable. We can interpret the lambda statistic by saying:

> Knowing whether or not someone regularly attends religious services has increased our ability to predict participation in volunteering by 16.5 per cent. By looking at the crosstab output, we can see that survey respondents who regularly attended religious services were also more likely to volunteer (66.2 per cent), compared to participants who did not regularly attend religious services (43.9 per cent).

## Generating Gamma

The SPSS procedures for computing gamma are similar to those for lambda, because both are based on bivariate crosstabulations. The Output/Viewer window (see Figure 13.2) also looks similar. The following procedure shows the steps for generating the gamma statistic for the research question, Does the presence of depressive symptoms affect one's reasons for living?

Crosstab

| | | | Q2 Q2. Depressive symptoms index | | | |
| --- | --- | --- | --- | --- | --- | --- |
| | | | 1 None | 2 Moderate | 3 High | Total |
| Q24 Q24. Reasons for living | 1 Little | Count | 30 | 58 | 63 | 151 |
| | | % within Q2 Q2. Depressive symptoms index | 6.5% | 12.7% | 28.4% | 13.1% |
| | 2 Moderate | Count | 183 | 226 | 104 | 513 |
| | | % within Q2 Q2. Depressive symptoms index | 39.9% | 49.5% | 46.8% | 45.1% |
| | 3 High | Count | 246 | 173 | 55 | 474 |
| | | % within Q2 Q2. Depressive symptoms index | 53.6% | 37.9% | 24.8% | 41.7% |
| Total | | Count | 459 | 457 | 222 | 1138 |
| | | % within Q2 Q2. Depressive symptoms index | 100.0% | 100.0% | 100.0% | 100.0% |

Symmetric Measures

| gamma statistic | | Value | Asymp. Std. Error[a] | Approx. T[b] | Approx. Sig. |
| --- | --- | --- | --- | --- | --- |
| Ordinal by Ordinal | Gamma | −.375 | .039 | −9.036 | .000 |
| N of valid Cases | | 1138 | | | |

a. Not assuming the null hypothesis
b. Using the asymptotic standard error assuming the null hypothesis

**FIGURE 13.2** Generating Crosstab and Gamma

## Procedure

1. Select Analyze from the SPSS menu bar.
2. Click Descriptive Statistics → Crosstabs.
   a. Move the dependent variable Q24 (*Reasons for living*) to the Row box.
   b. Move the independent variable Q2 (*Depressive symptoms*) to the Column box.
   c. Click on Cells → Column Percentages → Continue.
   d. Select the Statistics tab, then check Gamma.
      **Note:** if you wish to select other measures of association (but not necessarily PRE estimates) such as Somers' *d*, you can do so here.
3. Select Continue → OK.

In addition to a conventional percentage analysis achieved through the bivariate table, the above procedure will generate a table called Symmetric Measures in the Output/Viewer window (Figure 13.2). Under the column titled Value and the row titled Ordinal by Ordinal—Gamma, you will find the gamma coefficient/statistic, which is −0.375. For our purposes, you can disregard the other information in the Symmetric Measures table.

We would interpret this result by saying:

> Our gamma coefficient of −0.375 indicates that there is a negative relationship between *Depressive symptoms* and *Reasons for living*. That is, knowing the level of *Depressive symptoms* increases our ability to predict *Reasons for living* by 37.5 per cent.

## STEP 4 Practice

You now have the tools for calculating appropriate PRE measures of association for bivariate tables. You employ lambda when the variables are nominal and gamma when they are ordinal. You can perform these calculations either by hand or using computing software. To solidify your understanding you need to practise applying the statistical and software procedures.

This section provides you with opportunities to practise what you have learned about bivariate table analysis. The first set of questions uses hand calculations. The second set uses the SPSS procedures. For each set of questions:

1. Follow the procedural steps and complete the appropriate calculation (Set 1) or software application (Set 2).
2. Check your answers, using the Answer Key in the back of the text.
3. If your answer is incorrect, consult the Solutions section on the book's website. The Solutions provide a complete step-by-step analysis of how the answers are derived.

After you have completed the next section (Interpreting the Results), return to your calculations or output and *provide complete, written interpretations of each of the statistics you have generated.*

# Set 1: Hand-Calculation Practice Questions

1. Use Table 13.9 to answer the questions below.

**TABLE 13.9**  Math Preference by Faculty

| Math Preference | Faculty | |
| | Arts | Science |
| --- | --- | --- |
| Like | 45 | 87 |
| Dislike | 79 | 24 |

   a. How many prediction errors would you make using the marginal prediction rule?
   b. How many errors would you make using the relational prediction rule?
   c. Calculate lambda for this table.
   d. What does our obtained lambda statistic tell us about the bivariate relationship in this table?
   e. Remember lambda is asymmetric. Invert the variables in the table (make *Math preference* the IV, and *Faculty* the DV) and recalculate lambda. Does the obtained value for lambda change?

2. Use Table 13.10 to answer the following questions.
   a. How many prediction errors would you make using the marginal prediction rule?
   b. How many errors would you make using the relational prediction rule?
   c. Calculate lambda for this table.

**TABLE 13.10** Health Care Support by Nationality

| Support for Universal Health Care | Nationality | | Total |
| --- | --- | --- | --- |
| | **American** | **Canadian** | |
| Low | 55 44 | 14 11 | 69 |
| Medium | 47 38 | 44 35 | 91 |
| High | 23 18 | 72 55 | 95 |
| Total | 125 | 130 | 255 |

d. What does our obtained lambda statistic tell us about the bivariate relationship in this table?

e. How does this result compare to the result we got when we used the "percentage down, compare across" technique to examine this relationship in the Chapter 10 practice questions?

3. Use Table 13.11 to answer the following questions.

**TABLE 13.11** Life Satisfaction by Marital Status

| Life Satisfaction | Marital Status | | | Total |
| --- | --- | --- | --- | --- |
| | **Married** | **Divorced/ Separated** | **Single** | |
| Low | 33 | 22 | 33 | 88 |
| Medium | 44 | 55 | 67 | 166 |
| High | 72 | 64 | 59 | 195 |
| Total | 149 | 141 | 159 | 449 |

a. Calculate lambda for this table.

b. What does our obtained lambda statistic tell us about the bivariate relationship in this table?

4. Use Table 13.12 to answer the following questions.

**TABLE 13.12** Corporal Punishment Practice by Education

| Corporal Punishment | Highest Level of Education | | |
| --- | --- | --- | --- |
| | **< High School** | **High School** | **Post-Secondary** |
| Never | 24 41 | 63 59 | 86 75 |
| Sometimes | 22 38 | 33 31 | 23 20 |
| Often | 12 21 | 11 10 | 6 5 |
| | 58 | 107 | 115 |

a. How many same-ordered pairs are there?

b. How many opposite-ordered pairs are there?

c. What are the values of $E_1$ and $E_2$ for this table?

d. Calculate gamma for this table.

e. What does our obtained gamma statistic tell us about the bivariate relationship in this table?

f. How does this result compare to the result we got when we used the "percentage down, compare across" technique to examine this relationship in the Chapter 10 practice questions?

5. Use Table 13.13 to answer the following questions about a survey on quality of life and job satisfaction.

**TABLE 13.13** Quality of Life by Job Satisfaction

| | Job Satisfaction | | | |
|---|---|---|---|---|
| **Quality of Life** | **Very Unsatisfied** | **Unsatisfied** | **Satisfied** | **Very Satisfied** |
| Low | 48 | 43 | 21 | 10 |
| Moderate | 20 | 49 | 42 | 24 |
| High | 8 | 18 | 49 | 44 |

a. How many same-ordered pairs are there?
b. How many opposite-ordered pairs are there?
c. What are the values of $E_1$ and $E_2$ for this table?
d. Calculate gamma for this table.
e. What does our obtained gamma statistic tell us about the bivariate relationship in this table?
f. Invert the variables in the table (make *Quality of life* the IV, and *Job satisfaction* the DV) and recalculate gamma. Does the obtained value for gamma change?

6. The data in Table 13.14 is from a hypothetical study comparing North American and European driving preferences.

**TABLE 13.14** Transmission Preference by Continent of Residence

| | Continent of Residence | |
|---|---|---|
| **Experience using Transmission Type** | **North America** | **Europe** |
| Standard | 3 | 82 |
| Automatic | 102 | 9 |
| Both | 11 | 36 |

a. How many prediction errors would you make using the marginal prediction rule?
b. How many errors would you make using the relational prediction rule?
c. Calculate the appropriate PRE measure for this table.
d. What does our obtained PRE statistic tell us about the bivariate relationship in this table?

7. Table 13.15 contains hypothetical data from an international survey about government corruption. Respondents who indicated that they thought their government suffered from corruption were asked, in a follow-up question, how corrupt they felt

**TABLE 13.15** Perceptions of Government Corruption by Level of Income Inequality

| | Income Inequality | | |
|---|---|---|---|
| **Perceived Government Corruption** | **Low** | **Medium** | **High** |
| Low | 68 | 32 | 21 |
| Medium | 32 | 58 | 36 |
| High | 22 | 45 | 62 |

their government was. Researchers were interested in whether there was a connection between respondents' perceptions of government corruption and the degree of income inequality in their countries.

   a.   How many same-ordered pairs are there?
   b.   How many opposite-ordered pairs are there?
   c.   What are the values of $E_1$ and $E_2$ for this table?
   d.   Calculate the appropriate PRE measure for this table.
   e.   What does our obtained PRE statistic tell us about the bivariate relationship in this table?

8. Table 13.16 contains hypothetical data on the relationship between political ideology and income level.

**TABLE 13.16** Political Ideology by Income Level

| Political Ideology | Income | | |
|---|---|---|---|
| | Low | Medium | High |
| Left-wing | 25 | 18 | 9 |
| Centrist | 18 | 19 | 20 |
| Right-wing | 7 | 15 | 31 |

   a.   How many same-ordered pairs are there?
   b.   How many opposite-ordered pairs are there?
   c.   What are the values of $E_1$ and $E_2$ for this table?
   d.   Calculate the appropriate PRE measure for this table.
   e.   What does our obtained PRE statistic tell us about the bivariate relationship in this table?

9. Table 13.17 contains fictional data from a traffic court dealing with appeals from individuals who are seeking to have their fines reduced. Use the appropriate PRE measure to determine if there is a relationship between person's sex and the type of traffic offence they were charged with.

**TABLE 13.17** Type of Traffic Offence by Gender

| Type of Traffic Offence | Gender | |
|---|---|---|
| | Male | Female |
| Parking ticket | 22 | 26 |
| Speeding ticket | 32 | 15 |
| Not wearing a seat belt | 9 | 3 |
| Using cellphone | 18 | 24 |
| Failing to use turn signal | 26 | 20 |

   a.   How many prediction errors would you make using the marginal prediction rule?
   b.   How many errors would you make using the relational prediction rule?
   c.   Calculate the appropriate PRE measure for this table.
   d.   What does our obtained PRE statistic tell us about the bivariate relationship in this table?

# Set 2: SPSS Practice Questions

Working again with the Student Health and Well-Being Survey data set, use the Crosstabs function to obtain the appropriate PRE measure of association for each of the following pairs of variables. Interpret what the relevant PRE statistic tells us about each particular relationship.

1. Q25 *Sex* (IV) and Q19 *Volunteering* (DV)
2. Q27 *Born in Canada or not* (IV) and Q13B *Problems with school* (DV)
3. nQ35 *Frequency of religious service attendance* (IV) and Q19 *Volunteering* (DV)
4. nQ35 *Frequency of religious service attendance* (IV) and nQ1D *Satisfaction with mental health* (DV)
5. nQ1B *Satisfaction with level of exercise* (IV) and nQ1A *Satisfaction with overall level of physical health* (DV)
6. Q21D *Mutually fulfilling friendships* (IV) and nQ1D *Satisfaction with mental health* (DV)
7. Q32 *Ethnicity* (IV) and Q28 *Relationship status* (DV)
8. Q27 *Born in Canada* (IV) and Q19 *Volunteering* (DV)
9. Q36 *Political affiliation* (IV) and Q27 *Born in Canada* (DV)
10. Q10E *Told to stop or cut back on soft drug use* (IV) and Q5C *Trouble controlling temper* (DV)
11. Q5A *Argumentative* (IV) and Q23A *Called names or insulted* (DV)
12. Q10A *Work or school in the past year* (IV) and Q23D *Threatened or harassed* (DV)

## STEP 5  Interpreting the Results

Chapter 10 introduced you to bivariate tables and their interpretations. In that chapter you learned that the interpretation of bivariate table evidence required speaking to two issues— form and strength. Form describes the *kind of connection* that exists between the independent and dependent variables. In other words, form addresses the question, As the independent variable moves (changes) in a specified direction, how does the dependent variable change? Strength addresses a separate issue; namely, *How much difference* does changing the independent variable make to the dependent variable?

In Chapter 10 we introduced technical tools for determining both the form and strength of the connection between two variables in a table. Form was determined by using the analogy of changing distributions of persons in an apartment block. Strength was determined by applying the "percentage down, compare across" rule. Both of these tools worked, in the sense that they provided rough approximations of how the independent and dependent variables were connected. These tools had limitations, however. Specifically, the tools were subject to ambiguous interpretations. As Chapter 12 pointed out, PRE measures of association reduce ambiguity of interpretation and provide more reliable interpretations of bivariate relationships.

This chapter introduced you to two common PRE statistics, lambda and gamma. Lambda is used when both variables in a table are nominal; gamma is appropriate when the bivariate table contains ordinal variables. You now understand the logic of these measures and how to compute them either through hand calculations or computer generation. The final task involves interpretation.

# Interpreting Lambda Form

Lambda is an appropriate PRE statistic when both the independent and dependent variables are nominal. Nominal variables use categories that serve as names. In other words, the values of a nominal variable have no arithmetic properties; they are simply labels that identify that one category is different from another. It is important to emphasize this characteristic of lambda because it has implications for interpreting the form of the relationship in a table.

The fact that the values of nominal variables are only names means that *information about form* cannot *be built into* PRE *statistics (like lambda) that describe bivariate relationships between nominal variables.* As you shall see, this is not true of PRE statistics that describe variables at higher levels of measurement, but it is true for lambda, for the following simple reason: Since the values of nominal variables are merely names, how they are ordered is arbitrary. Imagine the columns of a bivariate table using the independent variable *Religion*, with categories *Christian, Jew, Muslim.* Whether the first column includes the Christians or Jews or Muslims is entirely arbitrary. And once one of these alternatives is selected for the first column, which of the remaining categories is listed in the second column is also an arbitrary choice. The same holds true for a nominal dependent variable, such as *Ethnicity.* How the categories of this variable are arranged in the rows of the table is also arbitrary.

The result of this arbitrariness resulting from the nature of nominal variables means that the form in any bivariate table using nominal variables is *table-specific. Table-specific* means that if the table was set up in a different way, the description of form would be different. Since there is no necessary, fixed way of setting up a bivariate table using nominal variables, form information cannot be built into the calculation of lambda. All PRE statistics must use general principles, so they can be applied to a wide range of different variables. General principles cannot capture the arbitrary features of nominal tables.

This is the rationale for why the computed result of lambda does not include information about the form of the relationship. This result has two practical applications. First, this situation does *not* mean that when lambda is computed on a bivariate table including two nominal variables that it is unnecessary to interpret the form of the relationship between the variables. The fact that lambda does not provide this information does not mean it is unimportant. Therefore, the second implication of this situation is that, for bivariate tables using nominal variables, *form must be identified and interpreted using the "apartment movement" technique describe in Chapter 10.*

# Interpreting Lambda Strength

Although lambda does not provide precise information about form, it does provide a clear interpretation of the strength of the relationship between nominal variables in a table. Lambda is one statistic in the PRE family, and therefore it has a conventional interpretation, in the sense that it is not idiosyncratic. All members of the PRE family of statistics share similar interpretations of strength.

Lambda is a proportional reduction in error (PRE) statistic. Its range of legitimate values varies between 0.0 and 1.0.[7] The results, typically expressed to two decimal points, are *proportions* that can be translated to *percentages* by multiplying by 100.

---

[7]If you calculate lambda and get a negative value or a result greater than 1.0, you have made a mistake—so try again!

All PRE statistics share a common interpretive framework regarding strength. As a member of this family, lambda is no exception. The strength interpretation begins with the phrase "You proportionately reduce your errors by $X$ per cent . . .," where $X$ is the lambda value multiplied by 100. So, for example, if the lambda value was 0.58, the interpretation would begin "You proportionately reduce your errors by 58 per cent . . ."

The remainder of the strength interpretation is based on what comparisons were made in the calculation of lambda. Specifically, lambda compared how well the dependent variable could be predicted under two conditions—first, using only information about the distribution of the dependent variable and, second, with knowledge of the independent variable. These comparisons generated $E_1$ and $E_2$. This comparison needs to be expressed as lambda's interpretation of strength.

In its general form, the interpretation of lambda states:

> You proportionately reduce your errors by $P$ [lambda value multiplied by 100] per cent when you know respondent's $X$ [independent variable] as opposed to not knowing $X$, when predicting their $Y$ [dependent variable].

Interpreting the strength of lambda for the specific result of Table 13.1 illustrates the application of this general interpretation. Table 13.1 reported the relationship between *Sex* (the independent variable) and *Religiosity* (the dependent variable) as lambda 0.11. The lambda result means that:

> You proportionately reduce your errors by 11 per cent when you know a respondent's *Sex*, as opposed to not knowing their *Sex*, when predicting the regularity of their religious practice (*Religiosity*).

Notice how this PRE statistic provides an unambiguous, precise interpretation of the relationship between the independent and dependent variables. This is an improvement over the vague assessments using the tools from Chapter 10. Given that lambda can range from zero (meaning that knowing the independent variable provides no improvement in predicting the dependent variable) to 1.0 (meaning that knowing the independent variable completely predicts dependent variable results), specific lambda results are now precisely located along this range.

## Lambda Anomaly

The tools for analyzing bivariate tables introduced in Chapter 10 serve the same purpose as those in this chapter. In all cases, these tools aim to help you understand the character of the connection (relationship) between the independent and dependent variables. So, if you analyze the same table using different tools, it is reasonable to expect that the results should be somewhat similar. Of course, the results won't be identical, since PRE interpretations are more precise than the percentage comparison tools. To confirm this point, here is a percentage analysis of the strength of the relationship for Table 13.1, which we know has a lambda value of 0.11. The percentages are in parentheses in Table 13.18.

The average conditional distribution (*Sex*) percentage differences in Table 13.14 is 24 per cent. Using the interpretive guideline table (Table 10.8), this result is interpreted as *sex* making a *modest* difference to religiosity. On a scale between zero and 1.0, the computed lambda value of 0.11 is not that large, so the percentage analysis interpretation of *modest* is in line with the lambda result. This comparison confirms that the same relationship (table) analyzed using different tools yields approximately similar results.

**TABLE 13.18** Percentage Analysis of Table 13.1

| Regular Religious Practice | Sex | | |
|---|---|---|---|
| | **Male** | **Female** | **Total** |
| Yes | 47 (36%) | 375 (60%) | 422 |
| No | 83 (64%) | 249 (40%) | 332 |
| Total | 130 | 624 | 754 |

Table 13.19, however, shows that this is not always the case. Computing lambda for this table results in a value of 0.00,[8] which leads to the conclusion that *Hair colour* makes no difference to dating experience. However, the average percentage differences between the conditional distributions reveals quite a substantial relationship between *Hair colour* and *Last date experience*.[9]

**TABLE 13.19** Hair Colour and Dating Experience

| Last Date Experience | Hair Colour | | |
|---|---|---|---|
| | **Blond** | **Not Blond** | **Total** |
| Fun | 60 (50%) | 0 (0%) | 60 |
| Not fun | 60 (50%) | 30 (100%) | 90 |

Table 13.19 depicts a situation in which lambda shows no relationship when in fact there is a relationship. This is an anomaly in using lambda called false zeros. **False zeros** occur when the computed value of lambda is zero when, in fact, there is a relationship between the variables. In this case, the computed lambda result (of zero) is false.

False zeros are a glitch in lambda that occurs for technical reasons. The important practical point is that you be aware of the false zero possibility and know how to manage it. Here is how you proceed. First, you must appreciate that every time lambda is zero the results are not necessarily false. False zeros are a possibility for lambda, not a certainty. So when the lambda result is zero, the challenge is to determine if this result is false or not. When this occurs you should examine the strength of the relationship using the percentage difference technique. If the percentage differences also show no relationship between the variables, then it is safe to conclude that the lambda of zero is *true*. However, if the percentage difference technique shows some relationship when the lambda result is zero, you should conclude that lambda is *false*. When lambda results in a false zero, you should not interpret it. Instead, you should revert to interpreting the percentage analysis results.

# Interpreting Gamma Form

Gamma is an appropriate PRE statistic when both variables in the table are at the ordinal level of measurement. For ordinal variables, the categories of the independent and dependent variables can be rank-ordered; they can be organized in terms of more or less of the property

---

[8]The lambda $(E_1 - E_2) \div E_1$ calculation = $(60 - 60) \div 60 = 0 \div 60 = 0$.
[9]The average percentage difference between the conditional distributions of the independent variable is 50 per cent.

being measured. This feature of ordinal variables means that, unlike lambda, the categories of the variable are not organized in an arbitrary way. There is a definite ordering to the way the values of an ordinal variable are listed.

A practical result of this point is that bivariate tables using ordinal variables must be set up in a specific way. It doesn't make sense to order the columns of an ordinal independent variable as Low, High, Medium, or the rows of an ordinal dependent variable as Strongly Agree, Neutral, Strongly Disagree, Agree, Disagree. The fact that the categories of ordinal variables are fixed means that information about form can be built into the gamma calculations.

An interpretation of form expresses the kind of connection that exists between the variables. Form states that as the independent variable changes in a specific way, the dependent variable is expected to change in a specified way. The categories of ordinal variables are organized in terms of more or less (increases or decreases) of the measured characteristic. Therefore, an interpretation of the form between two ordinal variables states that as the independent variable increases the dependent variable is expected to increase or decrease.

Legitimate gamma results range between −1.0 to +1.0. The sign of the gamma result provides information about the form of the relationship between the variables. If the sign of the gamma statistic is positive (+) this means that the variables are changing in the same way (either both increasing or both decreasing). If gamma has a negative sign (−), this means that the variables are changing in opposite ways (as one increases the other decreases). A gamma result of zero indicates that the table contains an equal number of same- and opposite-ordered pairs, which means that neither relational prediction rule predominates, and so no relationship exists.

There are two ways to get information about form from a computed gamma statistic. If you use the computational formula provided in Step 1, you will have to *add the appropriate sign to the gamma calculation*. The $(E_1 − E_2) \div E_1$ calculation of gamma does not provide form information. If you use this calculation formula you will have to add the correct sign after the fact. Knowing whether to add a positive or negative sign comes from looking at the ordering of the pairs. If there are more same-ordered pairs in the data set, then add a positive sign to the calculated value. If the data set contains more opposite-ordered pairs, then add a negative sign to the computed value.

The other way to get sign (i.e. form) information for gamma is to use an alternative calculation formula. However gamma is calculated, you need information about the number of pairs in the data set that have the same ordering ($N_s$), and the number of pairs that have opposite ordering ($N_o$). With these two pieces of information, gamma can also be calculated using the following formula:

$$\text{gamma} = \frac{N_s − N_o}{N_s + N_o}$$

Using this formula results in gamma having the correct sign information. In other words, using this formula will produce a value that is either positive or negative (or zero).

After using one of these alternative methods for obtaining form information about gamma, you interpret the form using the following general statements. If gamma has a positive sign, the interpretation is

As $X$ [independent variable] increases, the values of $Y$ [dependent variable] also increase.

If gamma has a negative sign, the interpretation is

As $X$ [independent variable] increases, the values of $Y$ [dependent variable] decrease.

For example, Table 13.6 reported the connection between *Reading competence* and *Arithmetic competence* among students. The table contains 12,560 same-ordered pairs and 4,144 opposite-ordered pairs. This is how $N_s$ and $N_o$ apply to the alternative calculation formula:

$$\text{gamma} = \frac{N_s - N_o}{N_s + N_o} = \frac{12{,}560 - 4{,}144}{12{,}560 + 4{,}144} = \frac{8{,}416}{16{,}704} = +0.50$$

In this case, the fact that the table contains more same-ordered pairs yields a result that has a positive sign. This sign provides the following interpretation about the form of the relationship:

As *Reading competence* increases, the values of *Arithmetic competence* also increase.

# Interpreting Gamma Strength

Strength reports on how much difference changing the independent variable makes to the dependent variable. For gamma, the strength information is present in the size of the computed value, *regardless of sign*. A gamma of −0.77 has the same strength as a gamma of +0.77. Regarding strength, it is the magnitude of the statistic (not its sign) that is relevant.

Gamma is a member of the PRE family of statistics. All measures of association in this family share a common interpretation format regarding strength. The generic statement about strength takes the following form:

You proportionately reduce your errors by $P$ [gamma value multiplied by 100] per cent when you know a respondent's $X$ [independent variable] as opposed to not knowing $X$, when predicting their $Y$ [dependent variable].

Applied to the gamma value for Table 13.6, it results in the following interpretation:

You proportionately reduce your errors by 50 per cent when you know a respondent's *Reading competence*, as opposed to not knowing this information, when predicting their *Arithmetic competence*.

As was the case with lambda, this PRE interpretation provides an unambiguous, precise interpretation of the relationship between the independent and dependent variables. Given that the strength of gamma increases as the values move away from zero (i.e. toward either +1.0 or −1.0), knowing that prediction errors are reduced by 55 per cent provides an exact understanding of the strength of the connection between *Reading competence* and *Arithmetic competence*.

## Gamma Weakness

Gamma is a complete PRE measure of association, in the sense that it provides, in one statistic, information about both the form and the strength of the relationship between the variables. Gamma does have a weakness, however, related to the issue of tied pairs. Understanding this issue requires returning to fundamentals, illustrated in Table 13.20.

**TABLE 13.20**   Volunteerism by Social Class

| Level of Volunteering | Social Class | | | Total |
|---|---|---|---|---|
| | **Low** | **Medium** | **High** | |
| Low | 110 | 70 | 40 | 220 |
| Medium | 80 | 100 | 100 | 280 |
| High | 70 | 80 | 150 | 300 |
| Total | 260 | 250 | 290 | 800 |

The data in Table 13.20 includes 800 respondents, which is a moderate size. If these variables were collected from a representative sample of adults in a Canadian city through telephone interviewing, the cost of data collection is easily in the tens of thousands of dollars. Collecting good quality information is not cheap; it takes considerable time, energy, and financial resources. The 800 cases in the sample contain 319,600 pairs of information.[10] The data table also contains 98,200 same-ordered pairs and 45,700 opposite-ordered pairs.[11]

The calculation of gamma utilizes the number of same- and opposite-ordered pairs. Therefore, gamma for Table 13.20 is based on 143,900 pairs of information.[12] However, the 143,900 pairs on which gamma is based represent only a small percentage of all the pairs of information contained in the table. This occurs because most of the pairs in the table are tied pairs; they have the same score on the independent variable, the dependent variable, or both variables. Gamma ignores tied pairs because they do not fit into its computational model. The result, for a table like 13.20, is that the measure of the relationship between the variables is based on only a small proportion of the expensive information contained in the table.

In short, because it ignores tied pairs, gamma wastes a lot of valuable information. There is nothing that can be done about this fact, other than to acknowledge it. Acknowledging this fact means recognizing that the measure of the relationship in a table using gamma is based on some, not all, of the information about that relationship contained in the table.

Gamma demonstrates the utility of PRE measures of association. The tools from Chapter 10 can be used to gain an approximate understanding of the form and strength of the relationship between variables in a table. But these procedures are cumbersome and yield ambiguous results. By contrast, in a single calculation, gamma provides information about both the form and strength of the connection between the ordinal variables in a table. The power of PRE statistics can be extended from tables using categorical variables to scatterplots containing continuous ones. The procedures for a detailed consideration of the relationship between quantitative variables in scatterplots are the subject of the next chapter.

---

[10]Using the $N (N - 1) \div 2$ calculation.

[11]$N_s$ and $N_o$ computed using the calculation rules discussed previously.

[12]From $N_s$ plus $N_o$.

# Qualitative Interpretations of Strength in PRE Measures

Lambda and gamma are PRE measures of association that provide precise interpretations of the strength of a relationship. Written interpretations of the results state exactly how knowing the independent variable reduces errors in predicting the dependent variable ("You proportionately reduce your errors by ____ per cent . . ."). Some audiences do not find this quantitative precision satisfying; they prefer a qualitative interpretation of the relationship. This requires some guidelines for giving the quantitative findings a qualitative assessment. For this purpose Table 13.21 (from Szafran, 2012:199) is helpful.

**TABLE 13.21**  Strength in PRE Measures

| If the absolute value of a PRE measure is: | The strength can be interpreted as: |
| --- | --- |
| 0.000 | No relationship |
| 0.001 to 0.199 | Weak |
| 0.200 to 0.399 | Moderate |
| 0.400 to 0.599 | Strong |
| 0.600 to 0.999 | Very strong |
| 1.000 | Perfect relationship |

# Chapter Summary

This chapter introduced the proportional reduction in error (PRE) family of statistics used in the analysis of bivariate tables. Having completed this chapter you should appreciate that:

- Lambda is the appropriate PRE statistic when both independent and dependent variables are nominal.
- The marginal prediction rule for lambda is "predict the mode of the dependent variable."
- The relational prediction rule for lambda is "predict the mode of each conditional distribution of the independent variable."
- $E_1$ for lambda is the number of cases that are not in the dependent variable mode.
- $E_2$ for lambda is the number of cases that are not in the modes of the independent variable's conditional distributions.
- Lambda is computed using the conventional $(E_1 - E_2) \div E_1$ calculation.
- The lambda statistic does not provide information about the form of the relationship.
- The form for two nominal variables must be determined using other techniques.
- The legitimate range for lambda is 0.0 to + 1.0.
- The strength of lambda is interpreted using a conventional PRE interpretive template.
- It is possible for lambda calculations to produce false zero results. This possibility must be tested using alternative measures of strength. If confirmed, false zero lambda results should not be interpreted.

- Lambda is asymmetric, so the value of the statistic will vary depending on which variable is designated as the independent variable. Following the convention of creating bivariate tables with the dependent variable across the top (columns) and dependent variable down the side (rows) helps guard against possible confusion.
- Gamma is an appropriate PRE statistic when both variables are ordinal.
- The calculation of gamma relies on using pairs of cases and comparing the ordering of pairs on both variables.
- The marginal prediction rule for gamma is "guess."
- $E_1$ for gamma is half the number of pairs for which predictions are made.
- The relational prediction rule is either "always predict same order" or "always predict opposite order."
- The relational prediction rule is selected as the greater number of the same- or opposite-ordered pairs. $E_2$ is the smaller of the number of same- or opposite-ordered pairs.
- The sign of gamma provides information about the form of the relationship.
- The magnitude of gamma provides information about the strength of the relationship.
- The form of gamma is expressed as either a direct or an inverse relationship.
- The strength of gamma is expressed in terms of a conventional PRE interpretation.
- Tied pairs are those that have the same score on the independent or dependent variable, or both.
- The calculation of gamma is based on untied pairs.

This chapter demonstrated how PRE measures of association generate precise interpretations of bivariate relationships expressed in tables. PRE measures also apply to bivariate relationships expressed in scatterplots. Bivariate analyses of higher order quantitative variables is the subject of the next chapter.

# 14

# Statistics for Continuous Connections

## Overview

Chapters 10 and 13 complemented one another. Chapter 10 introduced basic tools for examining the form and strength of bivariate relationships that appear in tables. Chapter 13 provided a more sophisticated means for examining these same properties through the calculation and interpretation of PRE measures of association.

In a parallel way, this chapter complements Chapter 11. Chapter 11 introduced bivariate connections in the form of scatterplots. Scatterplots graph the relationship between two quantitative variables, those at either the interval or ratio levels of measurement. Since variables at higher levels of measurement contain more information, scatterplot interpretations address three considerations, including form, extent, and precision. Form refers to the general shape of the connection between the independent and dependent variables. Extent addresses the impact one variable has on the other, while precision refers to how accurately the form and extent reports characterize the evidence.

In terms of the statistical selection checklist, the tools included in this chapter are:

## Statistical Selection Checklist

1. Does your problem centre on a descriptive or an inferential issue?

    ☑ Descriptive

    ☐ Inferential

2. How many variables are being analyzed simultaneously?

    ☐ One

    ☑ Two

    ☐ Three or more

3. What is the level of measurement of each variable?

    ☐ Nominal

    ☐ Ordinal

    ☑ Interval

    ☑ Ratio

In this chapter you will learn:

- how to calculate the values of the intercept and slope for a particular scatterplot;
- how to draw specific regression lines on a scatterplot;
- how to interpret the form, extent, and precision measures; and
- how to calculate and interpret the strength of a scatterplot relationship.

This chapter extends the basic form, extent, and precision ideas introduced earlier. It does so by providing precise tools for calculating and interpreting each of these properties on a scatterplot. All these tools are based on an understanding of linear relationships, which is where we begin.

## STEP 1 Understanding the Tools

# Regression Lines

## Preliminaries

Figure 14.1 presents a scatterplot of the connection between *Hours studied* (the independent variable) and *Grades* (the dependent variable). Both of these variables are ratio and the ordered pairs of these variables from a data set are plotted on the graph.

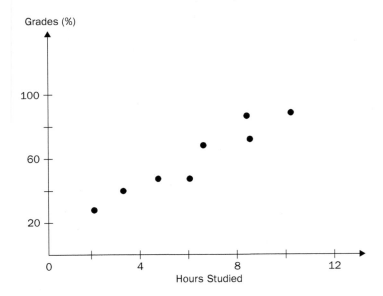

**FIGURE 14.1**  Hours Studied and Grades

The data points on this scatterplot represent evidence collected from a sample of students. Each of these points represents a specific fact. For instance, the dot located at $(x,y)$ represents the respondent in the data set who studied $x$ hours and obtained a grade of $y$. Similar interpretations can be made for every data point on the graph.

Common sense advocates "letting the facts speak for themselves." If you look at the evidence in Figure 14.1, what is it saying to you? The answer is probably nothing. The truth is that facts do not speak for themselves; understanding facts requires interpretation; and interpretation, in turn, requires analysis. The dots in Figure 14.1 are scattered; they are noisy. The task of analysis is to find the signal (meaning) in that noise. Performing such analysis is required for the facts to speak.

Chapter 11 introduced the regression line as the tool for identifying the signal in the scatterplot noise. The regression line is the straight line that best fits the data points. Best-fit is determined by minimizing the total distance between the data points and the line. For example, if two lines both come very close to the actual data points on a scatterplot, the one that has the least distance between its line and the data is called the best fit.

In scatterplot analysis, the best-fit regression line is actually called the **linear least squares regression line**. Several features about this label are instructive. First, the line being fit to the data points is linear, which signifies a *straight* line. Figure 14.2 illustrates the importance of this point.

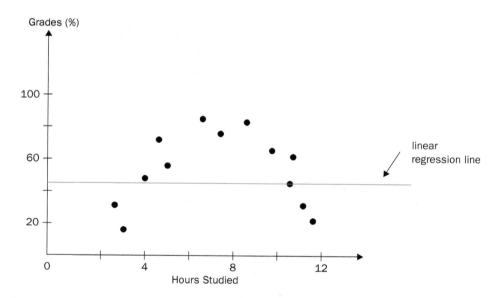

**FIGURE 14.2** Linear Fit to Curvilinear Pattern

There is a clear signal in the pattern of dots in Figure 14.2. The signal is that the variables are related in an inverted U-shaped way. As *Hours studied* begins increasing, it is associated with an increase in *Grades*. Then the pattern levels off, indicating that, beyond a certain point, more *Hours studied* has no effect on *Grades*. Then, at some later point, increasing *Hours studied* actually produces a reduction in *Grades*. In short, the form of the relationship between the variables in Figure 14.2 is curvilinear.

Standard regression analysis imposes the best-fit *straight* line on a scatterplot. Notice that when such a line is identified on Figure 14.2, it is flat—which signifies no relationship between the variables. In other words, the best-fit line on Figure 14.2 indicates that the variables are unrelated, when, in fact, they are clearly related (in a curvilinear way). The lesson from this illustration is important. Standard regression techniques identify and interpret the best-fit straight line. If the relationship is curvilinear, such techniques may report no relationship when there actually is a connection between the variables. So, as a precaution, researchers always visually examine a scatterplot to ensure that imposing a linear best-fit signal is appropriate.

The regression line is not only linear but it is also *least squares*. This feature describes how the distance between the data points (ordered pairs) and the line is determined. Figure 14.3 illustrates this point.

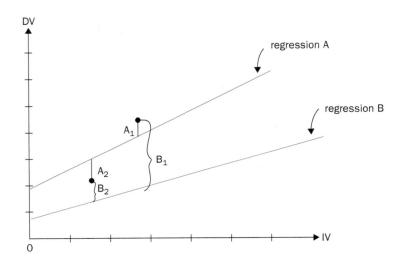

**FIGURE 14.3**  Least Squares Procedure

Figure 14.3 focuses on two data points. The distance of each point from alternative possible regression lines is noted. You could literally measure each of these distances with a ruler and then add them all up. If you performed such a measurement, you would see that the total of all these distances (or the total distance) would be smaller for regression line A than for B, as A is closer to the data points than alternative B. Of the alternatives, because line A has the least distance from the actual data points, it is the best-fit line. It is worth mentioning why the best-fit line is the "least squares" line. As the simple illustration in Figure 14.3 portrays, regression lines will have data points both above and below the line. Determining how *least* is measured requires adding up the distances of the data points from the line. Since data points are both above and below the line, some of the measured distances will be positive (i.e. above), while others will be negative (i.e. below). When these positives and negatives are simply added together, they cancel one another out, which produces undesirable results. To avoid this outcome, the measured distances from the line are squared, which always produces positive numbers.

## MATH TIPS

### How Do I Calculate a Square?

Squaring a number simply means multiplying one number by itself. How do I know I have to square a number? The square function is indicated by $a^2$ (a superscripted 2) beside a number. Here's an example:

$8^2$ means that you must multiply 8 by 8 (8 × 8). The result is 64.

Alternatively, most calculators have a squaring function which looks like this: $x^2$. Simply enter the number you wish to square into your calculator and press the $x^2$ button.

## Determining the Line

You now understand what the term *linear least squares regression* means. This label captures the features that determine the best-fit line. The next question asks, How is a line with these features determined? Answering this question begins with understanding the features of any straight line.

Figure 14.4 portrays three possible regression lines displaying the connection between the independent variable *Income* and the dependent variable *Happiness*. Examining different pairs of these lines reveals the central features of straight lines.

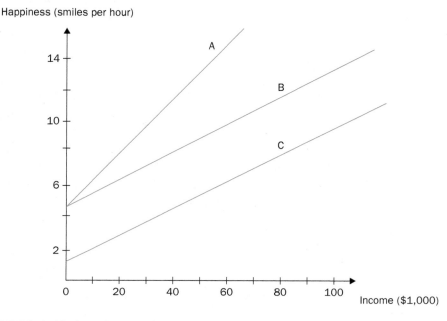

**FIGURE 14.4** Various Regression Lines

Look at lines A and B in Figure 14.4. What are their similarities and differences? Next, answer the same question comparing lines B and C. The answers to these questions are as follows: Lines A and B cross the vertical dependent variable axis at the same point, but have different slopes. Lines B and C cross the dependent variable axis at different points but have the same slope.

These fundamental comparisons reveal that any straight line connecting two variables is defined by two features. First, it has a specific intercept. The **intercept** is the point where the line crosses the axis of the dependent variable. Second, every straight line has a specific slope. The **slope** characterizes how much the dependent variable changes for a fixed change in the independent variable. In short, if you know the intercept and the slope for a line, you can draw it on a scatterplot. For example, if you are told that a fourth possible regression line relating *Income* and *Happiness* in Figure 14.4 has an intercept of 3 and a slope of 1,[1] you could draw it on the graph.

A regression line describes the relationship between an independent variable ($X$) and a dependent variable ($Y$). The best-fit line connecting these variables is characterized by a specific intercept ($a$) and a particular slope ($b$). Every straight line combines these features using the following formula:

$$Y = a + bX$$

This is the equation of a best-fit regression line.

This is the general formula that characterizes *every* straight line connecting *any* two variables. This formula states that if you begin with the independent variable score for a specific case ($x$), multiply that score by the value of the slope ($b$), and then add the value of the intercept ($a$), you will generate the best prediction of that case's dependent variable score ($y$). It is important to recognize that this general formula describes every linear regression line. The specific differences between different regression lines concern the variables involved ($x$ and $y$), the intercept ($a$), and the slope ($b$).

In short, determining the regression line between any two quantitative (interval/ratio) variables ($X$ and $Y$) in a data set requires figuring out what the specific intercept ($a$) and specific slope ($b$) should be for the general equation $Y = a + bX$. Determining the correct intercept and slope is a matter of calculation. The formulas that allow the computation of the slope and intercept of a regression line are applied to a data set. This is an important point to remember. You begin with the information about the variables contained in a data set. This is the same information used in determining the plotting of the ordered pairs used to graph the scatterplot. You then use this information to determine (compute) the intercept and slope of the best-fit regression line. Table 14.1 provides a sample data set on the variables *Income* and *Happiness* for five cases. *Income* is measured in units of $10,000; *Happiness* is measured on a 10-point scale of increasing happiness.

---

[1] A slope of 1 means that the angle of the line will be 45 degrees.

**TABLE 14.1** Income and Happiness Scores

| Case Number | Income | Happiness |
|---|---|---|
| 1 | 2 | 1 |
| 2 | 3 | 5 |
| 3 | 4.5 | 7 |
| 4 | 5 | 6 |
| 5 | 7 | 8 |
| 6 | 10 | 9.5 |

The data from these six cases can be plotted as ordered pairs on a scatterplot. Figure 14.5 displays the results.

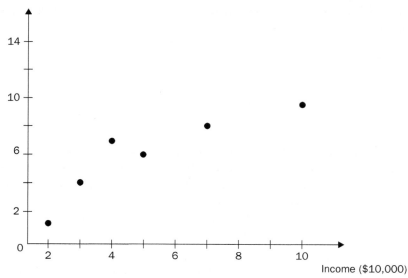

**FIGURE 14.5** Income and Happiness

The question becomes, What is the equation of the regression line that best fits (comes closest to) these data? We know that the formula's general shape will be:

$$\text{happiness} = a + b\,(\text{income})$$

Determining the specific values of the intercept ($a$) and slope ($b$) is a matter of calculation. The calculations are made as follows.

Begin by computing the slope ($b$) for the line that best fits the data. The slope is computed through the following formula:

$$b = \frac{N\Sigma XY - (\Sigma X)(\Sigma Y)}{N\Sigma X^2 - (\Sigma X)^2}$$

## MATH TIPS

### Slope Symbols

What do all these symbols mean? Let's work this out.

- $b$ refers to the slope of the regression line.
- $N$ refers to the sample size. How many cases (or individuals) do we have data for? We obtain this information by counting the number of rows of data. In the *Happiness* and *Income* example, there are six rows, therefore we have six people—so $N=6$.
- $\Sigma$ is the summation sign. Do you remember that we introduced this symbol to you in Chapter 8? It means take all the numbers and sum them.
- $X$ is the symbol used to indicate the independent variable.
- $Y$ is the symbol used to indicate the dependent variable.
- $XY$: Side by side, these symbols indicate that we should multiply these numbers together. $\Sigma XY$ means we should take the sum of all the cases of $X$ and $Y$ multiplied together—a process we explain below.
- $x^2$: The value of $X$ squared. This means we multiply $X$ by $X$.

## Calculating the Slope

1.  Some preliminary points to consider before we calculate the slope:
    a.  Ensure your table is organized in the correct manner: that individual cases are listed in the rows and the scores for the independent and dependent variable are listed in the columns.
    b.  Recall the formula for calculating the slope:

    $$b = \frac{N\Sigma XY - (\Sigma X)(\Sigma Y)}{N\Sigma X^2 - (\Sigma X)^2}$$

    c.  Know that despite the fact that the formula looks really big, we only need three sets of information: $N$, $Y$, and $X$.
2.  Calculate the sum of $X$, which is indicated mathematically as $\Sigma X$.
3.  Calculate the sum of $Y$, which is indicated mathematically as $\Sigma Y$.
4.  Calculate $XY$, which means multiplying $X$ by $Y$ for each individual row.
5.  Calculate $X^2$ by taking the square of each value for $X$.
6.  Enter the corresponding values for $X$, $Y$, and $\Sigma$ into the equation.
7.  Solve the numerator of the equation.
8.  Solve the denominator of the equation.
9.  Divide the numerator by the denominator.

Table 14.2 adds to the information from Table 14.1 the components required to solve the slope formula.

**TABLE 14.2** Income and Happiness Scores

| Case Number | Income (X) | Happiness (Y) | XY | X² |
|---|---|---|---|---|
| 1 | 2 | 1 | 2 | 4 |
| 2 | 3 | 5 | 15 | 9 |
| 3 | 4.5 | 7 | 31.5 | 20.25 |
| 4 | 5 | 6 | 30 | 25 |
| 5 | 7 | 8 | 56 | 49 |
| 6 | 10 | 9.5 | 95 | 100 |
| Σ | 31.5 | 36.5 | 229.5 | 207.25 |

The slope of the regression line linking *Income* and *Happiness* is now determined by inserting the values from Table 14.2 into the general formula:

$$
\begin{aligned}
b &= \frac{N\Sigma XY - (\Sigma X)(\Sigma Y)}{N\Sigma X^2 - (\Sigma X)^2} \\[2mm]
&= \frac{6(229.5) - (31.5)(36.5)}{6(207.25) - (31.5)^2} \\[2mm]
&= \frac{1{,}377 - 1{,}149.75}{1{,}243.5 - 992.25} \\[2mm]
&= \frac{227.25}{251.25} \\[2mm]
&= 0.90
\end{aligned}
$$

The slope of the relationship between *Income* and *Happiness* is 0.9. Slope describes how much the dependent variable changes for a unit change in the dependent variable. Slope is:

$$
\frac{\text{change in dependent variable}}{\text{change in independent variable}}
$$

The slope result of 0.9 is translatable into 0.9 ÷ 1, which parallels the equation above. The meaning of the slope of the relationship between *Income* and *Happiness* can now be stated clearly. A one-unit change in the *Income*[2] is associated with an expected nine-tenths of the unit change in *Happiness*.[3] Alternatively, each $10,000 increase in *Income* is associated with a nearly 10 per cent increase in *Happiness*.

Given the slope, the only remaining item required to complete the regression equation linking income and happiness is the intercept. The intercept is the point at which the regression

---

[2]Remember, each unit of income is $10,000. So this phrase could read, "Each $10,000 increase in *Income*. . ."

[3]*Happiness* is measured on a 10-point scale. Therefore, each 9/10 increase results in approximately a 10 per cent increase in *Happiness*.

line crosses the axis of the dependent variable. Since the intercept is a specific point, it can be identified as an ordered pair. Specifically, the intercept is the value of the dependent variable when the independent variable is zero. In any data set, the intercept is calculated using the following formula:

$$a = \overline{Y} - b\overline{X}$$

This formula states that the intercept is derived from three pieces of information, including the means of the independent ($\overline{X}$) and dependent ($\overline{Y}$) variables[4] as well as the value of the slope. The slope was just calculated and the means, using the data from Table 14.2, are 5.25 for *Income* and 6.08 for *Happiness*. Entering these values into the intercept equation yields the following result:

$$a = \overline{Y} - b\overline{X} = 6.08 - (0.9)\,5.25 = 6.08 - 4.73 = 1.35$$

The intercept is the point on the dependent-variable axis when the independent variable is zero. Therefore, this intercept can be interpreted as the expectation that someone with no income would have a happiness score (on a 10-point scale) of 1.35.

With the specific slope and the intercept values calculated, the regression line relating *Income* and *Happiness* is now determined. The specific equation for the best-fit line is:

$$\text{happiness} = 1.35 + (0.9)\,\text{income}$$

## Drawing the Regression Line

When scatterplots were introduced in Chapter 11, the best-fit regression line was an idea that was drawn on the graph freehand. The product was a rough approximation. The exact location of the regression line is now determined by the linear equation, and it can be drawn precisely on the scatterplot.

A line is really a set of points that are very, very close together. Drawing a straight line only requires connecting any two points on the line. Joining these two points, and extending them, produces all values on the line. In short, drawing the regression line on a scatterplot only requires knowing two points and joining them. The intercept is one point on the regression line. For the data set in Table 14.2, the intercept is the point defined by the ordered pair (0, 1.35). To draw the regression line on the scatterplot, all that's required is knowledge of one other point on the line.

The regression equation describes *all* the data points on the best-fit line. For example, you know the intercept is on the regression line. If you solve the *Income–Happiness* regression equation for an income value of zero, here is what you get:

$$\text{happiness} = 1.35 + (0.9)\,\text{income} = 1.35 + (0.9)\,0 = 1.35 + 0 = 1.35$$

---

[4]Remember, from univariate analysis, the mean is a measure of central tendency for variables at higher (interval or ratio) levels of measurement. Computing the intercept of the regression line uses the mean of both the independent variable ($\overline{X}$) and the dependent variable ($\overline{Y}$).

In other words, an *Income* score of zero is associated with a *Happiness* score of 1.35. The same arithmetic can be done for *any value* of the independent variable (*Income*). So, for an *Income* score of 5, the associated *Happiness* score is calculated as:

$$\text{happiness} = 1.35 + (0.9)\,\text{income} = 1.35 + (0.9)\,5 = 1.35 + 4.5 = 5.85$$

Now, two data points on the regression line are known. Both of these points can be described as ordered pairs and plotted on the scatterplot. For the *Income–Happiness* data set, the ordered pairs are (0, 1.35) and (5, 5.85). The first ordered pair is the intercept, the second the regression line point with an *Income* value of 5. If these two points are identified and joined on the scatterplot, the exact location of the best-fit regression line is revealed. Figure 14.6 shows the results.

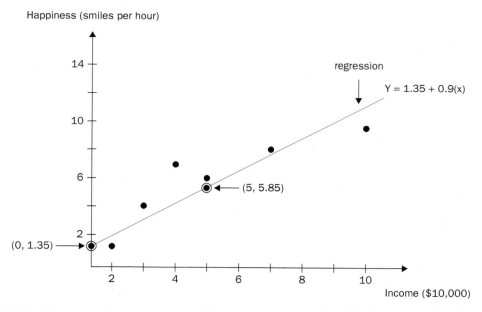

**FIGURE 14.6** Income and Happiness Scatterplot and Regression Line

# Form, Extent, and Precision

The tools for identifying and plotting the exact location of the best-fit regression line for a data set are simply refinements of the ideas introduced in Chapter 11. However, instead of roughly approximating the location of the regression line, its exact location is now specified through the regression line calculations. There is no room for argument on this point.

The identified regression line is undoubtedly the single straight line that comes closest to all the data points.

Interpreting scatterplots involves addressing three considerations—form, extent, and precision. The first two of these issues are derived from looking at the regression line. Remember that the regression line is the one that best fits the actual data points on a scatterplot. The third consideration, precision, addresses how well the regression line actually fits the data. By extension, precision informs us how accurate the form and extent descriptions (which rely on the regression line) are.

Precision refers to how scattered the data points are around the regression line. Chapter 11 displayed examples of scatterplots with different levels of precision. As the data points (ordered pairs) are more tightly clustered around the regression line, precision improves. Rather than speak vaguely about the level of precision (e.g. good, poor), exact levels of precision can be calculated for any scatterplot. The result is a measure called Pearson's *r*. **Pearson's *r*** is symbolized by a lower-case *r* and, in regression analysis, is commonly referred to as the correlation coefficient. This measure of precision ranges from −1.0 to +1.0. It has a value of zero when the ordered pairs on a scatterplot are random (signifying no relationship). As the magnitude of *r* (independent of its sign) increases, then precision increases. This correlation coefficient statistic is calculated using the following formula:

$$r = \frac{N\Sigma XY - (\Sigma X)(\Sigma Y)}{\sqrt{[N\Sigma X^2 - (\Sigma X)^2][N\Sigma Y^2 - (\Sigma Y)^2]}}$$

Most of the information required to calculate the correlation coefficient relies on information used in calculating the slope. However, some additional numbers are required from the data set. Table 14.3 adds the necessary additional information to the information already provided in Table 14.2.

**TABLE 14.3** Income and Happiness Information for Correlation Coefficient Calculation

| Case Number | Income (X) | Happiness (Y) | XY | X² | Y² |
|---|---|---|---|---|---|
| 1 | 2 | 1 | 2 | 4 | 1 |
| 2 | 3 | 5 | 15 | 9 | 25 |
| 3 | 4.5 | 7 | 31.5 | 20.25 | 49 |
| 4 | 5 | 6 | 30 | 25 | 36 |
| 5 | 7 | 8 | 56 | 49 | 64 |
| 6 | 10 | 9.5 | 95 | 100 | 90.25 |
| Σ | 31.5 | 36.5 | 229.5 | 207.25 | 265.25 |
| Σ² | 992.25 | 1,332.25 | | | |

The correlation coefficient can now be computed by applying the specific values from Table 14.3 to the equation.

$$
\begin{aligned}
r &= \frac{N\Sigma XY - (\Sigma X)(\Sigma Y)}{\sqrt{[N\Sigma X^2 - (\Sigma X)^2][N\Sigma Y^2 - (\Sigma Y)^2]}} \\
&= \frac{6(229.5) - 31.5(36.5)}{\sqrt{[6(207.25) - 992.25][6(265.25) - 1,332.25]}} \\
&= \frac{1,377 - 1,149.75}{\sqrt{[1,243.5 - 992.25][1,591.5 - 1,332.25]}} \\
&= \frac{227.25}{\sqrt{[251.25][259.25]}} \\
&= \frac{227.25}{\sqrt{65,136.56}} \\
&= \frac{227.25}{255.22} \\
&= 0.89
\end{aligned}
$$

This correlation coefficient calculation demonstrates that the regression line fits the data very well. If you look again at the clustering of the data points around the regression line in Figure 14.6, you will see that this is the case.

# PRE Interpretations

Chapter 12 introduced the proportional reduction in error (PRE) family of statistics. These statistics are measures of association; they report on the character and strength of the connection between two variables. The last chapter introduced the PRE measures appropriate for two nominal variables (lambda) and two ordinal variables (gamma). This chapter focuses on the bivariate situation of two interval or ratio variables. PRE measures (like all statistics) are tailored to specific levels of measurement, and there is a PRE measure designed for the relationship between two quantitative variables. This measure is called the **coefficient of determination** and is symbolized as $r^2$.

The coefficient of determination is a member of the PRE family of statistics, which are characterized by having marginal and relational prediction rules, ways of counting errors associated with these rules, and a standard calculation formula. The best way of understanding $r^2$ is by discussing its application of these common features.

## Marginal Prediction Rule and $E_1$

A marginal prediction rule is a way of predicting the dependent variable's scores of all cases in a data set based on only knowledge of the dependent variable's distribution. Scatterplots are based on quantitative variables, measured at either the interval or the ratio level of measurement. These variables are rich in information and, therefore, can employ sophisticated prediction rules.

Imagine you possess a data set on 500 individuals that, among other things, includes all scores on the variable *Income*. In marginal prediction rule terms, the question becomes, How can this information be used as a general rule for predicting each case's income? The answer is found in the statistical tools introduced in Chapter 5.

Chapter 5 discussed statistics that measure the central tendency of single variables. For interval or ratio variables, the most appropriate central tendency measure is the mean. The mean is a statistic that reports the typical or average score of a variable in a data set. *The mean is the marginal prediction rule for scatterplots.* In scatterplot analysis, if you only have information about the dependent variable's scores, and you want a general rule for predicting each case's dependent variable score, the best rule is to predict the mean. Figure 14.7 diagrams this procedure for a study in which the independent variable is *Education (years)* and the dependent variable is *Income*.

For clarity, Figure 14.7 only includes three data points. The marginal prediction rule states that, for each of these cases, the predicted dependent variable value is the mean. In this case it is $43,000.

None of the cases, however, has an income score of $43,000. Therefore, in predicting the mean ($43,000), each prediction contains some error. The amount of error using the marginal prediction rule is $E_1$. As portrayed in Figure 14.7, $E_1$ *is the total difference between the mean and the actual dependent variable scores.* $E_1$ is the sum of how different each actual dependent variable score (*y*) is from the mean of the dependent variable ($\bar{Y}$). Expressed arithmetically, this total is $\Sigma(y - \bar{Y})$.

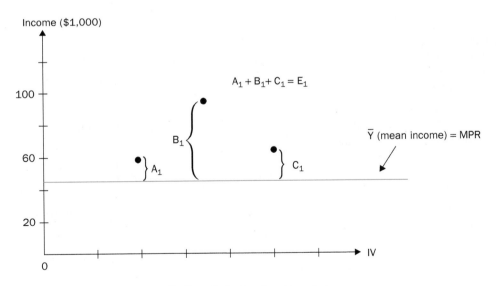

**FIGURE 14.7** Marginal Prediction Rule for Scatterplots

Because dependent variable scores occur both above and below the mean, adding them together will produce the uninformative total of zero. This issue is resolved by squaring the differences before adding them up.[5] So the actual calculation of $E_1$ for a scatterplot is $\Sigma(y - \bar{Y})^2$. This $E_1$ result is identified as the *total sums of squares (TSS)*.

## Relational Prediction Rule and $E_2$

In PRE measures, the marginal prediction rule and its results ($E_1$) are compared to predicting the dependent variable and its results having knowledge of each case's independent variable scores before predicting the dependent variable. The rule for predicting dependent variable scores based on independent variable values is the relational prediction rule and the errors using this rule is $E_2$.

For scatterplots, *the relational prediction rule is the regression line*. The formula for the best-fit regression line ($Y = a + bX$) is an equation for making predictions about the dependent variable ($Y$), given knowledge of an independent variable score ($X$). Since most relationships are imperfect, predictions using this relational prediction rule will contain some error. This error is $E_2$ and it is *the total difference between the regression line prediction and the actual dependent variable scores*. This is illustrated in Figure 14.8.

$E_2$ errors occur when the regression line prediction for a case ($Yp$) is different than the actual score ($y$). Again, because data points occur both below and above the regression line, these errors are squared. The arithmetic expression of $E_2$ becomes $\Sigma(y - Yp)^2$. These $E_2$ errors are the prediction mistakes that occur even after our best knowledge (of the independent variable) is used. In aggregate the $E_2$ total [$\Sigma(y - Yp)^2$] is identified as the *unexplained sums of squares (USS)*.

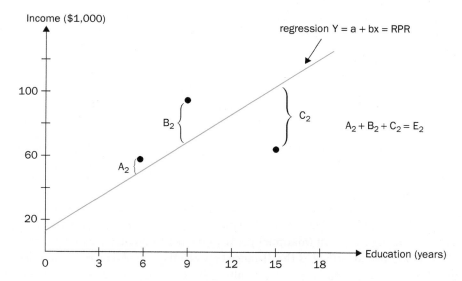

**FIGURE 14.8** Relational Prediction Rule for Scatterplots

---

[5]This is the same procedure for addressing the same issue in the earlier discussion of the least squares technique for identifying the best-fit regression line.

## PRE Concept

All statistics that are part of the family of PRE measures of association use the generic formula $(E_1 - E_2) \div E_1$. For scatterplots, the specific PRE measure of association is called the coefficient of determination. The calculation of this coefficient can be expressed in different ways.

The simplest expression simply inserts the $E_1$ and $E_2$ values into the equation, which results in the following:

$$\text{coefficient of determination} = \frac{\text{TSS} - \text{USS}}{\text{TSS}}$$

Figure 14.9 illustrates a useful translation of this formula.

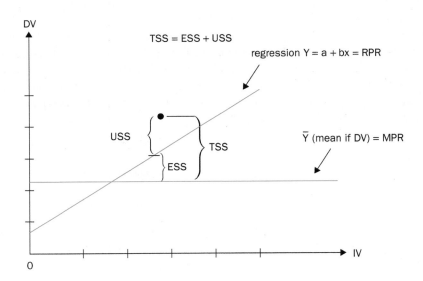

**FIGURE 14.9**  Sums of Squares

For clarity, Figure 14.9 includes only one data point. For this data point, the TSS is clearly identified as the distance from the actual data point to the mean. This distance is the error ($E_1$) resulting from using the mean (marginal prediction rule) to predict the location of the data point. The USS is also identified as the distance between the actual data point and the regression line. This distance is the error ($E_2$) resulting from using the regression equation (relational prediction rule) to predict the location of the data point. Notice that the USS is smaller than the TSS. In other words, using the regression line provides a better prediction (less error) than using the mean as a prediction rule.

As Figure 14.9 shows, the TSS distance can be divided into two components. One segment is the USS. The other segment is *the distance between the regression line prediction and the mean.* This distance is identified as the explained sums of squares (ESS). The ESS identifies the *improvement in prediction* by using the relational prediction rule (regression equation) over the marginal prediction rule (mean). Look carefully at Figure 14.9 and you will see that $ESS = TSS - USS$. Therefore, another way of thinking about the coefficient of determination is to see it as:

$$\text{coefficient of determination} = \frac{ESS}{TSS}$$

In other words, the PRE measure of association for scatterplots asks, Of the total differences (TSS), how much (what proportion) of them are explained (ESS)? The more the independent variable explains variation in (i.e. comes closer to correctly predicting) the dependent variable, the stronger the relationship.

## PRE Calculation

Understanding that the coefficient of determination is the proportion of total differences explained by the independent variable is quite different than actually calculating this measure of association. After all, researchers aren't going to get out their rulers and actually measure the TSS and ESS differences! Fortunately, the coefficient of determination is easily calculated, when you know its symbol is $r^2$. Obtaining the coefficient of determination is a matter of squaring the correlation coefficient Pearson's $r$.

## MATH TIPS

### Why are Differences from the Mean and Regression Line Squared?

We have already discussed how to square a term. In determining how well a regression line fits, why do we need to square differences? Remember when we were talking about calculating the total difference our scores are from the mean (for the marginal prediction rule) or the regression line (for the relational prediction rule). We need to square these differences because dependent variable scores occur both above and below these lines. If we simply added them together we would produce the uninformative total of zero. For example, if we think about the definition of the mean, there will be an equal number of scores above the mean and below the mean. Thus, if we add all the negative numbers to the positive numbers, then we would get a score of zero, which is not helpful in determining a PRE measure as all the measures would equal zero no matter the average distance from the mean. Squaring the difference removes this problem because when either positive or negative numbers are squared, the result is a positive number.

Pearson's $r$ is a measure of the precision of the regression line: how well it fits the data. It ranges from −1.0 to +1.0. Squaring this coefficient produces positive sums, which is why the range of the coefficient of determination can vary between 0.0 and 1.0.

## STEP 2 Learning the Calculations

# Calculating Pearson's Coefficient of Determination ($r^2$)

We have covered a lot of ground in this chapter. Let's step back a bit and review some of what we have learned.

The coefficient of determination is based on Pearson's $r$, which is determined by using this equation:

$$r = \frac{N\Sigma XY - (\Sigma X)(\Sigma Y)}{\sqrt{[N\Sigma X^2 - (\Sigma X)^2][N\Sigma Y^2 - (\Sigma Y)^2]}}$$

The coefficient of determination is simply $r^2$. So how do we calculate $r^2$ using this jumble of numbers and Greek symbols?

1. Recall that the calculation for Pearson's $r$ is

$$r = \frac{N\Sigma XY - (\Sigma X)(\Sigma Y)}{\sqrt{[N\Sigma X^2 - (\Sigma X)^2][N\Sigma Y^2 - (\Sigma Y)^2]}}$$

We will calculate Pearson's $r$ first, then calculate the coefficient of determination. To reinforce your knowledge of Pearson's $r$ and the coefficient of determination, let's use the following example. We would like to know whether or not your age has an effect on increasing your weight, because we just read an article in a fitness magazine that indicates that people gain 0.975 kilograms per year. We set up a table at the nearest convenience store and stop the first 10 people we meet. We ask them their date of birth and then weigh them (in kilograms). Table 14.4a shows the information we collect.

**TABLE 14.4a**　Age and Weight

| Case Number | Age (X) | Weight (Y) |
|:---:|:---:|:---:|
| 1 | 26 | 71 |
| 2 | 34 | 55 |
| 3 | 45 | 107 |
| 4 | 51 | 86 |
| 5 | 73 | 88 |
| 6 | 19 | 65 |
| 7 | 38 | 67 |
| 8 | 66 | 73 |
| 9 | 27 | 53 |
| 10 | 42 | 59 |

2.  Our next step is to calculate the numerator of the equation. As you can see, we need to calculate several numbers to do this. We need to know $N$ and the $\Sigma XY$ and also the $\Sigma X$ and $\Sigma Y$. Let's do this in small steps:

   a.  Remember that $N$ is the symbol for the total number of cases. In this example, we have 10 cases (corresponding to 10 unique rows containing the information for the 10 people we took measurements for). That is easy.

   b.  Now we need to calculate the $\Sigma XY$. That is easier than it looks. This part of the term means the sum of $X$ times $Y$. Using the table below, we can easily calculate this. For each case, we multiply $X$ by the value of $Y$. So for the first row, the value of $XY$ for case 1 is $26 \times 71$ which equals 1,846. Repeat this step for each of the remaining nine cases. You should obtain the result shown in Table 14.4b.

**TABLE 14.4b**  Age Multiplied by Weight

| Case Number | Age (X) | Weight (Y) | XY |
|---|---|---|---|
| 1 | 26 | 71 | 1,846 |
| 2 | 34 | 55 | 1,870 |
| 3 | 45 | 107 | 4,815 |
| 4 | 51 | 86 | 4,386 |
| 5 | 73 | 88 | 6,424 |
| 6 | 19 | 65 | 1,235 |
| 7 | 38 | 67 | 2,546 |
| 8 | 66 | 73 | 4,818 |
| 9 | 27 | 53 | 1,431 |
| 10 | 42 | 59 | 2,478 |

   c.  Next we need to calculate the sum for $XY$. That is simple, since we have already calculated $XY$, so all we need to do is add the values together, as shown in Table 14.4c. Note that we have added the last row to this table with the results. Can you obtain the same result?

**TABLE 14.4c**  Sum of Age Multiplied by Weight

| Case Number | Age (X) | Weight (Y) | XY |
|---|---|---|---|
| 1 | 26 | 71 | 1,846 |
| 2 | 34 | 55 | 1,870 |
| 3 | 45 | 107 | 4,815 |
| 4 | 51 | 86 | 4,386 |
| 5 | 73 | 88 | 6,424 |
| 6 | 19 | 65 | 1,235 |
| 7 | 38 | 67 | 2,546 |
| 8 | 66 | 73 | 4,818 |
| 9 | 27 | 53 | 1,431 |
| 10 | 42 | 59 | 2,478 |
| $\Sigma$ | | | 31,849 |

d.   Next we need to calculate the sum of $X$, which is depicted by $\Sigma X$ in the equation. All this means is that we need to add all the ages of our cases together (because *Age* is the independent variable, which is represented by $X$). We get the value of 421 when we add all the ages together.

e.   Now we need to calculate $\Sigma Y$, which is the same process as we used to calculate $\Sigma X$ in step d above. We simply add all the values of $Y$ together. For this example, $Y$ is *Weight*, as we are trying to predict *Weight* ($Y$) using *Age* ($X$). In this example, when we add all the weights of our 10 cases together, we get a value of 724. We have entered the values of $\Sigma X$ and $\Sigma Y$ to Table 14.4d.

**TABLE 14.4d**  Sums of Age and Weight

| Case Number | Age (X) | Weight (Y) | XY |
|---|---|---|---|
| 1 | 26 | 71 | 1,846 |
| 2 | 34 | 55 | 1,870 |
| 3 | 45 | 107 | 4,815 |
| 4 | 51 | 86 | 4,386 |
| 5 | 73 | 88 | 6,424 |
| 6 | 19 | 65 | 1,235 |
| 7 | 38 | 67 | 2,546 |
| 8 | 66 | 73 | 4,818 |
| 9 | 27 | 53 | 1,431 |
| 10 | 42 | 59 | 2,478 |
| $\Sigma$ | 421 | 724 | 31,849 |

f.   Now we have all the numbers we need to calculate the value of the numerator in our equation. All we need to do is to replace the equation symbols with the numbers we have just calculated. Recall that

$$
\begin{aligned}
N &= 10 \\
\Sigma XY &= 31,849 \\
\Sigma X &= 421 \\
\Sigma Y &= 724
\end{aligned}
$$

Let's plug these values into the numerator of the equation for $r$.

$$N\,\Sigma XY - (\Sigma X)(\Sigma Y)$$
$$10(31,849) - (421)(724)$$

Working through the steps in mathematical order we obtain the following results:

$$318,490 - 304,804$$
$$= 13,686$$

3. Now let's work on calculating the numbers we need to obtain the result for the denominator of the equation for $r$. The denominator for the equation looks like this

$$\left[N\Sigma X^2 - (\Sigma X)^2\right]\left[N\Sigma Y^2 - (\Sigma Y)^2\right]$$

a. Let's take this equation one symbol at a time. The first symbol is easy, $N$. We know that $N$ is 10.

b. The next part of the equation is $\Sigma X^2$. That means we want to take the sum of all the values of $X$ *after* they are squared. This means that we must square $X$ first and we must calculate the square of $X$ for all the cases in our sample. Thus, the value of $X^2$ for case 1 is $26 \times 26$, which is 676. Square $X$ for each of the remaining 9 cases to get the results shown in Table 14.4e.

**TABLE 14.4e** Age Squared

| Case Number | Age (X) | Weight (Y) | XY | X² |
|---|---|---|---|---|
| 1 | 26 | 71 | 1,846 | 676 |
| 2 | 34 | 55 | 1,870 | 1,156 |
| 3 | 45 | 107 | 4,815 | 2,025 |
| 4 | 51 | 86 | 4,386 | 2,601 |
| 5 | 73 | 88 | 6,424 | 5,329 |
| 6 | 19 | 65 | 1,235 | 361 |
| 7 | 38 | 67 | 2,546 | 1,444 |
| 8 | 66 | 73 | 4,818 | 4,356 |
| 9 | 27 | 53 | 1,431 | 729 |
| 10 | 42 | 59 | 2,478 | 1,764 |

c. Now we must add all the squares of $X$ together to obtain $\Sigma X^2$. We should obtain a total of 20,441.

d. Now we need to deal with the part of the equation symbolized by $(\Sigma X)^2$. Pay attention to your mathematical rules. Since the $\Sigma X$ part is in brackets, the equation is telling us that we must add all of the values of $X$ together before we square it. This means we can take 421 (which is the value of $\Sigma X$) which we calculated in step 2 and square it ($421 \times 421$) which is equal to 177,241.

e. Now let's calculate part of the equation that appears within the first set of square brackets: $N\Sigma X^2 - (\Sigma X)^2$. We plug in the numbers we have just calculated:

$$10(20,441) - 177,241$$
$$= 204,410 - 177,241$$
$$= 27,169$$

4. Now let's concentrate on calculating the information we require for the second part of the equation: $N\Sigma Y^2 - (\Sigma Y)^2$

   a. Calculating $N$ is easy! We already have that information. There are 10 cases.

   b. For the calculation of $\Sigma Y^2$, the procedure is the same as the one we used for $\Sigma X^2$. That means we want to take the sum of all the values of $Y$ *after* they are squared. This means that we must square each value of $Y$ first, and we must calculate the square of $Y$ for all the cases in our sample. Thus, the value of $Y^2$ for case 1 is $71 \times 71$, which is 5,041. Square $Y$ for each of the remaining nine cases to get the results shown in Table 14.4f.

**TABLE 14.4f** Age Squared and Weight Squared

| Case Number | Age (X) | Weight (Y) | XY | X² | Y² |
|---|---|---|---|---|---|
| 1 | 26 | 71 | 1,846 | 676 | 5,041 |
| 2 | 34 | 55 | 1,870 | 1,156 | 3,025 |
| 3 | 45 | 107 | 4,815 | 2,025 | 11,449 |
| 4 | 51 | 86 | 4,386 | 2,601 | 7,396 |
| 5 | 73 | 88 | 6,424 | 5,329 | 7,744 |
| 6 | 19 | 65 | 1,235 | 361 | 4,225 |
| 7 | 38 | 67 | 2,546 | 1,444 | 4,489 |
| 8 | 66 | 73 | 4,818 | 4,356 | 5,329 |
| 9 | 27 | 53 | 1,431 | 729 | 2,809 |
| 10 | 42 | 59 | 2,478 | 1,764 | 3,481 |
| Σ | 421 | 724 | 31,849 | | |

   c. Now we must add all the squares of $Y$ together to obtain $\Sigma Y^2$. We should obtain a total of 54,988.

   d. Now let's work on the $(\Sigma Y)^2$ part of the equation, just like we did in step 3d above. Pay attention to your mathematical rules. Remember the $\Sigma Y$ part is in brackets, and the equation is telling us that we must add all of the values of $Y$ together before we square it. This means we can take 724 (which is the value of $\Sigma Y$) and square it ($724 \times 724$), which is equal to 524,176.

   e. Finally we are at the point where we have all the numbers we need to calculate the denominator for this equation.

$$\left[ N\Sigma Y^2 - (\Sigma Y)^2 \right]$$
$$10(54,988) - (524,176)$$
$$= 549,880 - 524,176$$
$$= 25,704$$

5. Now that we have the numerator (13,686) and the denominator (25,704), all that's left is to do a simple division to get the answer:

$$= 13,686 \div 25,704$$
$$= 0.532$$

Now what does that mean? This number describes the slope of the line that is the relationship between *Age* and *Weight*. Slope describes how much the dependent variable (*Weight*) changes for a unit change in the independent variable (*Age*). The meaning of the slope of the relationship between *Age* and *Weight* can now be stated clearly. A one-unit change in *Age* is associated with an expected 0.532 of the unit change in *Weight*.

Now you can confidently calculate the slope of other pairs of dependent and independent variables that are measured using continuous level data.

## STEP 3 Using Computer Software

In Chapter 11, we went over the procedure for creating a scatterplot in IBM SPSS statistics software ("SPSS"), which is used when performing analyses between two variables that are measured at higher (i.e. interval and ratio) levels of measurement. You will recall that the scatterplot provides a visual representation of how the independent and dependent variables are related. The next step is to get an estimate of how well the line fits the actual data points, which is achieved by computing the Pearson's *r* statistic. For learning continuity, we will use the same variables as found in Chapter 11. Remember, as with all statistical analysis, to select all values that you do not want to include (e.g. *no response*) to *missing*.

# Pearson's *r* Correlation Coefficient

## Procedure

1. Select Analyze from the SPSS menu bar.
2. Go to Correlate → Bivariate.
   a. On the left side, scroll down the list of variables until you find your independent and dependent variables.
   b. Highlight the variable Q16F (*Psychological well-being—Self-acceptance*).
   c. Click the arrow button.
   d. Do this again for the second variable Q22 (*Satisfaction with life*).
3. Click OK.

This procedure will result in a table titled Correlations appearing in your Output/Viewer window (Figure 14.10). Notice that the coefficients in this table appear as a matrix. In other words, the calculations include both variables as columns and rows. For this reason, the Pearson's *r* correlation coefficient will be duplicated.

Within the Correlations table (Figure 14.10), examine the Pearson's *r* (correlation coefficient) for one combination of the independent (Q16F) and dependent variables (Q22). If we look in the cell in the upper-right corner, we observe that *r = 0.688*. For current purposes, you can ignore the information about statistical significance (*sig.*) and sample size (*N*), which are also included in the cells of the table.

**Correlations**

| | Pearson's r | Q16F Q16F. Psychological well-being- Self-acceptance | Q22 Q22. Satisfaction with life |
|---|---|---|---|
| Q16F Q16F. Psychological well-being - Self-acceptance | Pearson Correlation | 1 | .688 |
| | Sig. (2-tailed) | | .000 |
| | N | 1157 | 1096 |
| Q22 Q22. Satisfaction with life | Pearson Correlation | .688 | 1 |
| | Sig. (2-tailed) | .000 | |
| | N | 1096 | 1176 |

**. Correlation is significant at the 0.01 level (2-tailed).

**FIGURE 14.10**  Correlations Output

# Bivariate Linear Regression

Recall that the Pearson's r correlation coefficient is a statistic that measures the fit of a straight line on a scatterplot. The constants (intercept and slope) that define the line of best fit are produced through regression analysis. For example, in our example, research has shown that individuals who have higher levels of psychological well-being (of which self-acceptance is one of several components) report being more satisfied with their life. For our example, linear regression can be used to determine which line best fits the data displaying the connection between *Self-acceptance* (the independent variable) and *Satisfaction with life* (the dependent variable). The intercept and slope of such a best-fit regression line are determined using the following procedure.

## Procedure

1.  Select Analyze from the SPSS menu bar.
2.  Go to Regression → Linear.
    a.  Scroll down the list of variables on the left side to Q22 (*Satisfaction with life*) and click the arrow to move it to the Dependent Variable box.
    b.  Place Q16F (*Self-acceptance*) into the Independent Variable box.
    c.  Click OK.

Running this procedure should result in an output similar to that in Figure 14.11 (less the Variable Entered/Removed and ANOVA tables).

The regression output contains a lot of information that can be ignored for current purposes. Of importance are the four pieces of information noted in Figure 14.11.

In the table labelled Coefficients are the intercept and slope.

1.  The intercept of the regression line is found next to the label Constant.
2.  The slope of the regression line is found below the intercept in the column labelled B and the row labelled Q16F, which is our independent variable (*Self-acceptance*).
3.  The Pearson's r correlation coefficient is found in the Model Summary table in the column labelled R.

**Model Summary**

| Model | R | R Square | Adjusted R Square | Std. Error of the Estimate |
|---|---|---|---|---|
| 1 | .688ᵃ | .473 | .473 | .4338 |

a. Predictors: (Constant), Q16F Q16F. Psychological well-being-Self-acceptance

Pearson's r          Coefficient of determination          Intercept          Slope

**Coefficientsᵃ**

| Model | | Unstandardized Coefficients | | Standardized Coefficients | t | Sig. |
|---|---|---|---|---|---|---|
| | | B | Std. Error | Beta | | |
| 1 | (Constant) | 6.596 | .427 | | 15.463 | .000 |
| | Q16F Q16F. Psychological well-being - Self-acceptance | .841 | .027 | .688 | 31.366 | .000 |

a. Dependent Variable: Q22 Q22. Satisfaction with life

**FIGURE 14.11**  Regression Output

4. If you are interested in the PRE statistic that measures the strength of the connection between variables (the coefficient of determination), it is found in the column labelled R-Square.

There are several interpretations based on our output.

## Regression Equation

In terms of the regression equation ($Y = a + bX$), the slope ($b$) of the relationship between *Self-acceptance* and *Satisfaction with life* is 0.841, which tells us how much the dependent variable changes for a unit change in the independent variable.

Recall that $a$ is the intercept, which is the value of $Y$ when $X$ is zero. As such, it is expected in our survey that a respondent who scored zero on *Self-acceptance* would have a score of 6.596 for their *Satisfaction with life*, which is very low considering the maximum score on our dependent variable is 24 and the mean score is 15.13 with a standard deviation of 4.861.

We also know that $X$ is a given value from our data set, which represents one survey respondent. For example, in the IBM Data Editor in Data View under the column/variable ID we can go to ID #5. If we scroll across to Q16F and look at the cell for ID #5, we see that this student's score is 10.

Now we are ready to plug in some numbers, because we know $X$ (10) and the SPSS output has computed $a$ (intercept) and $b$ (slope) for us. Following the regression equation ($Y = a + bX \rightarrow Y = 6.596 + 0.841 \times 10$), we calculate that $Y$ equals 18.41. We can interpret our calculation by saying, "A student who scores 10 (very low) on *Self-acceptance* will score 18.41 on *Life satisfaction*."

Let's try another one. If we go to ID #1 and scroll across to Q16F, this student's score is 24. By calculating the regression equation, we determine that he or she would score 26.78, which can be interpreted by saying "A student who scores 24 (very high) on *Self-acceptance* will score 26.78 on *Life satisfaction*."

Finally, if we compare our two calculated scores, we see that the more self-acceptance students have the greater their overall life satisfaction is.

### Pearson's r and the Coefficient of Determination

We have learned that Pearson's *r* is a measure of precision. In particular, the larger the magnitude of Pearson's *r*, the better the regression line fits the data points. In our output, Pearson's *r* is 0.688, which can be rounded up to 0.70. Using the suggested interpretive descriptions described in Step 5, we can make the following interpretation: The Pearson's *r* is 0.70, which means the regression line describing the connection between *Self-acceptance* and *Life satisfaction* provides a very accurate description of the actual connection between these variables.

In Chapter 12, you learned about proportional reduction in error (PRE) measures, which report on the character and strength of the relationship between the independent and dependent variables. We have learned from this chapter that the PRE for quantitative variables (i.e. interval and ratio) is the coefficient of determination, or $r^2$. In our output, the $r^2$ was 0.473, which can be interpreted by saying, "Knowing a respondent's level of *Self-acceptance* (as opposed to not knowing his or her level of acceptance) proportionately reduces our errors in predicting *Life satisfaction* by 47.3 per cent."

## STEP 4 Practice

You now have the tools for calculating and interpreting the statistics related to the features of scatterplots. You can perform these analyses either by hand or using computing software. To solidify your understanding you need to practise applying the statistical and software procedures.

This section provides you with opportunities to practise what you have learned about scatterplot analysis. The first set of questions uses hand calculations. The second set uses the SPSS procedures. For each set of questions:

1. Follow the procedural steps and complete the appropriate calculation (Set 1) or software application (Set 2).
2. Check your answers, using the Answer Key in the back of the text.
3. If your answer is incorrect, consult the Solutions section on the course website. The Solutions provide a complete step-by-step analysis of how the answers are derived.

After you have completed the next section (Interpreting the Results), return to your calculations or output and *provide complete, written interpretations of each of the statistics you have generated.*

# Set 1: Hand-Calculation Practice Questions

1. Table 14.5 provides data on hours of cultural conversation per day in students' homes and reading achievement for five students. Use the data in this table to answer the following questions.
   a. Calculate the slope for the relationship between hours per week of culture-related conversation (books, movies, politics and current events, etc.) within the home and students' reading achievement from the data in Table 14.5.

**TABLE 14.5** Cultural Conversation and Reading Achievement

| Case # | Cultural Conversation (X) | Reading Achievement (Y) | XY | X² | Y² |
|---|---|---|---|---|---|
| 1 | 2 | 66 | 132 | 4 | 4,356 |
| 2 | 6 | 84 | 504 | 36 | 7,056 |
| 3 | 4 | 72 | 288 | 16 | 5,184 |
| 4 | 6 | 74 | 444 | 36 | 5,476 |
| 5 | 8 | 88 | 704 | 64 | 7,744 |
| $\Sigma$ | 26 | 384 | 2,072 | 156 | 29,816 |
| $\Sigma^2$ | 676 | 147,456 | | | |

   b.  Find the intercept for this regression line.

   c.  According to our regression equation, a student who reported participating in 10 hours of cultural conversation at home per week would have a reading achievement score of how much? What if they reported 3 hours per week?

   d.  Calculate Pearson's $r$ and the coefficient of determination for this relationship.

   e.  What do the slope, Pearson's $r$, and the coefficient of determination tell us about the form, extent, precision, and strength of this relationship?

2.  Table 14.6 provides data on hours of television viewing (per day) and body mass index (BMI) for a sample of six students. Use the data in this table to answer the following questions.

**TABLE 14.6** Television Viewing and BMI

| Case # | TV Viewing (X) | BMI (Y) |
|---|---|---|
| 1 | 2 | 21 |
| 2 | 2.5 | 24 |
| 3 | 1 | 17 |
| 4 | 5 | 29 |
| 5 | 3 | 19 |
| 6 | 3.5 | 26 |

   a.  Set up a calculation table and compute the slope for the relationship between hours per day of TV viewing and BMI.

   b.  Find the intercept for this regression line.

   c.  According to our regression equation, what would be the predicted BMI for someone who spent 4 hours per day viewing TV? 7 hours?

   d.  Calculate Pearson's $r$ and the coefficient of determination for this relationship.

   e.  What do the slope, Pearson's $r$, and the coefficient of determination tell us about the form, extent, precision, and strength of this relationship?

3.    Below is data for education and number of children for seven women.

**Years of Education**

| Marion | 15 |
| Tarla | 24 |
| Kathy | 12 |
| Chantal | 16 |
| Sook-Yin | 17 |
| Leslie | 20 |
| Jean | 10 |

**Number of Children**

| Marion | 2 |
| Tarla | 0 |
| Kathy | 4 |
| Chantal | 1 |
| Sook-Yin | 1 |
| Leslie | 2 |
| Jean | 3 |

a.    Use the data in these tables to construct a calculation table with *Number of children* as the dependent variable.

b.    Compute the slope for the relationship between *Years of education* and *Number of children*.

c.    Find the intercept for this regression line.

d.    According to our regression equation, how many children would a woman with 21 years of education be expected to have? 10 years?

e.    Calculate Pearson's *r* and the coefficient of determination for this relationship.

f.    What do the slope, Pearson's *r*, and the coefficient of determination tell us about the form, extent, precision, and strength of this relationship?

4.    In Chapter 10 we already established linearity of the relationship between life expectancy and country's health-care spending using a scatterplot. We are now interested in exploring the relationship further using the statistical tools learned in Chapter 14 to investigate the form, extent, precision, and strength of this association.

a.    Using the data in Table 14.7, set up a calculation table.

b.    Find the slope and intercept of a regression line, and construct the equation for this relationship's regression line.

c.    According to the prediction model you constructed, what is the predicted life expectancy in a country whose total health expenditure per capita is $5,000? $1,000?

d.    Compute Pearson's *r* and the coefficient of determination.

e.    Comment on the form, extent, precision, and strength of the relationship.

**TABLE 14.7** Country, Health Expenditure per Capita, and Life Expectancy

| Country | Total Health Expenditure per Capita (PPP int. $)** | Life Expectancy* |
|---|---|---|
| Australia | 3691.6 | 82 |
| Canada | 4520.0 | 82 |
| Chile | 1292.2 | 79 |
| Czech Republic | 1922.8 | 78 |
| Japan | 3174.3 | 83 |
| Mexico | 940.1 | 75 |
| Peru | 496.2 | 77 |
| Slovenia | 2518.9 | 80 |
| United Kingdom | 3321.7 | 80 |
| United States of America | 8607.9 | 79 |

5.  In this next exercise we will examine how the investigation of the relationship between life expectancy and country's total health expenditure per capita changes when we exclude the United States from our analysis. (**Tip**: You do not need to set up a new calculation table if you already constructed one for the previous exercise. Make sure, however, to recalculate the mean and sum values this time excluding the US from the analysis).

a.  Find the slope and intercept of a regression line, and construct the equation of this relationship's regression line.

b.  According to this second prediction model, what is the predicted life expectancy in a country whose total health expenditure per capita is $5,000? $1,000? How are these predictions different compared to those you obtained using the regression model which took the United States into account?

c.  Compute Pearson's *r* and the coefficient of determination.

d.  Comment on the form, extent, precision, and strength of the relationship.

e.  How have your conclusions about the relationship changed by excluding the United States from the analysis? Why?

# Set 2: SPSS Practice Questions

Working again with the Student Health and Well-Being Survey data set, use the Bivariate Correlate function to obtain the Pearson's *r* for the relationships between the following pairs of variables:

1.  a.  NQ24 *Reasons for living* and Q4 *Depression*
    b.  NQ24 *Reasons for living* and Q22 *Satisfaction with life*
    c.  Q4 *Depression* and Q17 *Self-esteem*
    d.  Q3A *Positive affect* and Q3B *Negative affect*

2.  a.  Use the Linear Regression function to obtain the slope, constant, and coefficient of determination for the effect of *Reasons for living* on *Depression*.
    b.  Write the regression equation for this relationship.
    c.  What do the slope, Pearson's *r*, and the coefficient of determination tell us about the form, extent, precision, and strength of this relationship?

3.  a.  Use the Linear Regression function to obtain the slope, constant, and coefficient of determination for the effect of *Reasons for living* on *Satisfaction with life*.
    b.  Write the regression equation for this relationship.
    c.  What do the slope, Pearson's *r*, and the coefficient of determination tell us about the form, extent, precision, and strength of this relationship?

4.  a.  Use the Linear Regression function to obtain the slope, constant, and coefficient of determination for the effect of *Depression* on *Self-esteem*.
    b.  Write the regression equation for this relationship.
    c.  What do the slope, Pearson's *r*, and the coefficient of determination tell us about the form, extent, precision, and strength of this relationship?

5.  a.  Use the Linear Regression function to obtain the slope, constant, and coefficient of determination for the effect of *Negative affect* on *Positive affect*.
    b.  Write the regression equation for this relationship.
    c.  What do the slope, Pearson's *r*, and the coefficient of determination tell us about the form, extent, precision, and strength of this relationship?

6.  Find the Pearson's *r*, slope, constant, and coefficient of determination for *Coping with stress by seeking social support* (Q14A) as the independent variable, and *Personal growth* (Q16C) as the dependent variable. What is the regression equation for this relationship? What do the slope, Pearson's *r*, and coefficient of determination show about the form, extent, precision, and strength of this relationship? What would be the predicted *Coping with stress* score for an individual with a *Personal growth* score of 10?

7.  Find the Pearson's *r*, slope, constant, and coefficient of determination for *Self-acceptance* (Q16F) as the independent variable, and *Coping with stress by avoidant behavior* (Q14C) as the dependent variable. What is the regression equation for this relationship? What do the slope, Pearson's *r*, and coefficient of determination show about the form, extent, precision, and strength of this relationship? What would be the predicted *Self-acceptance* score for an individual with a *Coping with stress by avoidant behavior* score of 7?

8.  Find the Pearson's *r*, slope, constant, and coefficient of determination for *Strength of religious faith* (Q18) as the independent variable, and *Positive relations with others* (Q16D) as the dependent variable. What is the regression equation for this relationship? What do the slope, Pearson's *r*, and coefficient of determination show about the form, extent, precision, and strength of this relationship? What would be the predicted *Strength of religious faith* for an individual with a *Positive relations with others* score of 17?

9.  Find the Pearson's *r*, slope, constant, and coefficient of determination for *Autonomy* (Q16A) as the independent variable, and *Coping with stress by planful problem*

*solving* (Q14B) as the dependent variable. What is the regression equation for this relationship? What do the slope, Pearson's *r*, and coefficient of determination show about the form, extent, precision, and strength of this relationship? What would be the predicted *Autonomy* score for an individual with a *Coping with stress by planful problem solving* score of 3?

## STEP 5 Interpreting the Results

Scatterplots graph the relationship between two quantitative variables whose measures come from a data set. The scatterplot produces a picture of the relationship. Statistical tools are used to interpret this picture and, in doing so, understand the nature of the connection between the variables. This chapter introduced the statistical tools for precisely measuring the features of a scatterplot relationship. The final task requires interpreting these statistical results. Here is how you interpret the form, extent, precision, and strength of the connection between quantitative variables.

# Interpreting the Form of Connections between Quantitative Variables

Form speaks to the kind of relationship that connects the two variables. Standard regression analysis imposes a straight line form on the scatterplot display. The first interpretive task is to double-check that it is a reasonable imposition. To check whether the relationship is roughly linear, examine the scatterplot display.

The simplest way to identify form is to *look at the sign of slope*.[6] If the slope has a positive sign, the relationship is direct. If the slope has a negative sign, the relationship is inverse. Direct and inverse have the same interpretations used previously. In general, if the form is direct, the interpretive statement is:

As *X* [independent variable] increases, then *Y* [dependent variable] also increases.

Example: As level of education increases, the amount of lifetime earnings also increases.

If the relationship is inverse (indicating that the variables move in opposite directions), the interpretive statement is:

As *X* [independent variable] increases, then *Y* [dependent variable] decreases.

Example: As the number of hours of paid employment increases, students' grades decrease.

---

[6]The slope is the value of *b* in the linear regression equation.

# Interpreting the Extent of Connections between Quantitative Variables

Extent refers to the impact of changing the independent variable on the dependent variable. As impact increases, the independent variable makes a greater difference to the dependent variable. The easiest way to assess extent is to *look at the magnitude of the slope*. The larger the absolute value of the slope (i.e. independent of its sign), the greater the extent.

Slope is interpreted as the amount of change in the dependent variable for a one-unit change in the independent variable. In general, the interpretation of slope is:

> Changing $X$ [independent variable] by $C$ amount [size of a one-unit change in the independent variable] produces $D$ change [absolute value of the slope] in $Y$ [dependent variable].

> Example ($b = 12,000$): Changing level of education by one year produces a $12,000 change in income.[7]

# Interpreting the Precision of Connections between Quantitative Variables

The interpretations of form and extent come from information contained in the regression equation, which defines the best-fit line. Precision is indicated by the *magnitude of Pearson's r*.[8] The larger the magnitude of Pearson's *r*, the better the regression line fits the data points. A correlation coefficient of zero indicates the scatterplot is a random display which no regression line fits. When *r* is 1.0, it indicates that the regression line perfectly fits (describes) the pattern in the data. Other than at these extremes, there is no standard interpretation of the correlation coefficient. To give you sense of precision, Figure 14.12 illustrates the scatterplot pattern for a variety of correlation coefficients.

To assist your interpretation of precision, we use the suggested cut-off descriptions for the magnitude of Pearson's *r* that appear in Table 14.8.

---

[7] The statement "*Increasing* level of education by one year produces a $12,000 *increase* in income" is an interpretation statement that *combines* both form and extent information.

[8] Pearson's *r* ranges from −1.0 to +1.0. The *sign* of this coefficient provides information about *form*.

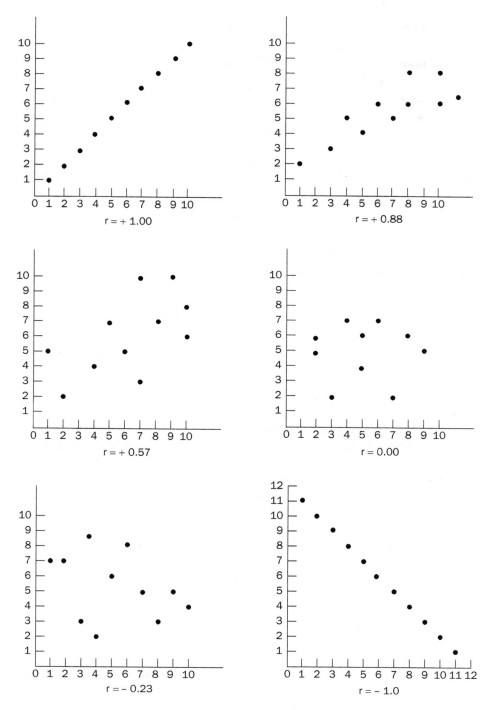

**FIGURE 14.12** Scatterplots for Various Values of *r*

**TABLE 14.8**  Magnitude of Pearson's *r*

| *r* Magnitude | Interpretation |
|---|---|
| > 0.70 | very accurate |
| 0.40 – 0.69 | reasonable |
| 0.30 – 0.39 | moderate |
| 0.20 – 0.29 | weak |
| < 0.19 | poor |

Using these suggested interpretive descriptions, the standard format for precision is:

The regression line describing the connection between *X* and *Y* [independent and dependent variables] provides a(n) [appropriate descriptor] description of the actual connection between these variables.

Example ($b = -0.88$): The regression line describing the connection between *Education* and *Income* provides a very accurate description of the actual connection between these variables.

# Interpreting the Strength of Connections between Quantitative Variables

When two variables are presented in tables, the interpretation of their relationship makes statements about both form and strength. Scatterplots display variables at higher levels of measurement. These variables contain more information and, therefore, more sophisticated interpretive criteria (form, extent, precision). Still, for these variables, it is often useful to speak to the strength of the connection between the variables.

In scatterplots, strength is a measure that *combines* the properties of extent and precision.[9] Measures of strength are interpreted in terms of prediction capacity. Stronger relationships between two variables result in improved ability to predict dependent variable values. Measures in the proportional reduction in error (PRE) family are statistics that measure strength. The coefficient of determination ($r^2$) is a PRE statistic and, therefore, uses the following convention PRE interpretation:

You proportionately reduce your errors by *P* [multiply the $r^2$ value by 100] per cent when you know the respondent's *X* [independent variable] as opposed to not knowing *X*, when predicting their *Y* [dependent variable].

Example ($r^2 = 0.73$): You proportionately reduce your errors by 73 per cent when you know a respondent's level of education, as opposed to not knowing their level of education, when predicting their income.

---

[9]Regression lines that have steeper slopes (greater extent) and are better-fit (more precision) indicate relationships between variables with greater strength.

# Chapter Summary

This chapter introduced the statistical tools for understanding relationships between quantitative variables displayed as scatterplots. These tools refined the basic ideas discussed in Chapter 11 by showing how to precisely calculate, apply, and interpret the features of a scatterplot. Having completed this chapter you should appreciate that:

- Regression lines impose straight lines on scatterplots.
- If relationships are curvilinear, using linear techniques leads to misunderstanding.
- The best-fit regression line is determined by minimizing the squared differences from the line.
- All straight lines are defined by the equation $Y = a + bX$.
- If the slope and intercept for a data set are known, the regression line can be described.
- Slopes and intercepts are calculated using the ordered-pairs evidence from a data set.
- The slope identifies the amount of change in the dependent variable for a fixed change in the independent variable.
- The intercept identifies the point where the regression line crosses the dependent variable axis.
- The completed linear regression formula can be used to draw the line on a scatterplot.
- Form information is obtained from the sign of the regression line slope.
- Extent information is obtained from the magnitude of the slope.
- Precision information is obtained by calculating Pearson's $r$.
- The strength of a scatterplot relationship combines information about extent and precision.
- Strength is interpreted in proportional reduction in error terms from the coefficient of determination statistic.

The chapter completes the discussion of bivariate analysis tools. You now have the statistics to calculate and interpret bivariate relationships for two variables at any level of measurement. Part IV introduces various tools for examining connections that involve more than two variables.

# PART IV
## Multivariate Analysis

# 15

# Taking Additional Variables into Account

## Overview

The last two chapters covered a lot of ground in the statistical maze and added several tools to your collection. Taking account of the statistical tools you picked up earlier, your toolkit is getting rather full. To review, in Part II (Chapters 4 through 8) you obtained the tools for conducting univariate (single-variable) statistical descriptions. In Part III, Chapters 9 through 14, you added statistical concepts, tools, and techniques conducting bivariate descriptions for variables at various levels of measurement. All the tools at your command in terms of the statistical selection checklist now include:

## Statistical Selection Checklist

1. Does your problem centre on a descriptive or an inferential issue?

   ☑ Descriptive

   ❏ Inferential

2. How many variables are being analyzed simultaneously?

   ☑ One

   ☑ Two

   ❏ Three or more

3. What is the level of measurement of each variable?

   ☑ Nominal

   ☑ Ordinal

   ☑ Interval

   ☑ Ratio

No wonder your toolbag is getting heavy! You have all the tools for conducting descriptive, univariate analysis for variables at every level of measurement, and tools for conducting descriptive, bivariate analysis for every kind of variable. Stated differently, you can conduct descriptive data analysis for every situation but one—the situation in which three or more variables are being analyzed simultaneously. And the path we go on next takes you into that region of the statistical maze.

As we have done previously, before going on to a new region it is worth pausing to get your bearings, which is what this chapter does. This chapter is fairly short and introduces you to multivariate analysis by:

- emphasizing the importance of context to understanding bivariate relationships;
- explaining the four most common types of contextual effects; and
- connecting multivariate analysis considerations to causal analysis.

Since this chapter discusses concepts and principles rather than specific statistical tools, it does not follow the five-step model. Instead, it focuses on the ideas required to understand the tools discussed in upcoming chapters. The story begins with appreciating the importance of context.

# Key Ideas in Multivariate Analysis

## The Importance of Context

Your statistical toolkit contains several tools for examining bivariate relationships. Chapters 9 through 11 taught you the fundamental concepts and tools for describing bivariate connections, while Chapters 12 through 14 showed you how to report these relationships in proportional reduction in error (PRE) terms. As useful as these bivariate analysis tools are, they neglect an important consideration—that of context.

Relationships between people rarely exist in isolation, and neither do relationships between variables. Relationships, either human or scientific, usually exist in some context. And context matters. Imagine two university students meeting for the first time in a bar on Saturday night. At closing time, one declares their undying love for the other. A naive recipient of this declaration takes it at face value. A prudent recipient understands the context of the declaration and imagines, "But what will they say when they are sober?" So it is with all relationships. The connection between the independent variable *Family income* and the dependent variable *Educational attainment* differs depending on whether the connection is measured in countries that have tuition systems or those that have "free" higher education. The connection between *Sex* and *Happiness* differs by age; the relationship between *Ethnic diversity* and *Social integration*, by national social policy.

## The Goal of Science

Science is one way of knowing about yourself and the environments embedding your experience. It is not the only way. Long before the introduction of science, people used other methods of knowing, including authority, tradition, meditation, revelation, and the like. These non-scientific methods, of course, continue to be in use today.

The proclaimed superiority of science as a way of knowing is rooted in this method's ability to describe and explain the way things really are. The prestige of science rests on distinguishing true claims from false ones. Science is imperfect in this regard, but the rise in its status in recent centuries is a testament to its superiority over other methods.

Since science's prestige and credibility rest on distinguishing true from false claims, it is a very conservative method. Science is particularly guarded about making one type of error; it does all it can to avoid declaring something is true when it is actually false. If science claims a particular drug cures multiple sclerosis when in fact it does not, the credibility of science suffers. If science claims that increasing the consumption of leafy green vegetables reduces inflammation when it doesn't, the value of the stock of science declines.

Given science's acute interest in avoiding false claims, the method has developed statistical tools for taking context into account. Its techniques aim to develop authentic statements. In this case *authentic* means that appearance matches reality. When something is authentic, "what you see is what you get." People trust authentic things because there is little risk of being fooled by them. When appearances match reality, we can plan our responses accordingly. When reality differs from appearance, we are at risk of acting foolishly. Our anticipated responses are less effective because we have misread our surroundings.

Contextual considerations shape appearances. Since the credibility of science as a method of knowing rests on making accurate statements about reality, the methods of science must take context into account.

## Statistical Accounting of Context

Your statistical toolkit already contains instruments for describing the character of the relationship between two variables. This is what bivariate analysis does. When variables are assessed at lower levels of measurement, bivariate relationships appear in the form of tables. You know how to determine and describe the form and the strength of the relationship expressed in tables—using percentage analysis or PRE measures of association like lambda and gamma. Scatterplots display the relationship between variables at higher levels of measurement, and you have the tools for describing these connections in terms of form, extent, precision, and strength.

In short, bivariate statistics assess the relationship between two variables *in isolation*. In other words, bivariate tools treat the relationship as if it occurred in a bubble or vacuum, as illustrated in Figure 15.1.

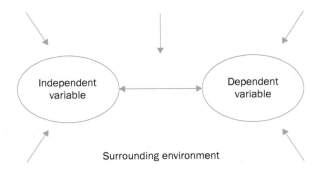

**FIGURE 15.1**   Bivariate *Ceteris Paribus*

Science uses the following fancy term to describe this situation—**ceteris paribus**—which means "all other things being equal." This statement is really a disclaimer. Like all disclaimers, this one tries to limit risk by stating that the bivariate description of the relationship between the two variables is accurate, *given the assumption that everything else stays the same.*

This assumption may be reasonable under the strictest of controlled laboratory conditions, but it is doubtful in other settings. It is rarely (if ever) the case in real life that relationships between people or variables occur in a bubble, uncontaminated by other influences. Consider the case of a couple who, during a marriage ceremony, recite traditional marriage vows that declare their undying commitment to one another "for better or for worse, for richer, for poorer, in sickness and in health." About 40 per cent of these couples in Canada will end their relationship in divorce. These divorces will not occur because the couple lied to each other on their wedding day but, rather, because other things (contextual considerations) did not stay the same. When times get much worse, when people get substantially richer or poorer, when health declines, relationships are affected.

In short, science's concern for making accurate statements about reality requires that it develop tools that take into account other factors that might affect bivariate relationships. In terms of statistical analysis, this requires expanding from bivariate analysis of relationships to trivariate or multivariate analysis. Trivariate analysis (obviously) takes three variables into account, which includes the original independent and dependent variables that constitute the bivariate relationship plus the effects of one additional (third) variable. Multivariate analysis expands the consideration of additional variables from one (third) variable to others (fourth, fifth, etc.). Trivariate analysis, then, is the simplest form of multivariate analysis. Figure 15.2 diagrams the consideration of how third, fourth, and additional variables may affect an original bivariate relationship.

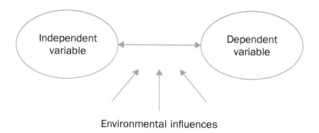

**FIGURE 15.2** Multivariate Model

## The Logic of Trivariate Analysis

Testing whether a third (or fourth or fifth) variable is operating to affect the relationship between an independent and a dependent variable involves comparison. Specifically, the logic involves examining and comparing the original relationship between the independent and dependent variable relationship under two conditions: (1) when the third variable is *allowed to change*, and (2) when the third variable is *held constant*.

Remember that variables only produce their effects when they are operating, that is, when they are changing. So the comparison between the two conditions boils down to comparing the independent–dependent variable relationship when the third variable is operating (condition 1) versus when it is not (condition 2).

These two conditions are diagrammed in Figure 15.3. Note that, in condition 1, the third variable is present and exerting its influence on the independent–dependent variable connection. In condition 2, the third variable is absent. This absence occurs because the variable is not operating (i.e. is held constant).

**FIGURE 15.3**  Trivariate Analysis Comparisons

The next question is, How does comparing these conditions inform us whether or not the third variable (the context) is affecting the original independent–dependent variable relationship? The answer is as follows. Alternative 1: If the nature of the connection between the independent and dependent variables *stays the same under the two conditions*, then the third variable has no effect. Alternative 2: If the nature of the independent–dependent variable connection *is different between the two conditions*, the third variable is having an effect.

You can appreciate the logic of these comparisons through the following scenario. Imagine you are in a romantic relationship; you being person X, your partner being person Y. You care very much for your partner but are suspicious that some other person (Z) might be affecting your relationship. It has crossed your mind that your partner doesn't really love you, as they claim. Instead, you imagine that your partner's declaration of undying love is really just the advice of person Z, who is suggesting such declarations as a way of keeping you in the dark about the true Y–Z romance. One way to resolve this concern is to remove person Z from the picture.[1] If your partner continues to declare their undying love for you even after Z has left the scene, then your suspicions about Z's influence were wrong. Alternatively, if Y's behaviour toward you changes after Z's removal, something was up.

## Effects on the Original Relationship

The comparison of the original independent–dependent variable relationship under the two conditions—one where a third variable is allowed to vary versus one where it is not—lets you determine *if* the third variable is having an effect. This still leaves open the question of *what kind of effect* the third variable may be having. Return to the romantic scenario above. Learning that person Z is having an effect on the X–Y connection does not tell you what kind of influence they have. Their presence could have detracted from your partnership, but it could also have enhanced it, or influenced it under some conditions and not others.

In theory, it is possible to specify a very large number of possible effects a third variable could have on an original independent–dependent variable relationship. In practice, however, researchers typically look for a limited number of possibilities. These are as follows.

---

[1]Let your imagination decide what "remove person Z from the picture" might mean. Perhaps they could be sent to a remote island with no access to communication, or worse. . . .

## Authentic Relationships

Figure 15.4 diagrams the connections between the variables when the original relationship is authentic.

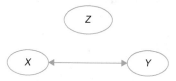

**FIGURE 15.4**  Authentic Pattern

The links in this diagram explain what an authentic connection means. The first thing to observe is that there is a connection (<—>) between the independent and dependent variables, $X$ and $Y$. Also notice that there is no connection between the third variable ($Z$) and the independent and dependent variables. This lack of connection between the third variable and the others indicates that it has no influence. Without a path to affect the $X$–$Y$ connection, $Z$ cannot make any difference to it.

Another way of stating the influence of $Z$ on $X$–$Y$ is as follows. The observed connection between $X$ and $Y$ remains the same, whether $Z$ is allowed to vary or whether it is held constant. In the real world, this is the test for authenticity; what you see (condition 1) is what you get (confirmed in condition 2). When connections are **authentic**, the way they appear is consistent. Context does not matter.

In everyday life some relationships are authentic; some friends and lovers do remain loyal no matter what the context. But when relationships are tested, authenticity is not always confirmed. There are other possibilities.

## Spurious Relationships

Figure 15.5 diagrams the connections between the variables when the original relationship is spurious.

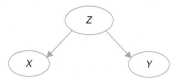

**FIGURE 15.5**  Spurious Pattern

*Spurious* is the opposite of *authentic*. Spurious relationships are illusory. In an authentic relationship, what appears to be a genuine connection between the independent and dependent variables holds up (is replicated) under different conditions. In a spurious relationship what appears to be an independent–dependent variable connection disappears when tested.

Look at the diagram of the spurious pattern in Figure 15.5. Notice that there is actually no connection between the independent and dependent variables. There is, however, a connection between the third variable ($Z$) and both of these variables. Keep in mind that variables can only have an effect when (1) they are allowed to vary (change) and (2) when there is a path of connection. You can now discern the pattern of spurious relationships under two conditions.

In condition 1, the third variable is allowed to vary. The arrow on the left in Figure 15.5 indicates that when $Z$ is allowed to vary it has a systematic effect on $X$; that is, $Z$ changing causes a systematic change in $X$. Likewise, the $Z$–$Y$ arrow indicates that $Z$ affects $Y$. So, when you look at the connection between $X$ and $Y$, when $Z$ is varying (condition 1), what is observed? The answer is that $X$ and $Y$ vary systematically; when one changes, the other changes in a predictable way. In other words, under condition 1 it *appears* as if $X$ and $Y$ are connected. What unfolds under condition 2? In this condition $Z$ is held constant, which means that it is not able to have any effects. You can see the results on the $X$–$Y$ connection under this condition if you cover up $Z$ and the arrows following from it in Figure 15.5. Here you see that there is no connection between $X$ and $Y$.

In short, a spurious relationship displays the following pattern. An $X$–$Y$ relationship is apparent when $Z$ is operating (condition 1) but this apparent $X$–$Y$ connection disappears when $Z$ is held constant (condition 2). In short, the test of authenticity failed; the apparent $X$–$Y$ relationship is illusory; it appears to be one-way when, in fact, it is not.

### Intervening Relationships

In a spurious relationship, the contextual variable ($Z$) produced the changes in $X$ and $Y$ that made them appear related. This, however, is not the only way in which appearances can be deceiving. Figure 15.6 diagrams another pattern, when the original connection between the variables is intervening.

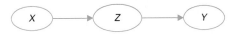

**FIGURE 15.6**  Intervening Pattern

Following the arrows lets you see the intervening pattern. In the first condition, $Z$ is varying so that $X$ is affecting $Z$ which, in turn, is affecting $Y$. So, when you only pay attention to the connection of $X$ and $Y$ while $Z$ is allowed to operate (condition 1), what you see is a systematic connection—$X$ and $Y$ are changing together in a reliable way. If you remove the effects of $Z$ (literally, cover up $Z$ and its arrows in Figure 15.6), there is no existing connection between $X$ and $Y$. In short, the pattern exhibited by an **intervening** relationship is the apparent relationship that exists when $Z$ is operating (condition 1) but disappears when $Z$ and its effects are controlled (condition 2).

In social experience, this pattern occurs when a translator intervenes to facilitate communication. Imagine two people, one of whom speaks only English; the other, only Swahili. Relying only on verbal communication, these parties cannot have a relationship. The English speaker can't understand a word of what the Swahili speaker is saying, and vice versa. Now imagine an English–Swahili translator intervenes. The translator's conduct (variation) allows a relationship to exist. The English speaker would share her thoughts to the translator, who would pass them along in Swahili. The process would then be reversed. In this example, the operation of the translator makes the relationship between the original parties possible. And if the translator stops operating, the relationship (communication) disappears.

Notice that the empirical pattern of the intervening relationship (condition 1, some relationship; condition 2, no relationship) is the same pattern displayed in spurious relationships. Clearly, however, these patterns are not the same. Look at the diagrams in Figures 15.5 and 15.6. How do they differ? Well, in both figures the $Z$–$Y$ connection is the same; when $Z$ operates it affects $Y$. The difference between the spurious and intervening models resides in the $X$–$Z$ connection. In the spurious case, $Z$ affects $X$; in the intervening case, $X$ affects $Z$. This is a difference in the sequencing or time-ordering of the $X$ and $Z$ variables. In short, although the empirical patterns of spurious and intervening relationships are similar, their temporal patterns differ. This matter will be elaborated later. For now, it is sufficient that you appreciate that distinguishing spurious from intervening relationships requires additional evidence that clarifies the sequencing of the $X$–$Z$ connection.

### Interactive Relationships

Figure 15.7 diagrams the model for interactive relationships.

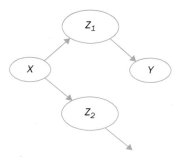

**FIGURE 15.7**  Interactive Relationships

Understanding this figure requires paying particular attention to the $Z$ variable. In Figures 15.4, 15.5, and 15.6 there is a variable labelled $Z$. In Figure 15.7 the $Z$ variable has a subscript 1 or 2, as in $Z_1$ and $Z_2$. These subscripts designate that the contextual variable ($Z$) does not have uniform effects. In other words, the effect of $Z$ on the $X$–$Y$ relationship depends on the particular value of the $Z$ variable, which is the sign of **interaction**. In the diagram, if $Z$ has a value of 1, then one kind of connection between $X$ and $Y$ appears. Alternatively, if $Z$ has a value of 2, then a different $X$–$Y$ connection emerges.

This understanding is clarified by the arrows in Figure 15.7. When $X$ is combined (interacts with) the $Z_1$ value, changing $X$ leads to a systematic connection to $Y$. On the other hand, if $X$ interacts with the $Z_2$ value, then $X$ has no effect on $Y$.

Think about the connection between attending a party (the independent variable) and having fun (the dependent variable). Your experience probably demonstrates that this connection is not a uniform one. Sometimes going to a party leads to lots of fun; on other occasions, going to a party yields no fun. This puzzle is clarified by appreciating that the $X$–$Y$ connection in this case is conditioned by context. One contextual ($Z$) variable is who you are with. When your date is someone you like ($Z_1$), then the $X$–$Y$ connection is probably strong. On the other hand if your date is someone you dislike ($Z_2$), then attending a party probably does not produce any fun.[2]

## Models as Ideal Types

The famous sociologist Max Weber taught us about the importance of ideal types. **Ideal types** are abstract models (ideals) that help us see reality more clearly. An ideal type, by definition, is pure and clear. In contrast, the real world is mostly messy and confusing.

---

[2]You may have experienced how other contextual variables affect the *Party attendance–Fun* relationship. Think about how this $X$–$Y$ connection changes as the amount of alcohol consumed ($Z$) changes.

Comparing an imaginary ideal type with real-world cases helps us see reality more sharply. The ideal type emphasizes salient, specific considerations and disregards others. In doing so, it helps to see what we are looking for in real-world cases. The features of an ideal type of blue jay are different than the ideal-type features of a cardinal. Although both are beautiful birds, knowing the central features of each lets us distinguish them at backyard feeders with a glance. Notice that such rapid identification occurs even though the actual blue jays and actual cardinals are both different from the ideal type and different in detail from other members of their group.

The authentic, spurious, intervening, and interaction models are ideal types. In reality hardly any relationships are unaffected by other considerations (i.e. are perfectly authentic). The same holds true of the other patterns. But these models are crucial as ideal types for helping us sort out the messiness of real-world relationships. The research task is to compare the actual case (evidence) to each of the ideal types and decide *which comes closest* to describing the actual pattern.[3] For example, we tell others whom we care for that we love them, even though our experience of love is less than perfect (ideal). What matters is that the pattern of our interaction is better described as loving than it is by other descriptions (types) of relationships.

## Models and Causal Criteria

Earlier in this chapter we emphasized that the credibility of science, as a method of knowing, rests on being accurate in its statements of what is real. Early chapters in this book discussed techniques for describing the realistic features of single variables. Then we introduced tools for characterizing the relationships between two variables (bivariate understanding). This chapter began the journey for elaborating bivariate relationships to take additional (contextual) variables into account.

In its quest for description, science gives special attention to identifying causal connections between variables. At this point it is worth reviewing what scientists mean when they report that one variable causes another, and connecting the four multivariate models to this discussion.

In order to determine that an independent variable ($X$) is the cause of a dependent variable ($Y$), science requires that the relationship must pass at least three tests. These tests include sequence, association, and non-spuriousness.

The **sequence** criterion is a test of time-ordering. Specifically, for sequencing to be evident, science must demonstrate that changes in the independent variable occurred *prior to* changes in the dependent variable. If the dependent variable ($Y$) changes before the independent variable ($X$) changes, then clearly $X$ cannot be the cause of changes in $Y$. The information for the sequence test cannot be determined through data analysis. Information about time-ordering must be gained from some other source. Sometimes that source is the methodological design of the investigation. For example, classical experiments are controlled so that the introduction of the independent variable must occur before detecting changes in the dependent variable. Sometimes sequences can be inferred through logic. Clearly the independent variable *Childhood nurturing* occurs prior to observed differences in a dependent variable such as *Academic*

---

[3]Authentic, spurious, intervening, and interactive relationships do not exhaust the list of ideal types, but they do cover the most common ones. On occasion it is possible to detect real-world examples of suppressor and distorter models. In suppressor relationships, the third variable's operation is keeping the $X$–$Y$ relationship from detection. In distorter relationships, the operation of the third variable reverses the form of the $X$–$Y$ connection.

*achievement*.[4] For present purposes, the point is that passing the sequence test for causal imputation must rely on information outside of that obtained through data analysis.

The second causal criterion is association. **Association** means that the independent and dependent variables change systematically. In other words, when one changes there is a reliable, predictable change in the other. Unlike sequence, the association is established through statistical analysis. In fact, this is precisely what bivariate statistical analysis tools do; they establish whether or not an association exists between the variables under consideration.

**Non-spuriousness** is the third causal criterion. This test is conducted through the multivariate analysis techniques discussed in this chapter and the following ones. Taken literally, *non-spurious* means "not phony or inauthentic," which is an awkward, double-negative way of saying that non-spurious relationships are authentic ones. In short, passing this third test for causation relies on using the statistical tests discussed in this section and confirming that the apparent association (confirmed by the second causal criterion) is a genuine one.

In summary, since multivariate analysis begins with an original bivariate relationship (condition 1) and proceeds to explore the effects of context on that original relationship (comparing conditions 1 and 2), these techniques are central to establishing causal connections between variables. Outside of the sequence criterion, which needs to be established by other means, multivariate analysis conducts the tests necessary for determining whether a connection between two variables is a causal one.

# Two Closing Considerations

This chapter introduced the main ideas behind multivariate analysis. The following two chapters will show you the specific techniques for conducting the elaborations on an original bivariate relationship and determining which model most closely approximates the evidence. Before turning to these techniques, two additional conceptual considerations are worth noting. The first concerns the language used in multivariate analysis; the second relates to the selection of contextual variables.

## Relationship Order

Multivariate techniques are employed to elaborate on original bivariate relationships to determine the effects of contextual variables. Multivariate analyses often refer to the order of the relationships under consideration. For example, researchers often talk of first-order or second-order relationships. To follow the discussion, it is important to understand what researchers mean when they describe relationship order.

In multivariate analysis, the order of a relationship is determined by the number of *control (contextual) variables* that are included in the analysis. Remember that multivariate analysis begins with an original relationship, which is the bivariate relationship between the independent and dependent variables. This original, bivariate relationship is unelaborated; it takes no contextual variables into account. For this reason, the original bivariate relationship that

---

[4]Sometimes, however, sequencing is not so obvious and the appropriate test of sequence is not so clear. Think of the connection between the variables *Dropping out of high school* and *Drug abuse*. There is a systematic relationship between these variables, but which came prior to the other?

begins a multivariate analysis is called a **zero-order relationship**. Its order is zero because that is the number of control (contextual) variables that are included at that point in the analysis.

From an original, bivariate relationship, multivariate analysis proceeds to take contextual variables into account to determine their influence on the original relationship. In the illustrations provided earlier, the control variable is identified as *Z*. When an *X–Y* relationship controls for the effects of *Z*, this is called a **first-order relationship**, because it takes one control (contextual) variable into account. For example, the relationship between *Studying* (independent variable) and *Grades* (dependent variable) *among only women* is a first-order relationship because the connection between the independent and dependent variables controls for *Sex*. Likewise, the *Studying–Grades* relationship *among only men* is a first-order relationship.

Multivariate techniques are not restricted, however, to taking only one contextual variable into account at a time. For example, it is possible to examine the *Studying–Grades* connection among only *Women* ( first control variable) who are *Single* (second control variable). This relationship constitutes a **second-order relationship**, because two control variables (*Sex* and *Relationship status*) are being accounted for.[5] As the influence of more control variables are taken into account simultaneously, the description of the order of the relationships increases. Notice that order only increases as the number of control variables considered *simultaneously* increases. If the *Studying–Grades* connection is examined first among women, and then among single persons, the order remains as first order. Sequential consideration of multiple control variables keeps the order as first order. Simultaneous consideration of multiple control variables elevates the order to high levels.

## Selecting Contextual Variables

Multivariate analysis techniques focus on determining the effects of contextual variables on an original bivariate relationship. The analysis involves examining the bivariate relationship under conditions where the contextual variable is allowed to operate and when it is held constant (controlled). For this reason, contextual variables in multivariate analysis are identified as *control* variables.

To use multivariate analysis, at least one contextual (control) variable needs to be employed. As just noted, the simultaneous inclusion of more contextual variables adds to the complexity of the analysis by increasing the order of the relationships. The practical question, however, is deciding which contextual variables to include in the analysis.

This is a practical question because, in the everyday world, everything is a potential control variable. This is the case because relationships are situated in some real-world context and, in the real world, everything is connected. Clearly, a multivariate analysis cannot include all other variables in the world, which leads to the selection issue.

In practice, a multivariate analysis should take into account all *theoretically relevant* contextual variables. This provision sharply narrows the field of consideration. Notice that *theoretically relevant* is different than *interesting*. If a multivariate analysis were obliged to take all interesting considerations into account, it would be an endless task. After a researcher

---

[5]Note how quickly the addition of contextual variables affects the number of tables. Assuming that Sex has only two categories (*male, female*) and *Relationship status* has only two categories (*single, attached*), then four second-order relationships appear between the independent and dependent variables. These are single men, single women, attached men, and attached women.

completed an analysis, someone could always say, "I am interested" in how that relationship is affected by astrological sign, or coffee consumption, or whatever variable came to mind.

Restricting the multivariate analysis to theoretically relevant contextual variables makes producers and consumers of scientific analysis do serious intellectual work. They have to seriously think about the relationships under examination and make a theoretically informed argument for how they would expect contextual variables to influence the evidence. When this argument is made, researchers are under an obligation to treat these contextual considerations seriously, because there is now good reason to expect they are having an influence.

# Chapter Summary

This chapter introduced you to multivariate analysis. It did so by focusing on ideas rather than technical tools. These ideas, however, are central to understanding the use of the statistical techniques presented in the following two chapters. Having completed this chapter you should appreciate that:

- Both human and statistical relationships occur in a context and context matters.
- Science is particularly interested in clarifying the effects of context.
- In statistical analysis, context refers to the effects of variables other than the original independent and dependent variables.
- The simplest version of multivariate analysis is the trivariate case, which includes consideration of a single contextual variable.
- The logic of multivariate analysis involves examining and comparing the original bivariate relationship under two conditions: (1) when the third variable is allowed to operate (vary) and (2) when the third variable's effects are taken out or held constant (controlled).
- The identification of a third variable's effects is aided through the use of ideal-type models.
- An authentic pattern exists when the bivariate relationship is replicated under the two conditions.
- A spurious pattern is potentially present when the original relationship disappears under the controlled condition.
- An intervening pattern is potentially present when the original relationship disappears under the controlled condition.
- The distinction between spurious and intervening patterns requires information about the time-ordering (sequence) of the variables.
- An interaction pattern is present when different values of the control variable produce differing effects.
- The number of control variables simultaneously included in the analysis determines the order of the relationship.
- The selection criterion for including contextual variables in a multivariate analysis is theoretical relevance.

With these ideas in place, we can now turn to their application in the form of data analysis tools. Chapter 16 introduces the multivariate statistical techniques employed when the variables are at lower levels of measurement. Chapter 17 discusses multivariate techniques for interval and ratio variables.

# 16

# The Elaboration Model

## Overview

You are now ready to journey into the multivariate analysis region of the statistical maze. The last chapter emphasized that a full understanding of any bivariate relationship requires a consideration of the context of the relationship. When a bivariate relationship is considered in isolation of its context, appearances may not match reality, and the researcher risks being fooled. In a simple bivariate relationship, additional variables are operating. When these additional variables operate (vary) they can affect the character of the original bivariate relationship. Examining the potential effects of contextual variables requires systematic examination.

The systematic examination of the potential effects of a third variable on an original bivariate relationship involves examining the independent–dependent variable connection under two conditions. In condition 1, the bivariate relationship is described when a third (contextual) variable is operating. In condition 2, the independent–dependent variable connection is described with the third variable held constant. In effect, these two conditions amount to comparing the independent–dependent variable connection when the third variable is operating (condition 1) and when it is not operating (condition 2). As described in Chapter 15 the results of the comparison of these two conditions gives clues into whether the original bivariate relationship is authentic, spurious, intervening, or interacting. This chapter explores how this detection system operates on actual evidence.

This chapter examines multivariate techniques for variables at lower levels of measurement, which means the evidence is presented in tables. In terms of the statistical selection checklist, this chapter covers:

## Statistical Selection Checklist

1. Does your problem centre on a descriptive or an inferential issue?

    ☑ Descriptive

    ☐ Inferential

2. How many variables are being analyzed simultaneously?

    ☐ One

    ☐ Two

    ☑ Three or more

3. What is the level of measurement of each variable?

☑ Nominal

☑ Ordinal

☐ Interval

☐ Ratio

In this chapter you will learn:

* how original and partial tables are constructed;
* the steps in the logic of the elaboration model; and
* how to employ the elaboration model to identify contextual effects.

The journey toward these goals begins with a consideration of tabular evidence.

## STEP 1 Understanding the Tools

# Kinds of Tables

## Table Analysis Review

Multivariate analysis of nominal and ordinal variables begins with a table that contains the original bivariate relationship between the independent and dependent variables. It then proceeds to analyze this original relationship by taking into account the effects of at least one contextual variable. Since this form of multivariate analysis involves so much table analysis, it is worth reviewing how tables are analyzed. This review will use Table 16.1.

This table expresses the bivariate connection between an independent and a dependent variable. By convention the independent variable (*Sex*) is listed across the top of the table, and the dependent variable (*Voting pattern*) is found on the left side. Given this configuration, the attributes of the independent variable (*male, female*) are the columns of the table, while the attributes of the dependent variable (*conservative, liberal*) are the table's rows.

The margins of the table include numbers that make up the frequency distributions of the independent and dependent variables. The Total column is a frequency distribution of the dependent variable, stating that the sample of 510 respondents included 230 who voted for a conservative party and 280 who voted for a liberal one. The Total row is a frequency distribution of the independent variable, stating that the sample included 200 males and 310 females.

**TABLE 16.1** Original Bivariate Table

| Voting Pattern | Sex | | Total |
| --- | --- | --- | --- |
| | Male | Female | |
| Conservative | 140 | 90 | 230 |
| Liberal | 60 | 220 | 280 |
| Total | 200 | 310 | 510 |

The numbers inside the table are the joint frequencies which display the connection between the independent and dependent variables. For example, the 220 frequency represents the joining of a particular attribute of *Sex* (*female*) with a specific voting pattern (*liberal*). An analysis of the joint frequencies in a table reveals what (if any) connection (relationship) exists between the independent and dependent variables.

For data displayed in tables, the analysis of the evidence needs to address two issues: the form and the strength of the relationship. Form describes the character of the connection between the variables. In doing so, form reports what changes in the dependent variable are expected to occur through changes in the independent variable. Strength addresses a different issue: it characterizes how much difference changing the independent variable makes to the dependent variable.

Chapter 10 presented various tools for determining the form and strength of a relationship expressed in a table. One technique is to use percentage analysis techniques. Table 16.2 shows the results of applying percentage analysis to Table 16.1.

**TABLE 16.2** Percentage Analysis of Original Table

| | Sex | | |
| Voting Pattern | Male | Female | Total |
| --- | --- | --- | --- |
| Conservative | 140 (70%) | 90 (29%) | 230 |
| Liberal | 60 (30%) | 220 (71%) | 280 |
| Total | 200 | 310 | 510 |

Percentage difference = 41%

As gender moves from male (B) to female (C), voting patterns shift (move down) from more conservative (A) to more liberal (C)

Notice that the columns (not the rows) have been percentaged. Using the form techniques from Chapter 10, it is clear that the connection is as follows: Women have more liberal voting patterns than males. The strength of this relationship is 41 per cent, which means that *Sex* makes a moderate difference to *Voting pattern*.

Alternatively, using the techniques in Chapter 13, the relationship between *Sex* and *Voting pattern* could be assessed using a proportional reduction in error (PRE) statistic. For this table, lambda would be an appropriate measure. The lambda calculation is 0.35, which means that you proportionally reduce the errors in predicting the voting pattern by 35 per cent when a respondent's sex is known, as compared to not having such knowledge.

In summary, you have the tools for describing and interpreting the relationship between independent and dependent variables when this connection is expressed in a table. As the previous chapter cautioned, the apparent relationship in a bivariate table may not be authentic. Sometimes appearances do not match reality. The following sections show you how to test whether or not the bivariate relationship expressed in a table is affected by contextual variables and, if so, how.

## Original and Partial Tables

The end of the last chapter noted that the type of connection between an independent and a dependent variable can be expressed as the order of the table. A zero-order table is one in which the independent–dependent variable connection is expressed with no consideration of how contextual variables are operating. Zero-order tables express simple bivariate connections and are also known as original tables. Table 16.1, for example, is an original (zero-order) table expressing the connection between *Sex* and *Voting pattern*.

In an original table, contextual variables are operating (varying). When these variables are operating they may be affecting the original independent–dependent variable connection. Look, for example, at the 140 in the upper-left cell of Table 16.1. This joint frequency claims that the data set contains 140 males who voted conservative. Imagine that these 140 males were actually standing in the hallway. If we examined them, would they all be the same height? Probably not. Have the same intelligence? Unlikely. Share the same religion? Improbable. In short, within these 140 respondents, there is extensive variation in height, intelligence, religion, and a multitude of other variables. The same would hold true if you carefully examined the 90 female conservatives, or the 60 male liberals, or even the 220 females who voted for a liberal party. In short, within Table 16.1, height, intelligence, religion, and numerous other attributes are varying. And because they are varying, these variables may be affecting the original *Sex–Voting pattern* relationship that we observed.

You will recall that exploring the possible effects of contextual variables involves examining the independent–dependent variable connection under two conditions. The first condition occurs when the contextual variable under examination is allowed to vary (operate). As we have just seen, this condition is met in an original (zero-order) bivariate table. The second testing condition for assessing the influence of contextual variables on the original bivariate connection occurs when the third variable is controlled (held constant). This condition is met in first-order tables.

First-order tables are partial tables. Partial tables get their name for the following reason. We have demonstrated that the *Sex–Voting pattern* connection reported in Table 16.1 includes the operation of many other variables, such as height, intelligence, and religion. *Union membership* is another contextual variable in operation. Among the male conservatives in Table 16.1 some will be union members, others will not be. The same is true of the female conservatives, as well as the male and female liberals. If we suspect that *Union membership* may be affecting the *Sex–Voting pattern* connection, we need to examine the original bivariate connection under conditions where *Union membership* is held constant (condition 2). Table 16.3 provides an example.

**TABLE 16.3** Original and Partial Tables for Sex, Voting Pattern, and Union Membership

**TABLE 16.3a** Original (Zero-Order) Table

| Voting Pattern | Sex | | Total |
| --- | --- | --- | --- |
| | Male | Female | |
| Conservative | 140 | 90 | 230 |
| Liberal | 60 | 220 | 280 |
| Total | 200 | 310 | 510 |

The partial tables 16.3b and 16.3c are named because they are parts of the original table. If you placed the Union Members partial table on top of the Non-Unionized partial table they would add up to the original table connecting *Sex* and *Voting pattern*. For example, the values of the upper-left cells of the partial tables (54 and 86) add up to the value of the upper-left cell in the original table (140). The same is true for all other joint frequencies and marginal distributions.

**TABLE 16.3b**  Partial (First-Order) Table: Union Members

| | Sex | | |
|---|---|---|---|
| **Voting Pattern** | **Male** | **Female** | **Total** |
| Conservative | 54 | 41 | 95 |
| Liberal | 21 | 89 | 110 |
| Total | 75 | 130 | 205 |

**TABLE 16.3c**  Partial (First-Order) Table: Non-Unionized

| | Sex | | |
|---|---|---|---|
| **Voting Pattern** | **Male** | **Female** | **Total** |
| Conservative | 86 | 49 | 135 |
| Liberal | 39 | 131 | 170 |
| Total | 125 | 180 | 305 |

Partial tables are first order, which means that the effects of one control (contextual) variable are removed from the original bivariate relationship. This means that, in the partial tables 16.3b and 16.3c, *Union membership* is having no effect on the connection between *Sex* and *Voting pattern*. This is clear for the following reason. Look at the 54 male conservatives in the upper-left cell of Table 16.3b. While they may differ in height, intelligence, and religion, they are *all the same* with respect to *Union membership*; all 54 are union members. The same is true for the 21 male liberals, the 41 female conservatives, and the 89 female liberals. In short, every respondent in Table 16.3b is a union member. So, when we examine the relationship between *Sex* and *Voting pattern* in this table, *Union membership* has no effect because it is held constant. The same situation exists in Table 16.3c, except that *Union membership* is held constant by making sure the table contains all non-union members.

In summary, an original table reports on the bivariate connection between the independent and dependent variables when a third (contextual) variable is operating. The partial tables report on the bivariate relationship where the effects of the third variable are controlled. When these contextual effects are controlled their influence is removed. In terms of the conditions discussed in the last chapter, the original table presents condition 1 (contextual variable operating), while the partial tables present condition 2 (contextual effect removed).

Before discussing how an analysis of original and partial tables proceeds to determine whether the original relationship is authentic, spurious, intervening, or interactive, two additional points about partial tables are worth noting. First, the process of decomposing an original, zero-order table into partial (first-order) tables is called **elaboration**. The act of elaboration involves exploring a subject more carefully, which is just what partial tables do with respect to an original table. They explore the original bivariate connection more carefully and completely to determine what, if any, contextual effects are operating. Second, since partial tables control (hold constant) the effects of a third variable, elaboration is always going to result in as many partial tables as there are attributes of the third (control) variable. In the earlier example, two partial tables were created because *Union membership* is a variable that has two attributes (*union member* and *non-unionized*). If the *Sex–Voting pattern* connection was elaborated for the effects of religion, the result may be several partial tables (e.g. *Christian, Jewish, Muslim, Buddhist, Hindu*).

Now that you understand how original and partial tables are produced, you can learn how they are analyzed in order to determine which of the ideal type patterns (authentic, spurious, intervening, interactive) best describes the effects of contextual variables on the original relationship.

# The Logic of Elaboration

Elaboration analysis begins with two sets of tables: an original table that contains the connection between the independent and dependent variables when other (contextual) variables are operating, and partial tables that examine the independent–dependent variable connection when a contextual variable's effects are removed. These tables are the evidence for the two conditions that must be compared to determine the effects of a third variable on an original bivariate relationship.

Analyzing the evidence in the original and partial tables to determine the effects of a contextual variable involves the following four steps:

1.   Analyze the original and partial tables separately.
2.   Compare the partial tables to each other.
3.   Compare the partial tables to the original table.
4.   Draw a conclusion about the pattern in the relationships.

Understanding the meaning and application of each of these steps is best understood through application to the following example.

## 1. Analyze the Original and Partial Tables Separately

Table 16.3 contains the original and partial tables for the relationship between *Sex* and *Voting pattern*, controlling for the third variable *Union membership*. Notice that both the original and partial tables display the same independent–dependent variable relationship: they both display the connection between *Sex* and *Voting pattern*. The only difference between the original and partial tables is that *Union membership* varies in the original table and is controlled (held constant) in the partial tables. This first step in the elaboration model is to analyze these tables separately. Since both the original and partial tables are looking at the relationship between the same independent and dependent variables, this step translates into conducting a series of *bivariate* analyses. To do this you employ the bivariate tools in your statistical toolkit. For example, you could use percentage analysis techniques to determine the form and strength of the various connections, or you could use PRE measures of association to do the same thing.

Applying percentage analysis to the original and partial tables in Table 16.3 yields the results shown in Table 16.4.

## 2. Compare the Partial Tables to Each Other

This step focuses on only the partial tables. Remember that there will be as many partial tables as there are attributes of the third (control) variable. Table 16.4 contains two partial tables because the control variable, *Union membership*, has two attributes: a respondent is either a member of a union or not.

**TABLE 16.4** Percentage Analysis of Table 16.3

**TABLE 16.4a**  Original (Zero-Order) Table

| | Sex | | |
|---|---|---|---|
| **Voting Pattern** | **Male** | **Female** | **Total** |
| Conservative | 70% | 29% | 220 |
| Liberal | 30% | 71% | 280 |
| Total (N) | 200 | 310 | 510 |

Percentage difference = 41%

**TABLE 16.4b**  Partial (First-Order) Table: Union Members

| | Sex | | |
|---|---|---|---|
| **Voting Pattern** | **Male** | **Female** | **Total** |
| Conservative | 72% | 32% | 95 |
| Liberal | 28% | 68% | 110 |
| Total (N) | 75 | 130 | 205 |

Percentage difference = 40%

**TABLE 16.4c**  Partial (First-Order) Table: Non-Unionized

| | Sex | | |
|---|---|---|---|
| **Voting Pattern** | **Male** | **Female** | **Total** |
| Conservative | 69% | 27% | 135 |
| Liberal | 31% | 73% | 170 |
| Total | 125 | 180 | 305 |

Percentage difference = 42%

Comparing the partial tables to each other means comparing the results of the *analyzed* partial tables to each other. In other words, you examine the results from the first step and compare these conclusions. Our example contains two partial tables. Therefore, applying this second step means comparing the form and strength of the union members partial table to the form and strength of the non-unionized partial table. Look at the results of the analysis of these two partial tables (from step 1). Here you see that, for union members, females displayed more liberal voting patterns than males (form) and that the connection between *Sex* and *Voting pattern* in this partial table is moderate (strength). For the non-unionized partial table, the form is that females display more liberal voting patterns than males and the strength (42 per cent) is moderate.

When the partial tables in Table 16.4 are compared to one another they show *essentially the same pattern*. In other words, the form and strength of the connection between *Sex* and *Voting pattern* is similar in both partial tables.

## 3. Compare the Partial Tables to the Original Table

Like step 2, this step involves comparing the results from the *analyzed* tables to each other. In other words, compare the interpretation (from step 1) of the original table to the interpretation of the first partial table, then compare the original to the second partial table. Continue this comparison process until the original table results have been compared to each of the partial tables.

In our example, the original table findings are compared, first, to the Union Members partial table and then, second, to the Non-Unionized partial table. Performing this comparison yields the following results. In the original table the form indicates that females display more liberal voting patterns than males. The strength of this connection is moderate (41 per cent difference). In the Union Members partial table, the form and strength are similar to that of the original. The same is true when comparing the original table to the Non-Unionized partial table; they display similar form and strength patterns. In summary, in Table 16.4, both partial tables display the same connection between *Sex* and *Voting pattern* as found in the original table.

## 4. Draw a Conclusion about the Pattern in the Relationships

The second and third steps compare the results produced in the first step. The first step involved a series of bivariate analyses. The goal of elaboration, however, is to determine what's occurring among three variables (trivariate analysis). This final step attains that goal by examining the results of steps two and three. It does so by looking for patterns.

Chapter 15 described four models characterizing what effect, if any, a third variable might have on an original bivariate relationship. These included the authentic, spurious, intervening, and interaction patterns. Each of these patterns was described as an ideal type. In practical application, the task is to determine which model is most closely approximated by the evidence. The previous three steps provide the evidence; this final step shows how to interpret this evidence and draw a conclusion about which model is operating. Each of the four models shows a specific evidence pattern. Your task is to choose the one that best fits the actual evidence for the variables involved in the analysis. Here are the patterns.

### Evidence of an Authentic Relationship

An authentic relationship between an independent and a dependent variable is one that is unaffected by the third (contextual) variable. In other words, in an authentic relationship it does not matter whether the third variable is allowed to operate (vary) or whether it is controlled (held constant). In an authentic relationship, the pattern detected in the original relationship is *replicated* in the partial tables. Replication occurs in the evidence as follows:

- From step 2, the partial tables are the *same* (or similar) to each other.
- From step 3, the partial tables are the *same* (or similar) to the original table.

### Evidence of a Spurious Relationship

A spurious original relationship is a phony one; what appears to be a relationship is actually non-existent. In a spurious relationship, the operation of the third (contextual) variable makes all the difference to whether there appears to be a relationship between the independent and dependent variables. In this sense, the third variable *explains* the appearance of the original connection. Spurious relationships exist in the following pattern of evidence:

- From step 2, the partial tables are the *same* (or similar) to each other, and show no (or little) relationship.
- From step 3, the partial tables are *different* from the original, with the partials displaying no relationship and the original showing some relationship.

### Evidence of an Intervening Relationship

In an intervening relationship, the third variable comes between the operation of the independent and dependent variables. In doing so the third variable translates or *interprets* the effects of the independent variable on the dependent variable. Like all translators, in the intervening model the third variable makes all the difference to whether the independent and dependent variables can affect one another. Interpretation is evident in the following pattern:

- From step 2, the partial tables are the *same* (or similar) to each other and show no (or little) relationship.
- From step 3, the partial tables are *different* from the original, with the partials displaying no relationship and the original showing some relationship.

Notice that the evidence pattern of the intervening relationship is the same as that in a spurious relationship. When the evidence shows this pattern, a researcher can determine that the relationship is either spurious or intervening (and not one of the other types) but cannot determine which it is. In other words, spurious and intervening patterns cannot be distinguished on the basis of the empirical data alone. As noted in Chapter 15, making this determination requires additional information about the sequencing (time-ordering) of the variables.

### Evidence of an Interaction Relationship

In an interaction relationship the effects of the third (contextual) variable are not straightforward. In the interaction case, it is not that the third variable has an effect (e.g. spurious or intervening) or has no effect (e.g. authentic). In interaction the third variable has different effects on the original independent–dependent relationship depending on the amount (value) of it that is present. In this way interaction relationships *specify* how particular amounts of the contextual variable impact the original bivariate relationship. Specification is evidenced by the following data pattern:

- From step 2, the partial tables are *different* from each other.
- From step 3, the partial tables are *different* from the original.

Two points about the interaction pattern are worth highlighting. First, if you look at the authentic, spurious, and intervening pattern evidence from step 2, you will notice that in these three cases the partial tables are always the same (or similar). This is not so for interaction, where the partial table results are dissimilar. Second, when the partial tables are compared to the original in an interaction pattern, the results are also dissimilar. This partial–original table comparison can result in many kinds of dissimilarity. Often one partial table will display a stronger relationship than the original; and another partial, a weaker relationship than the original. Sometimes the partial tables have different forms from one another. The possibilities are wide-ranging, since there are numerous ways in which a third variable can affect the original relationship.

Table 16.5 summarizes the possible patterns. Notice that some of the descriptive labels have synonyms that are used in research reports. For example, authentic relationships are sometimes identified as replicated ones. The control variable possibilities include antecedent and intervening. *Antecedent* means the control variable changes before the independent or dependent variables change. *Intervening* means that the control variable changes after the independent variable but before the dependent variable changes.

**TABLE 16.5** Relationship Patterns

| Partials Compared with Original | Control Variable | |
|---|---|---|
| | **Antecedent** | **Intervening** |
| Same | authentic/replication | authentic/replication |
| Less or none | spurious/explanation | intervening/interpretation |
| Split* | interaction/specification | interaction/specification |

*one partial is the same or greater, and the other is less or none

Now that you are familiar with the patterns that define the various kinds of elaboration models, let's apply them to the evidence in Table 16.4. Here is what we learned from steps 2 and 3: The partial tables are similar to one another (step 2) and the partial tables are similar to the original (step 3). Now the question is, Which of the four ideal-type models do these patterns most closely approximate? The answer is the authentic model. In other words, the original table findings are replicated in the partial tables. In the original table, which contains both union members and non-unionized respondents, females display more liberal voting patterns than males. When the effects of union membership/non-membership are controlled, as they are in the partial tables, females continue to show more liberal voting patterns than males. From this evidence we can conclude that *Union membership* does not make any difference to the connection between *Sex* and *Voting pattern*.

## STEP 2 Learning the Calculations

# Applying Elaboration Models

Let's use the example in Table 16.4 to understand elaboration a bit better. We'll work with two examples. The first example is a spurious relationship and the second example will outline an interaction relationship. Remember that there are four steps to applying the elaboration model:

1. Analyze the original and partial tables separately (recall that you can use percentage differences or PRE measures for nominal and ordinal level variables. We will use percentage differences for these examples).
2. Compare the partial tables to each other.

3. Compare the partial tables to the original.
4. Draw a conclusion about the pattern in the relationships.

1. The original data from Table 16.4 is replicated in Table 16.6a. We will change the variables and the numbers in the partial tables so that we can provide appropriate examples of the spurious and interaction relationships, but we won't change the values in the original (zero-order) table.

**TABLE 16.6**  Original Table Percentage Analysis

**TABLE 16.6a**  Original (Zero-Order) Table

| Voting Pattern | Sex | | Total |
| --- | --- | --- | --- |
| | Male | Female | |
| Conservative | 70% | 27% | 220 |
| Liberal | 30% | 73% | 280 |
| Total (N) | 200 | 300 | 500 |

Percentage difference = 43%

Instead of looking at the influence of *Union membership* (which does not have an effect on the relationship between *Sex* and *Voting pattern*), let's use another control variable to examine the spurious relationship. Let's look at *Religious affiliation* as measured in two categories: *religious* and *not religious*, in Table 16.6b.

**TABLE 16.6b**  Partial (First-Order) Table: Religious Canadians

| Voting Pattern | Sex | | Total |
| --- | --- | --- | --- |
| | Male | Female | |
| Conservative | 65% | 63% | 92 |
| Liberal | 35% | 37% | 108 |
| Total (N) | 75 | 125 | 200 |

Percentage difference = 2%

**TABLE 16.6c**  Partial (First-Order) Table: Non-religious Canadians

| Voting Pattern | Sex | | Total |
| --- | --- | --- | --- |
| | Male | Female | |
| Conservative | 59% | 64% | 128 |
| Liberal | 41% | 36% | 172 |
| Total (N) | 125 | 175 | 300 |

Percentage difference = 5%

2. Now that we have calculated the percentage differences for each table (see Chapter 10 if you need a review of this procedure), we can now compare the results of the partial tables. Table 16.6b (religious Canadians) yielded a result of 2 per cent while Table 16.6c (non-religious Canadians) yielded a result of 5 per cent. That's not much of a difference; the effect of *Religious affiliation* is small. In sum, *Religious affiliation* has no influence on the relationship between *Sex* and *Voting pattern*.

3. We now need to compare the results of the analysis of the partials with the results we observed in the zero-order table. In the zero-order table, the percentage analysis revealed that there is a 43 per cent difference in the way the sexes vote. Yet when we controlled for *Religious affiliation*, we found that it only had a 2 per cent or 5 per cent influence. There is a big difference between what we observed in the zero-order and partial tables.

4. Now we need to identify the relationship. Remember that we have four choices: authentic, spurious, intervening, or interaction. Given that the analysis of the partial tables is the same (one partial has a 2 per cent effect and the other has a 5 per cent effect), and that the results of the partial tables are significantly different than those from the zero-order table, which indicate a moderate relationship at 43 per cent, we can identify the relationship as intervening.[1]

Let's try another example, this time identifying an interaction relationship. We'll use the same example of voting behaviour and sex—but we will introduce a different control variable. Here's the data from Table 16.6 once again.

**TABLE 16.7a** Original (Zero-Order) Table Percentage Analysis

| Voting Pattern | Sex | | Total |
| --- | --- | --- | --- |
| | Male | Female | |
| Conservative | 70% | 27% | 220 |
| Liberal | 30% | 73% | 280 |
| Total (N) | 200 | 300 | 500 |

Percentage difference = 43%

For this example, let's look at the influence of *Race*. In this example, we will measure *Race* as *white* or *visible minority* (i.e. not white).

Recall that in an interaction relationship, two conditions must exist. First, the results of the percentage analysis comparison between the partial (first-order) tables are different from one another, and, second, when they are compared to the results of the zero-order table, those relationships are different.

So, you know the drill, let's calculate the percentage differences for the partials.

[1] The pattern of evidence indicates the relationship could be either spurious or intervening. As Table 16.6 reminds us, the difference between these alternatives centres on the sequence of changes. Which of the following narratives sounds more plausible? The intervening alternative suggests that there are Sex differences in *Religious affiliation* and that *Religious affiliation* affects *Voting patterns*. The spurious alternative suggests that *Religious affiliation* affects *Sex* and *Voting patterns*. This is implausible, since it indicates that a change in religious commitment produces a change in sex!

**TABLE 16.7b** Partial (First-Order) Table: White Canadians

| | Sex | | |
|---|---|---|---|
| Voting Pattern | Male | Female | Total |
| Conservative | **75%** | **73%** | 92 |
| Liberal | **25%** | **27%** | 108 |
| Total (N) | 75 | 125 | 200 |

Percentage difference = 2%

**TABLE 16.7c** Partial (First-Order) Table: Visible Minority Canadians

| | Sex | | |
|---|---|---|---|
| Voting Pattern | Male | Female | Total |
| Conservative | **65%** | **8%** | 128 |
| Liberal | **35%** | **92%** | 172 |
| Total (N) | 125 | 175 | 300 |

Percentage difference = 57%

So for step 2, we must compare the results of the two partial (first-order) tables. In Table 16.7b, the influence of sex on voting patterns for white Canadians is 2 per cent, yet for visible minority Canadians in Table 16.7c, the influence of sex is 57 per cent, a big difference. It means that visible minority women are significantly more likely to vote for a liberal party than both visible minority men and white female and males. In step 3, we need to compare the partial tables with the results of the zero-order table. Recall that the difference between men and women in the zero-order table is 43 per cent. Given that the difference by sex for white voters is small (2 per cent) but the sex difference for visible minority Canadians is large, we can say that there is an effect of race on the relationship between sex and voting patterns.

But what is the nature of this relationship? In step 4, we identify the nature of this relationship. Recall that when the partial tables are different (in this case, a 2 per cent difference for white voters versus a 57 per cent difference for visible minority voters) and when these relationships are different from the zero-order table (43 per cent), then we can identify this relationship as an interaction.

That wasn't so hard, was it?

## STEP 3 Using Computer Software

In previous chapters, we covered IBM SPSS statistics software ("SPSS") procedures for generating tables that displayed the relationship between two variables, the independent and dependent variables. However, in order to determine whether or not the (zero-order) relationship is being affected by other variables, we need to expand our analysis to a trivariate or multivariate analysis. One such method is the elaboration model.

Not surprisingly, generating trivariate tables is similar to their bivariate counterparts, with a couple of additional steps.

# Generating a Bivariate Table

## Procedure

Similar to other statistical procedures, before beginning, run frequency distributions in order to determine if any values in your independent, dependent, or control variables need to be set to *missing*. See Chapter 3 for instructions on how to set desired values to *missing*. To generate an original, bivariate table, proceed as follows:

1. Under Analysis in the SPSS Menu, click Descriptive Statistics and then Crosstabs.
2. Move your independent variable from the left side to the right Column(s) box. In the example illustrated below, the independent variable is "Q13D. In the past 6 months: Excessive partying?"
3. Select your dependent variable and move it to the Row(s) box on the right-hand side (e.g. "Q13B. In the past 6 months: Problems with school?").
4. Next, select the Cells box (upper-right corner).
   a. In the Percentages box, check Column. By selecting the column per cent, you will be able to analyze the output by following the rule "percentage down and compare across."
   b. Click on the Continue box.
5. Select Continue and OK to execute the Crosstab command.

From the output shown in Figure 16.1, you can see that students who engaged in excessive partying were more likely to indicate that they were having problems in school (51.8 per cent compared to 44 per cent) which, according to the per cent analysis table (Table 10.8), is a small relationship.

Q13B Q13B. In the past 6 months have you: Problems with school * Q13D Q13D. In the past 6 months have you: Excessively been partying Crosstabulation

|  |  |  | Q13D Q13D. In the past 6 months have you: Excessively been partying | | Total |
|---|---|---|---|---|---|
|  |  |  | 1 Yes | 2 No |  |
| Q13B Q13B. In the past 6 months have you: Problems with school | 1 Yes | Count | 177 | 396 | 573 |
|  |  | % within Q13D Q13D. In the past 6 months have you: Excessively been partying | 51.8% | 44.0% | 46.1% |
|  | 2 No | Count | 165 | 504 | 669 |
|  |  | % within Q13D Q13D. In the past 6 months have you: Excessively been partying | 48.2% | 56.0% | 53.9% |
| Total |  | Count | 342 | 900 | 1242 |
|  |  | % within Q13D Q13D. In the past 6 months have you: Excessively been partying | 100.0% | 100.0% | 100.0% |

per cent analysis
7.8%
small relationship

**FIGURE 16.1** Original Table

# Generating Trivariate Partial Tables

Generating trivariate crosstabs (i.e. partial tables) requires one extra step beyond those used to generate a bivariate table.

## Procedure

1. Select Analyze from the menu bar.
2. Select Descriptive Statistics → Crosstabs.
   a. Move the independent variable, Q13D, to the Column box.
   b. Move the dependent variable, Q13B, to the Row box.
   c. Move the control variable, Q25. *Respondent sex*, to the Layer 1 of 1 box.
   d. Select Cells → check Column Percentages.
3. Select Continue → OK.

As shown in Figure 16.2, the strength of the relationship between Q13D (*Excessive partying*) and Q13B (*Problems with school*) for *females* is 11.7 per cent, which is still a small association. However, the strength of the relationship for *males* is 1.4 per cent, which indicates that there is no association between excessive partying and problems with school.

**Q13B Q13B. In the past 6 months have you: Problems with school * Q13D Q13D. In the past 6 months have you: Excessively been partying * Q25 Q25. What is your sex Crosstabulation**

| Q25 Q25. What is your sex | | | | Q13D Q13D. In the past 6 months have you: Excessively been partying | | Total |
| | | | | 1 Yes | 2 No | |
| 1 Female | Q13B Q13B. In the past 6 months have you: Problems with school | 1 Yes | Count | 115 | 249 | 364 |
| | | | % within Q13D Q13D. In the past 6 months have you: Excessively been partying | 55.8% | 44.1% | 47.3% |
| | | 2 No | Count | 91 | 315 | 406 |
| | | | % within Q13D Q13D. In the past 6 months have you: Excessively been partying | 44.2% | 55.9% | 52.7% |
| | Total | | Count | 206 | 564 | 770 |
| | | | % within Q13D Q13D. In the past 6 months have you: Excessively been partying | 100.0% | 100.0% | 100.0% |
| 2 Male | Q13B Q13B. In the past 6 months have you: Problems with school | 1 Yes | Count | 62 | 146 | 208 |
| | | | % within Q13D Q13D. In the past 6 months have you: Excessively been partying | 45.9% | 44.5% | 44.9% |
| | | 2 No | Count | 73 | 182 | 255 |
| | | | % within Q13D Q13D. In the past 6 months have you: Excessively been partying | 54.1% | 55.5% | 55.1% |
| | Total | | Count | 135 | 328 | 463 |
| | | | % within Q13D Q13D. In the past 6 months have you: Excessively been partying | 100.0% | 100.0% | 100.0% |

11.7%
small relationship

1.4%
no relationship

**FIGURE 16.2** Partial (Trivariate) Tables

The elaboration model for our trivariate analysis is an interaction relationship, meaning that the initial (zero-order) bivariate association between excessive partying and problems with school affects female students but not male students, since there is no relationship for males who party in excessive ways and have problems with school.

## STEP 4 Practice

You now have the tools for using the elaboration model to determine the effects of contextual variables on an original bivariate relationship. You can conduct this analysis either by hand or using computing software. To solidify your understanding you need to practise applying the statistical and software procedures.

This section provides you opportunities to practise what you have learned about the elaboration model. The first set of questions uses hand calculations. The second set uses the SPSS procedures. For each set of questions:

1. Follow the procedural steps and complete the appropriate calculation (Set 1) or software application (Set 2).
2. Check your answers, using the Answer Key in the back of the text.
3. If your answer is incorrect, consult the Solutions section on the book's website. The Solutions provide a complete step-by-step analysis of how the answers are derived.

After you have completed the next section (Interpreting the Results), return to your calculations or output and *provide complete, written interpretations of each of the statistics you have generated.*

# Set 1: Hand-Calculation Practice Questions

1. Use the data in Table 16.8 to answer the following questions.

**TABLE 16.8** Sex, Gun Registry Support, and Residence Data

| Case Number | Sex | Support Gun Registry? | Rural or Urban Resident |
|---|---|---|---|
| 1 | M | Y | R |
| 2 | F | Y | U |
| 3 | F | Y | U |
| 4 | M | N | R |
| 5 | M | N | U |
| 6 | F | Y | R |
| 7 | F | Y | R |
| 8 | M | Y | U |
| 9 | M | N | U |

*(continued)*

**TABLE 16.8** continued

| Case Number | Sex | Support Gun Registry? | Rural or Urban Resident |
|---|---|---|---|
| 10 | M | N | U |
| 11 | F | N | R |
| 12 | F | N | R |
| 13 | M | Y | U |
| 14 | F | Y | U |
| 15 | F | Y | U |
| 16 | F | N | R |
| 17 | F | N | R |
| 18 | F | N | R |
| 19 | F | N | R |
| 20 | F | Y | U |
| 21 | M | N | R |
| 22 | M | N | R |
| 23 | F | Y | U |
| 24 | F | Y | U |
| 25 | F | Y | U |
| 26 | F | N | U |
| 27 | F | N | U |
| 28 | M | N | R |
| 29 | M | Y | U |
| 30 | M | Y | U |

a. Use data from the Table 16.8 to create a zero-order bivariate table for the relationship between *Sex* (IV) and *Support for the national gun registry* (DV).

b. Use the same data set to create partial tables with *Rural/Urban residence* as the control variable.

c. Use the table percentaging technique to determine whether there is a zero-order relationship between *Sex* and *Support for the gun registry*. If there is, how do we interpret it? (**Hint**: see Table 10.8.)

d. Do the same for each of the partial tables.

e. Does the control variable influence the original relationship?

f. Is the control variable antecedent (affects *X*) or intervening (affects *Y*)?

g. According to the logic of elaboration, which pattern of relationship best describes this data?

2. Substantial evidence exists to support the claim that married people tend to have better mental health than single people. Table 16.9 contains data from a fictional survey where 30 individuals provided information about their marital status (married or single), self-rated mental well-being (good or poor), and their sex. In the following exercises you will use the elaboration model from Chapter 16 to investigate the relationship.

**TABLE 16.9** Marital Status, Well-Being, and Sex

| Case Number | Marital Status (married/single) | Mental Well-being (good/poor) | Sex (male/female) |
|---|---|---|---|
| 1 | M | G | M |
| 2 | M | P | F |
| 3 | M | G | F |
| 4 | S | G | F |
| 5 | S | P | M |
| 6 | M | G | M |
| 7 | S | P | M |
| 8 | M | G | F |
| 9 | M | P | M |
| 10 | S | G | F |
| 11 | S | G | F |
| 12 | S | P | M |
| 13 | S | P | F |
| 14 | M | G | M |
| 15 | M | G | M |
| 16 | S | G | M |
| 17 | M | G | F |
| 18 | S | G | M |
| 19 | S | P | M |
| 20 | M | P | F |
| 21 | M | G | F |
| 22 | M | G | M |
| 23 | S | P | M |
| 24 | S | P | F |
| 25 | M | G | F |
| 26 | S | G | F |
| 27 | M | P | M |
| 28 | M | G | M |
| 29 | S | G | M |
| 30 | S | G | F |

a. Use data from Table 16.9 to create a zero-order bivariate table for the relationship between *Marital status* (IV) and *Mental well-being* (DV).
b. Use the same data set to create partial tables with *Sex* as the control variable.
c. Use the table percentaging technique to determine whether there is a zero-order relationship between *Marital status* and *Mental well-being*. If there is, how do we interpret it? (**Hint**: see Table 10.8.)
d. Do the same for each of the partial tables.
e. Does the control variable influence the original relationship?

f.   According to the logic of elaboration, which pattern of relationship best describes this data?

g.   What do you conclude about the original bivariate relationship from the results of the elaboration model? Is the original relationship authentic?

# Set 2: SPSS Practice Questions

1.   a.   Use the Crosstabs function to create a bivariate table of the relationship between NQ35 *Regular religious service attendance* (IV) and Q9 *Consumption of drugs or not* (DV).

b.   Use the Crosstabs function to create partial tables controlling for whether the respondent is satisfied with their mental health (NQ1D).

c.   Use the table percentaging technique to determine whether there is a zero-order relationship between religious service attendance and drug use. If there is, how do we interpret it?

d.   Do the same for each of the partial tables.

e.   Does level of satisfaction with mental health influence the original relationship?

f.   Is the control variable antecedent (affects $X$) or intervening (affects $Y$)?

g.   According to the logic of elaboration which pattern of relationship best describes this data?

2.   a.   Use the crosstabs function to create a bivariate table of the relationship between nQ1B *Satisfaction with level of exercise* (IV) and nQ1A *Satisfaction with physical health* (DV).

b.   Use the Crosstabs function to create partial tables controlling for *Sex of respondent* (Q25).

c.   Use the table percentaging technique to determine whether there is a zero-order relationship between *Satisfaction with level of exercise* and *Satisfaction with physical health*. If there is, how do we interpret it?

d.   Do the same for each of the partial tables.

e.   Does the control variable influence the original relationship?

f.   Is the control variable antecedent (affects $X$) or intervening (affects $Y$)?

g.   According to the logic of elaboration which pattern of relationship best describes this data?

3.   a.   Create a bivariate table of the relationship between *Problems with school in the last six months* (Q13B) as the dependent variable and *Alcohol consumed in the past year* (Q7) as the independent variable. How do we interpret the relationship between these variables?

b.   Create partial tables controlling for the variable *Moved to a new home/apartment/residence in the past six months* (Q13A). Does the control variable influence the original relationship? If yes, how so?

4.   a.   Create a bivariate table of the relationship between *Volunteering* (Q19) as the dependent variable and *Born in Canada* (Q27) as the independent variable. How do we interpret the relationship between these variables?

b.   Create partial tables controlling for *Financial setback* (Q13E). Does the control variable influence the original relationship? If yes, how so?

## STEP 5 Interpreting the Results

The techniques in this chapter have focused on how to *identify* which ideal-type pattern (authentic, spurious, intervening, interactive) best approximates the empirical evidence. Once the best pattern is identified, it still must be *interpreted*.

Interpretation centres on stating in simple and direct language what the identified relationship means. For the ordinary person, stating that a relationship is spurious is not very helpful, since they don't know what that means. The same is true of the other patterns. The interpretive task is to provide a clear understanding of what the multivariate analysis means.

The interpretive task begins with recalling the purpose of multivariate analysis. Multivariate analysis is intended to identify what effect, if any, a contextual variable has on the bivariate relationship between an independent and a dependent variable. The interpretation must reflect this purpose. Interpretation occurs after the model that best approximates the evidence has been selected (i.e. authentic, spurious, intervening, interactive). Here are generic interpretations and examples of each model:

Authentic: This analysis shows that the original relationship between $X$ [independent variable] and $Y$ [dependent variable] is unaffected by the third variable $Z$ [control variable].

Example: The analysis shows that the original relationship between *Height* and *Happiness* is unaffected by *Peanut butter consumption*.

Spurious: This analysis shows that the original apparent relationship between $X$ [independent variable] and $Y$ [dependent variable] is phony. In other words, there is no actual connection between $X$ [independent variable] and $Y$ [dependent variable]. The apparent relationship between these variables is actually caused by the shared influence of a third variable $Z$ [control variable].

Example: The analysis shows that the original apparent relationship between *Height* and *Happiness* is phony, since both variables are affected by the variable *Sex*.

Intervening: This analysis shows that there is no direct connection between $X$ [independent variable] and $Y$ [dependent variable]. Although these variables appear to be directly related, in fact their connection is actually mediated by the intervention of a third variable $Z$ [control variable].

Example: The analysis shows no direct connection between *Height* and *Happiness*. Although these variables appear directly related, their connection is mediated by the intervention of *Income*.

Interactive: This analysis shows that the connection between $X$ [independent variable] and $Y$ [dependent variable] is complicated by the effects of $Z$ [control variable]. Under condition A [the appropriate attribute of the control variable] the connection between the independent and dependent variables is B [describe the connection]. By contrast, under condition C [the appropriate attribute of the control variable] the connection between the independent and dependent variables is D [describe the connection].

Example: The analysis shows that the connection between *Height* and *Happiness* is complicated by the effects of nutrition. When nutrition is good, there is a strong, direct connection between *Height* and *Happiness*. When nutrition is poor, there is a weak, direct connection between *Height* and *Happiness*.

# Chapter Summary

This chapter applied the ideas of the previous chapter to the situation where the variables are at lower (nominal and ordinal) levels of measurement. The relationships between variables at lower levels of measurement are displayed in the form of tables. Consequently, the multivariate analysis procedures covered in this chapter centred on reading and interpreting tabular patterns. In this chapter you learned that:

- Summarizing tables requires determining the form and strength of the relationship between the independent and dependent variables.
- Original tables describe the apparent connection between independent and dependent variables where the effects of additional (contextual) variables are operating.
- Partial tables describe an original relationship under conditions where one (or more) contextual variables is held constant (controlled).
- Partial tables are created by disaggregating an original table into component parts.
- Determining how a contextual variable is affecting an original bivariate relationship requires the use of the elaboration model.
- Elaborating on a relationship begins by summarizing the form and strength of the relationship in both the original and partial tables.
- After summarization, the elaboration model compares the partial table results to one another and the partial table results to the original.
- The pattern of the partial–partial and partial–original table comparisons reveals which multivariate model best describes how context is operating.
- In authentic relationships the partial tables replicate the original relationship.
- In spurious and intervening patterns, the partial tables reveal no relationship between the independent and dependent variables while the original relationship shows some connection.
- Spurious and intervening patterns must be distinguished by supplementary information about the time-ordering (sequence) of the variables.
- In interaction patterns, the effects of the contextual variable change depending on the particular amount (value) of the third variable.

This chapter's tools demonstrated how multivariate analysis is performed on tables which include variables at lower levels of measurement. The next chapter reveals how multivariate determinations are made with variables at the interval and ratio level.

# 17

# Multiple Regression

## Overview

Your journey into the region of the statistical maze dealing with multiple variables continues in this chapter. Chapter 15 demonstrated that context can have important effects on relationships. In research, third (control) variables measure context. Therefore, measuring the effects of context on an original bivariate relationship is a matter of determining what effects control variables have. Chapter 16 provided the tools for making such a determination when the variables were at lower (nominal and ordinal) levels of measurement. This chapter provides the tools for examining the effects of contextual variables when the variables are at higher (interval and ratio) levels of measurement.

In terms of the statistical selection checklist, this is where the contents of this chapter are situated:

## Statistical Selection Checklist

1.  Does your problem centre on a descriptive or an inferential issue?

    ☑ Descriptive

    ❑ Inferential

2.  How many variables are being analyzed simultaneously?

    ❑ One

    ❑ Two

    ☑ Three or more

3.  What is the level of measurement of each variable?

    ❑ Nominal

    ❑ Ordinal

    ☑ Interval

    ☑ Ratio

In this chapter you will learn about:

- the place and importance of partial correlation coefficients;
- the meaning and interpretation of partial regression coefficients;
- how multiple regression operates; and
- how to interpret multiple correlation coefficients.

Your understanding of these issues begins with reviewing some fundamentals.

## STEP 1 Understanding the Tools

# Key Considerations

## Connection to the Elaboration Model

The logic of multiple regression parallels the logic of elaboration used when the independent, dependent, and control variables are at lower levels of measurement. In both cases the original independent–dependent variable connection is examined, first, when contextual variables are operating and, second, when contextual effects are controlled. Multiple regression, however, has an important advantage over elaboration analysis using tables that is worth noting. Chapter 16 demonstrated how elaboration analysis begins with an original (bivariate) table and then proceeds to construct partial tables. The partial tables express the original bivariate connection under conditions where the third variable's influence is held constant. In partial tables, the context effects are literally held constant; when you examine a partial table, all cases have the same attribute of the control variable.

Literally seeing how third variable effects are controlled has the advantage of helping you understand what is occurring in elaboration analysis. Everything is transparent. The original table is decomposed into the partial tables and you can see, by adding the partial tables together, how you can recreate the original table. This technique carries an important disadvantage, however. When an original table is decomposed into partial tables, the parts contain fewer cases than the whole (original). This process strictly limits how many control variables can be considered simultaneously, because you quickly run out of cases. As the parts are made smaller and smaller, their reduced size rapidly renders them meaningless.

Using multiple regression to examine the contextual effects does not cause this problem, because the effects of contextual variables are statistically (rather than literally) controlled. As a result, multiple regression can easily determine the separate effects of multiple control variables. The result is that we can examine the original relationship between an independent and a dependent variable in more detail because we control the effects of several variables simultaneously.

## What Multiple Regression Can Do

Because it can consider several variables simultaneously, multiple regression analysis includes a powerful set of statistical tools. Before identifying these tools, it is worth clarifying how multiple regression extends the simple bivariate regression techniques you learned in Chapters 11 and 14. The difference is illustrated in Figure 17.1.

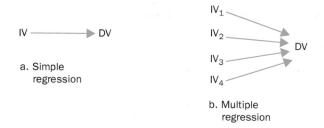

**FIGURE 17.1** Simple and Multiple Regression

Figure 17.1a illustrates the simple bivariate regression you are familiar with from previous chapters. In this analysis, the task is to describe how the independent variable is related to the dependent variable *under the condition where other variables are allowed to operate.* For example, in Figure 17.1a, the independent variable might be *Hours worked*; and the dependent variable, *Income*. A simple bivariate analysis would determine how working additional hours affects income.

Figure 17.1b illustrates that the situation may be somewhat more complicated than the Figure 17.1a diagram suggests. Figure 17.1b illustrates that the dependent variable might be affected by more than one independent variable and, moreover, that these different independent variables may affect each other. Return to the Figure 17.1a example, which examined how *Hours worked* affected the dependent variable *Income*. Undoubtedly, *Hours worked* is one determinant of *Income*; people who work more hours generally make more money. But is it the only determinant? Hardly. A person who works an additional hour at a fast-food outlet does make additional income, but it is not the same as a corporate lawyer working an additional hour. Occupation also affects income; persons in occupations with higher prestige generally make more income than those in less prestigious positions. A similar case can be made for the effects of education on income.

The simple bivariate model (Figure 17.1a) examines the *Hours worked–Income* connection without taking into account the operation of *Occupational prestige* and *Education*. The multivariate analysis (Figure 17.1b) takes the simultaneous influence of these various independent variables into account. It can determine, for example, how much effect *Hours worked* has on *Income*, separate from the effects of occupation and education. This is the kind of complexity that multivariate analysis can handle. It extends simple bivariate regression to take account of how several independent variables affect a dependent variable.

With this general appreciation in mind, we can specify the kinds of things that multiple regression can accomplish. These include identifying:

- which independent variables have the strongest relationship with the dependent variable;
- how much impact each independent variable has on the dependent variable;
- how the set of independent variables jointly affects the dependent variable; and
- how well the set of independent variables explains the dependent variable.

Remember that multivariate regression analysis is simply an extension of simple bivariate regression analysis. It is an analogue of how elaboration analysis using partial tables is an extension of bivariate table analysis. Since multivariate analysis builds upon simple bivariate regression, it is worth reviewing the key ideas of simple regression.

## Review of Simple Linear Regression and Correlation

Simple regression describes the relationship between two variables when both variables are measured at the interval or ratio level of measurement. In short, regression describes data presented in scatterplots. Linear regression and correlation, introduced in Chapters 11 and 14, centred on several key concepts and principles. Specific statistical tools measure the key concepts and principles. Here is a quick summary:

- The evidence on a scatterplot is the actual data collected by researchers. The task of simple regression is to determine the straight line that best approximates this evidence.
- The general form of a straight line is captured in the equation:

$$Y = a + b(X)$$

  where $X$ is the independent variable, $Y$ is the dependent variable, $a$ is the intercept, and $b$ the slope.

- Using the scatterplot evidence displaying the relationship between the specific independent ($X$) and dependent ($Y$) variables, formulas are solved to determine the values of the intercept's location ($a$) and the best-fit line's slope ($b$).
- The resulting specific regression line can be written as an equation:

$$income = 22,000 + 10,000(years\ schooling)$$

  It can then be drawn on the scatterplot.

- The regression line can predict dependent variable scores from specific independent variable scores:

  An individual with 15 years of schooling should expect an income of $172,000.

- In addition to prediction, regression lines contain two important pieces of information. First, regression lines tell you about the *form* (structure) of the relationship between the variables. If the slope of the regression equation has a positive (+) sign, this tells you that the variables change in the same direction; as one increases (or decreases), the other also increases (or decreases). If the slope of the regression equation has a negative (–) sign, this indicates that the variables change in opposite directions (as one increases the other decreases). Second, the regression equation provides information about the *impact* of the independent variable on the dependent variable. Impact is determined by looking at the size of the slope in the regression equation. This number tells us how many units of change in the dependent variable are associated with a one-unit change in the independent variable.
- Scatterplots present the actual data, while the regression line provides the best linear approximation of this evidence. Since interpretations of form and extent (impact) come from examining the best-fit regression line, another important piece of information concerns *precision*. Precision asks how well the best-fit regression line (and, by extension, the interpretation of form and extent) approximates the actual evidence. Precision is captured by the computation of the correlation coefficient ($r$).
- The concept of the *strength* of the connection between the independent and dependent variables combines information about impact and precision. Stronger connections

have greater impact and precision, since these ingredients allow us to see that the independent variable makes more difference to the dependent variable. Strength is measured by computing the coefficient of determination ($r^2$).

The computations and interpretations of simple regression and correlation analysis inform us about the relationship between an independent and a dependent variable under conditions where other contextual variables are operating. Multiple regression simply extends the basic correlation and regression ideas to remove the effects of whatever contextual variables are taken into account. In doing so, multiple regression accomplishes the goals listed earlier. We now turn to the statistical tools associated with accomplishing each of these objectives.

# Four Multiple Regression Tools

## Partial Correlation

Chapter 16 introduced the elaboration model, which is the procedure for determining the influence of a contextual variable on a simple bivariate relationship when the evidence is in the form of tables. Elaboration analysis compares two kinds of tables: the original table expressing the connection between the independent and dependent variables, and the partial tables which display this connection when a contextual variable's effects are controlled (partialled out). Comparing the original and partial tables provides an understanding of what effect the third variable has on the original relationship.

Multiple regression analysis includes a similar statistical outcome. In simple bivariate regression, the strength of the connection between the independent and dependent variables is measured through the coefficient of determination ($r^2$). This measure is based on the simple correlation coefficient ($r$). In other words, the correlation coefficient provides an indicator of the strength of the relationship between an independent and dependent variable, under the condition where other (contextual) variables are operating.

In contrast to simple bivariate regression, multiple regression analysis includes several independent variables. For each of these independent variables, a simple correlation coefficient ($r$) can be computed. This statistic is interpreted the same way in multiple regression as it is in simple regression. It indicates how strong/precise the connection is between each independent variable and the dependent variable, where contextual variables are allowed to operate.

Multiple regression analysis provides another correlation coefficient as well: this is a **partial correlation** coefficient. A partial correlation coefficient indicates the strength of the connection between an independent variable and the dependent variable *when the effects of all other independent variables in the model are removed (controlled).* The partial correlations for each independent variable in the model can be used to determine which independent variable has the strongest relationship with the dependent variable.

To illustrate how partial correlations provide this result, let's return to the example in Figure 17.1b, where the independent variables *Hours worked, Occupational prestige,* and *Years of education* are used to predict *Income*. Table 17.1 presents the partial correlations between each of these independent variables and the dependent variable, *Income*.

Remember that each of these partial correlations indicates the strength/precision of the connection between the independent variable and *Income* when the effects of the other two independent variables are taken out. In other words, the contaminating effects of context are

**TABLE 17.1**  Partial Correlations

| Independent Variables | Partial Correlation |
|---|---|
| Hours worked | 0.35 |
| Years of education | 0.19 |
| Occupational prestige | 0.17 |

removed. When we compare these three partial correlations, it is clear that *Hours worked* has the strongest isolated effect on *Income*, followed by *Years of education* and then *Occupational prestige*.

The computation of partial correlations is complicated, so we will not show you the formula for calculating this statistic. However, it is important for you to know how a conventional representation of a partial correlation is interpreted. Imagine we give the variables in our example the following numbers: 1 = *Hours worked*, 2 = *Years of education*, 3 = *Occupational prestige*, and 4 = *Income*. Using these numbers as abbreviations, here is how each of the partial correlations in Table 17.1 is formally reported:

$$r_{41.23} = 0.35$$

$$r_{42.13} = 0.19$$

$$r_{43.12} = 0.17$$

Notice that, for each partial correlation, the symbol $r$ is followed by a subscript of digits. The first digit in the subscript indicates the dependent variable; in these examples it is 4, which stands for *Income*. The second digit identifies the specific independent variable. Then there is a dot or period. The digits after the dot specify which independent variables are controlled in the partial correlation. So, for example, $r_{41.23}$ = 0.35 indicates the strength of the connection between *Income* (4) and *Hours worked* (1) controlling for the effects of *Years of education* (2) and *Occupational prestige* (3).

In summary, partial correlations are one important statistic that results from a multiple regression analysis. Partial correlation results are typically produced through computer analysis, so what you need to focus on is how to interpret the results. Partial correlations indicate the strength of the connection between an independent variable and the dependent variable, when the effects of the other independent variables are removed. Therefore, by comparing partial correlations you can determine which independent variables have the strongest relationship with the dependent variable.

## Partial Regression Coefficients

A second important result of multiple regression analysis determines how much impact each independent variable has on the dependent variable. In simple bivariate regression, the impact of a single independent variable on the dependent variable is determined by the value of the slope in the regression equation. Multivariate analysis builds upon this idea and allows the comparison of the slopes of the different independent variables. Understanding this idea requires extending the bivariate linear regression equation.

In simple regression, the equation for a best-fit regression line that approximates the actual data points is determined using a series of formulas. The general form of the equation is:

$$Y = a + b(X)$$

Multiple regression extends this equation by including several independent variables in the equation, and associated with each independent variable is a specific, calculated slope. Here is the general form of the multiple regression best-fit equation:

$$Y = a + b_1(X_1) + b_2(X_2) + b_3(X_3)$$

This equation includes three independent variables identified by subscripts ($X_1, X_2, X_3$), but the general equation can be expanded to include as many independent variables as the research question requires. Associated with each independent variable is a slope ($b$), with each one identified by a subscript. Notice that $Y$ still represents the dependent variable; and $a$, the intercept.

When a multiple regression equation is determined for a set of data, the slope of each independent variable provides valuable information. Table 17.2 provides an example.

**TABLE 17.2** Partial Regression Coefficients

| Independent Variables | Partial Correlation | Partial Regression Coefficients |
|---|---|---|
| Hours worked last month | 0.35 | 655 |
| Years of education | 0.19 | 4,000 |
| Occupational prestige | 0.17 | 906 |

The partial regression coefficients are the slopes associated with the various independent variables in the multiple regression equation ($b_1, b_2, b_3$). These slopes report the impact of each particular independent variable *when the effects of all the other independent variables are held constant.* For example, the $b_1$ value is 655, since *Hours worked* is the first independent variable. This partial slope tells us that each additional hour worked in the last month increased a respondent's annual income by $655 among persons with the same years of education and occupational prestige. Similarly the partial regression coefficient for *Years of education* ($b_2$) tells us that each additional year of education contributes an additional $4,000 to annual income among those with occupations of the same prestige who work the same number of hours.

In short, the interpretation of partial regression coefficients follows the same form as the interpretation of simple bivariate slopes, with the important addition that the effects of the other independent variables are held constant. Using these partial slopes you can describe the unique impact of each independent variable on the dependent variable. What you *cannot* do with partial regression coefficients (slopes) is compare them to determine which independent variable is having the largest impact on the dependent variable. Such comparisons are inappropriate because the different independent variables are measured in different units. A comparison of $b_1$ and $b_2$ in the previous example helps clarify this important point.

The $b_1$ partial slope tells you that the unique effect of each additional hour of work contributes $655 to annual income, while the $b_2$ partial slope identifies the unique impact of each year of schooling as $4,000. From these numbers you cannot conclude that the impact of schooling on income is greater than the impact of working. The $655 properly means "an additional $655 *per* hour *of work*" while the $4,000 means "an additional $4,000 *per* year *of education.*" *Hours worked* and *Years of education* are not measured in the same units, so they are not directly comparable, for the same reason you cannot directly compare apples and oranges.

Fortunately, however, there is a multiple regression statistic that allows comparison of the relative impact of each independent variable on the dependent variable. The technique uses the same tools you learned in Chapter 8 to compare variables measured in different units. The technique is standardization. In Chapter 8 you learned that if you wanted to compare one person's test grade to another person's weekly peanut butter consumption, these measures must be converted to a common base. Clearly these two measures are not directly comparable since test grades are in percentage units and peanut butter consumption is in spoonfuls. These scores become comparable, however, if each is converted to standard deviation units.[1]

The same idea applies to the partial regression coefficients in a multiple regression analysis. These slopes are initially in their original units; they are unstandardized scores. For comparison purposes, the slopes can be converted into standard deviation units, or standard scores. These standardized slopes are called Beta scores (or Beta weights). Beta scores have the following characteristics:

- They range between -1.0 and + 1.0.
- The sign indicates the form of the relationship.
- Zero indicates that the independent variable has no impact on the dependent variable.
- The interpretation of impact is in standard deviation units.

Table 17.3 provides Beta weights for the multiple regression example.

**TABLE 17.3**  Beta–Standardized Partial Regression Coefficients

| Independent Variables | Partial Correlation | Partial Regression Coefficients | Beta Scores |
|---|---|---|---|
| Hours worked last month | 0.35 | 106 | 0.27 |
| Years of education | 0.19 | 4,000 | 0.25 |
| Occupational prestige | 0.17 | 906 | 0.12 |

Because they are standardized, the Beta scores are directly comparable. Comparing Beta weights allows us to determine which independent variable has the largest impact on the dependent variable. In Table 17.3, *Hours worked* has a larger impact on *Income* than either of the two other independent variables. Its impact is slightly larger than the effect of *Years of education* and more than twice as large as the impact of *Occupational prestige*.

Interpreting Beta scores requires remembering that the impact of the independent variable on the dependent variable has been standardized and that the independent variable's effect has controlled for the impact of the other variables. Therefore, the 0.27 Beta in Table 17.3 is interpreted as follows:

> Holding both *Years of education* and *Occupational prestige* constant, a one-standard-deviation-unit change in *Hours worked* results in a change of 0.27 of a standard deviation in *Income*.

---

[1]See Chapter 8 for how to convert ordinary measurements into standard (*z*-) scores.

This interpretation, of course, sounds awkward because both the independent and dependent variable are expressed in standard scores. For this reason, standardized partial slopes (Betas) are typically used for the purpose of determining the relative impact of the independent variables on the dependent variable. Stating a particular independent variable's impact is usually expressed in unstandardized terms, using the partial regression coefficients.

## Multiple Regression

Recall from Chapter 1 that one test of scientific understanding is improved prediction. Ignorance is problematic because this state of affairs restricts our anticipation of and planning for the future. Through improved understanding, ignorance declines; the better our knowledge of some topic, the better our ability to forecast future events.

Scientific knowledge and prediction rests on the detection of patterns in empirical evidence. Patterns identify the underlying trend, and once this trend is clarified predictions can be made. Physicians detect trends in physical health and make their prognoses about a patient's recovery; economists detect trends in unemployment and predict future rates; astronomers detect a meteor's trend and anticipate whether or not emergency preparations need to be made on earth. All science-based activities use the understanding of existing patterns to anticipate future events.

In multiple regression, the state of affairs being predicted is a dependent variable score. Imagine a person standing before you in a large cardboard box that provides no clue about the person's occupation, work habits, education level, or anything else. If you were required to predict this person's income, your guess would likely be filled with error. However, if *Hours worked*, *Occupational prestige*, and *Years of education* are systematically related to *Income* (i.e. if these independent variables display a systematic pattern with *Income*), then knowing the individual's scores on these independent variables should allow you to better predict their income. The multiple regression equation provides a description of how the independent variables are systematically related to the dependent variable and, therefore, can be used for prediction.

Earlier we introduced the general form of the multiple regression equation:

$$Y = a + b_1(X_1) + b_2(X_2) + b_3(X_3)$$

The multiple regression equation reports how the set of independent variables jointly influences the dependent variable. Look at the equation carefully. It says that to predict a particular person's dependent variable score ($Y$) you do the following: take the intercept and add to it the person's score on the first independent variable ($X_1$) multiplied by the appropriate partial slope ($b_1$), and then add the person's score in the second independent variable ($X_2$) multiplied by the appropriate partial slope ($b_2$), followed by repeating the same procedure for every other independent variable.

Here is an illustration of how this procedure applies to the variables and evidence provided in Table 17.3. If the intercept value is 2,313, then the specific regression equation connection *Income* (the dependent variable) to *Hours worked*, *Years of education*, and *Occupational prestige* is the following:

$$\text{income} = 2{,}313 + 106(\text{hours worked}) + 4{,}000(\text{years of education}) + 906(\text{occupational prestige})$$

This equation defines the general pattern between the independent variables and *Income*. If we wanted to predict the *Income* score of a person who worked 160 hours per month, had 15 years of education, and an occupational prestige score of 10, we simple add the specific values of the independent variables to the equation:

$$\text{income} = 2{,}313 + 106(160) + 4{,}000(15) + 906(10)$$

As we solve this equation, the predicted income of a person with these characteristics is:

$$\$88{,}333 = 2{,}313 + 16{,}960 + 60{,}000 + 9{,}060.$$

In summary, a third use of multiple regression is to show how several variables come together to predict dependent variable scores. This goal is accomplished by inserting the partial regression coefficients into the general multiple regression equation (along with the intercept value) and solving the equation for specific independent variable characteristics (values or scores for each $X$).

## Multiple Correlation

A multiple regression equation describes the best-fitting trend between a set of independent variables and a dependent variable of interest. You have just seen how multiple regression information can be used to predict respondents' scores. The scores are a researcher's best predictions, given her current understanding of the topic. But knowing that a researcher is making her best prediction does not tell you how good that prediction is. Physicians' prognoses are sometimes wrong; so are economists' forecasts about future states of the economy; astronomers' anticipations about meteor trajectories can also be mistaken.

In short, establishing how good predictions are is a separate issue from the tools used to make the predictions. A final statistical tool that multiple regression provides is an assessment of how well the set of independent variables explains the dependent variable. Understanding this tool is assisted by returning to the simple bivariate regression situation discussed in Chapter 14.

Simple bivariate regression uses only a single variable to make predictions about a dependent variable. It does so by identifying the best-fitting regression line describing the patterned relationship between the variables. From this regression line, interpretations of the relationship's form, extent, and precision are made. Form comes from interpreting the sign of the slope; extent, from the magnitude of the slope; and precision, from the correlation coefficient ($r$).

Simple bivariate regression also includes another tool, one that measures the strength of the connection between the independent and dependent variables. This measure of strength is a proportional reduction in error (PRE) statistic called the coefficient of determination. This coefficient is symbolized as $r^2$.

Multiple regression uses an analogous statistic to measure the strength of the connection between the *set of independent variables* and the dependent variable. This statistic is called the **coefficient of multiple determination** and is symbolized as $R^2$. The coefficient of multiple determination is a proportional reduction in error (PRE) statistic and, accordingly, has a standardized interpretation. $R^2$ describes how well a set of independent variables explains a dependent variable by reporting how much of the total variation in dependent variable scores is explained by the set of independent variables.

For any specific multiple regression, the coefficient of multiple determination provides a measure of how well the set of independent variables predicts the dependent variable. $R^2$ can vary between zero and $+ 1.0$. In proportional reduction in error terms, an $R^2$ of zero means that the set of independent variables has no systematic connection to the dependent variable; they explain none of its variance. At the other extreme an $R^2$ of $+ 1.0$ means that the set of independent variables explains all of the variation in the dependent variable. Scores between these extremes indicate how well any specific set of variables explains the dependent variable outcomes in a set of scores. For example, if the coefficient of multiple determination for the *Income* regression equation illustrated earlier was $R^2 = 0.68$, this would mean that 68 per cent of the variation in *Income* scores can be explained by knowledge of respondents' *Hours worked*, *Years of education*, and *Occupational prestige*. Alternatively, this also means that 32 per cent of differences in *Income* scores remains unexplained; other variables, not taken into account in the analysis, remain operating and having important effects.

## Identifying Effects of Third Variables

Chapter 16 introduced four ideal-type models to describe what effect a third variable (measured at lower levels of measurement) might be having on an original bivariate relationship. In the authentic model, the contextual variable has no effect on the original relationship. Consequently, the original relationship between the independent and dependent variables is replicated when the third variable's effects are controlled. In the spurious model, the original relationship disappears when the control variable is held constant. The same disappearance occurs for the intervening model, although the sequence of the third variable's effects is different. Finally, in the interaction model, the original relationship changes depending on which specific value of the third variable is held constant.

The multiple regression techniques in this chapter have the same goal as the elaboration model techniques introduced in the last chapter. In both instances, the goal is to determine what effect contextual variables are having on the original relationship between the independent and dependent variables. Different techniques are involved because the variables are at different levels of measurement, and therefore have different properties that can be incorporated into the analysis.

Still, some commonalities between the outcomes of the elaboration analysis and multivariate analysis are notable. Let us begin with identifying authentic relationships. In tables, an authentic relationship is one that displays the same relationship in the original bivariate table and the partial tables. In authentic relationships the connection observed when the contextual variable is operating is replicated when the contextual variable's effects are removed. Remember that in partial tables the effects of a contextual variable are literally controlled. By contrast, in multiple regression analysis, the effects of all contextual variables are statistically controlled.

Given this situation, how are authentic relationships recognized in multiple regression? Answering this question turns on remembering that models are identified by comparing the independent–dependent variable relationship under two conditions: first, where the context variables are operating and, second, where contextual effects are removed. Here is how you see these conditions in multiple regression. The original relationship between any independent variable and the dependent variable (i.e. the condition where contextual effects are operating) is characterized by the simple correlation coefficient between the variables. The second condition, where the effects of contextual variables on this relationship are removed, is captured in the partial correlation coefficient between the independent and dependent variables. In short, an authentic connection in a multivariate analysis is one in which the simple and partial correlation coefficients are the same (or similar). This similarity indicates that whether the contextual variables are allowed to operate or whether they are controlled makes no difference to the bivariate connection.

In tables, spurious and intervening relationships are identified when the original bivariate connection disappears as contextual effects are removed. In multiple regression, this situation is evident when a substantial relationship in the simple correlation coefficient between the independent and dependent variables disappears (or is substantially reduced) in the partial correlation coefficient. This dramatic reduction in the coefficient size tells us that the context variables make a lot of difference to the observed connection between the original variables. As always, how this reduction is occurring, either through spurious or intervening mechanisms, is a matter that must be decided based on additional evidence about the sequencing of the variables.

Interaction effects can also be detected using multiple regression, but this involves a set of operations that are more complicated and beyond our current scope. The important general point to note is that multiple regression is analogous to the elaboration model considered in Chapter 16. Since the variables in multiple regression are at higher levels of measurement, the analyses can be more sophisticated. But the general objective remains the same: to determine how additional variables are affecting observed connections between an independent and a dependent variable.

## STEP 2 Learning the Calculations

# Locating the Information for Calculating Multiple Regression

Thankfully, we are not going to calculate multiple regression by hand! In this section, we are going to learn how to take information from an IBM® SPSS® statistics software ("SPSS") output and use it to identify the information we need to produce for the multiple regression equation, including the partial regression partial correlations, coefficient of multiple determination, and partial slopes. We are going to learn four strategies for using the SPSS output:

1. Where do I find the intercept on the printout? Where does it go in the equation?
2. Where are the partial slopes located? Where do they go in the equation?
3. Where do I find the coefficient of multiple determination? How do I interpret it?
4. How do we use the numbers we have generated from SPSS output to make predictions about individuals?

Let's begin with an example based on the Student Health and Well-Being Survey data set. First, we are interested in simply determining whether or not *Age* has an influence on *Satisfaction with life*. If there is an influence, how strong is the impact of *Age* on *Satisfaction with life* and what is the direction of that relationship? Figure 17.2 provides the printout from SPSS which you shall learn in step 3.

**Model Summary**

| Model | R | R Square | Adjusted R Square | Std. Error of the Estimate |
|---|---|---|---|---|
| 1 | .107ᵃ | .011 | .011 | 5.959 |

a. Predictors: (Constant), age Respondent Age

**Coefficientsa**

| Model | | Unstandardized Coefficients | | Standardized Coefficients | t | Sig. |
|---|---|---|---|---|---|---|
| | | B | Std. Error | Beta | | |
| 1 | (Constant) | 22.938 | 1.003 | | 22.864 | .000 |
| | age Respondent Age | −.179 | .049 | −.107 | −3.665 | .000 |

a. Dependent Variable: Q22 Q22. Satisfaction with life

**FIGURE 17.2** Simple Regression Output

Don't let its looks scare you! All the information you need to identify the numbers needed for your equation is located in these two tables. Recall the form of multiple regression equations as outlined above. In our example, however, we only have one independent variable and thus no $X_2$ or $X_3$.

$$Y = a + b_1(X_1) + b_2(X_2) + b_3(X_3)$$

## Locating the Intercept

Let's deal with the intercept first. Where do I find it? Recall that $a$ is the notation for intercept in the equation. Unfortunately, there is no $a$ labelled on the SPSS printout. It is located under *(Constant)* in the last table. We'll underline the appropriate number in Figure 17.3.

**Coefficientsa**

| Model | Constant = a | Unstandardized Coefficients | | Standardized Coefficients | t | Sig. |
|---|---|---|---|---|---|---|
| | | B | Std. Error | Beta | | |
| 1 | (Constant) | 22.938 | 1.003 | | 22.864 | .000 |
| | age Respondent Age | −.179 | .049 | −.107 | −3.665 | .000 |

a. Dependent Variable: Q22 Q22. Satisfaction with life

**FIGURE 17.3** Locating the Intercept on Output

In this case, the intercept is 22.94. Now that we have located the intercept, we need to place it into the equation.

$$Y = 22.94 + b_1(X_1) + b_2(X_2) + b_3(X_3)$$

## Locating the Partial Slopes

Our next step is to use the printout to locate the partial slopes in the SPSS printout and drop them into the equation. Recall that these are the numbers that tell us the effects of the independent variable when the values of all the other independent variables are held constant. In this case, the partial slopes are indicated by the letter *b* in the equation. The subscripts (numbers that are slightly beneath the writing line), indicate each independent variable. For instance, $b_1$ is the first independent variable—which in this case is *Age*. This is followed by $b_2$, which is the second independent variable, and so on for the other independent variables.

Luckily for us, the independent variables are clearly located in the SPSS output using their variable name. Let's look at that nasty printout once again, in Figure 17.4!

**Coefficients<sup>a</sup>**

| Model | | Unstandardized Coefficients | | Standardized Coefficients | t | Sig. |
|---|---|---|---|---|---|---|
| | | B | Std. Error | Beta | | |
| 1 | (Constant) | 22.938 | 1.003 | | 22.864 | .000 |
| | age Respondent Age | −.179 | .049 | −.107 | −3.665 | .000 |

a. Dependent Variable: Q22 Q22. Satisfaction with life

**FIGURE 17.4**  Locating the Partial Slope on Output

Right beneath the intercept is the listing of all the independent variables in the SPSS output. We can easily locate $b_1$, as it is the first variable in the list and is indicated by the label *Respondent Age*. Now which number do we use? That's easy to find as it is the only one! It's the column labelled B. In this case, the value of $b_1$ is -0.179. We can now drop that number into the appropriate place in the equation.

$$Y = 22.94 + -.179(X_1) + b_2(X_2) + b_3(X_3)$$

We will repeat the process for the other variables in the equation later, which will give us the partial regression coefficients. But for now, let's now look for the Beta. Remember that these coefficients are standardized (converted to a common base—standard deviation units) so that we can compare the relative influence of the different independent variables on the dependent variable. This will allow us to determine the independent variables with the greatest influence on the dependent variable and those with lesser influence.

The standardized slopes (Betas) are located in the position on the output identified in Figure 17.5.

**Coefficients<sup>a</sup>**

| Model | | Unstandardized Coefficients | | Standardized Coefficients | t | Sig. |
|---|---|---|---|---|---|---|
| | | B | Std. Error | Beta | | |
| 1 | (Constant) | 22.938 | 1.003 | | 22.864 | .000 |
| | age Respondent Age | −.179 | .049 | −.107 | −3.665 | .000 |

a. Dependent Variable: Q22 Q22. Satisfaction with life

**FIGURE 17.5**  Locating the Standardized Slope on Output

Luckily for us, these too are clearly labelled under the column entitled Beta. Let's locate the Beta for *Age*. It has a value of -0.107, which, since there is only one IV, is the Pearson's correlation coefficient (*r*) between *Age* and *Satisfaction with life*. We can interpret the *r* by saying there is a poor negative relationship between *Age* and *Satisfaction with life* (see Chapter 14 for review).

## Additional Variables

Let's compare that result to the value of Beta for some other independent variables—Q14B (*Coping with stress through planful problem solving*) and Q17 (*Rosenberg self-esteem scale*). The output of the computer analysis containing these three independent variables (*Age, Coping style, Self-esteem*) is contained in Figure 17.6. This output is identified as multiple regression, because it identifies the effects of several independent variables on the dependent variable (*Satisfaction with life*).

**Model Summary**

| Model | R | R Square | Adjusted R Square | Std. Error of the Estimate |
|---|---|---|---|---|
| 1 | .628a | .394 | .393 | 4.671 |

a. Predictors: (Constant), Q17 Q17. Rosenberg self-esteem Scale, age Respondent Age, Q14B Q14B. Coping with stress by planful problem solving

**Coefficientsa**

| Model | | Unstandardized Coefficients | | Standardized Coefficients | t | Sig. |
|---|---|---|---|---|---|---|
| | | B | Std. Error | Beta | | |
| 1 | (Constant) | 7.538 | 1.031 | | 7.309 | .000 |
| | age Respondent Age | −.193 | .040 | −.115 | −4.829 | .000 |
| | Q14B Q14B. Coping with sress by planful problem solving | .071 | .049 | .036 | 1.453 | .146 |
| | Q17 Q17. Rosenberg self-esteem Scale | .346 | .014 | .611 | 24.690 | .000 |

a. Dependent Variable: Q22 Q22. Satisfaction with life

**FIGURE 17.6**  Multiple Regression Output

Once again, don't panic! We will discuss what all these numbers mean in sequence.

Since we left off with Betas in the previous section, we will start here. Look at the output and identify which standardized coefficient (Beta) is the largest. Which one is second largest? Which one is smallest? Betas allow us to compare the effects of each independent variable separately so we can determine which variable has the strongest influence on the dependent variable. The answers are found in Figure 17.7.

**Coefficientsa**

| Model | | Unstandardized Coefficients | | Standardized Coefficients | t | Sig. |
|---|---|---|---|---|---|---|
| | | B | Std. Error | Beta | | |
| 1 | (Constant) | 7.538 | 1.031 | | 7.309 | .000 |
| | age Respondent Age | −.193 | .040 | #2 −.115 | −4.829 | .000 |
| | Q14B Q14B. Coping with stress by planful problem solving | .071 | .049 | #3 .036 | 1.453 | .146 |
| | Q17 Q17. Rosenberg self-esteem Scale | .346 | .014 | #1 .611 | 24.690 | .000 |

a. Dependent Variable: Q22 Q22. Satisfaction with life

**FIGURE 17.7**  Rank Ordering Independent Variable Effects

In this case, Q17 (*Self-esteem*) has the greatest influence, as its Beta is 0.611. The variable with the second highest influence is *Age*, with a Beta of -0.115. The variable with the least influence is Q14B (*Coping with stress by planful problem solving*), with a Beta of 0.036.

## Locating the Coefficient of Multiple Determination

Locating the coefficient of multiple determination is just as easy. This number appears at the top of the printout and is clearly labelled as R Square. Let's take a look at the printout provided in Figure 17.8.

**Model Summary**

| Model | R | R Square | Adjusted R Square | Std. Error of the Estimate |
|---|---|---|---|---|
| 1 | .628[a] | .394 | .393 | 4.671 |

a. Predictors: (Constant), Q17 Q17. Rosenberg self-esteem Scale, age Respondent Age, Q14B Q14B. Coping with stress by planful problem solving

**FIGURE 17.8**  Locating the Coefficient of Multiple Determination

R Square in this example is 0.394 and is identified by the circle. This number does not go into the equation because it tells you how well the numbers in your equation measure the variation in the dependent variable. Put another way, 0.394 tells you how well all the variables in the equation, taken together, predict or determine the dependent variable. To evaluate 0.394, take a look at Table 13.21. In this case, the effect of all the variables in the dependent variable is *moderate*. We can interpret this number like a percentage. In other words, all the independent variables in this equation predict 39.4 per cent of the variation in *Satisfaction with life*, or, in PRE terms, reduce error in predicting *Satisfaction with life* by 39.4 per cent.

## How Do We Make Predictions Using the Equation?

As mentioned earlier, in addition to enabling us to identify the form, strength, and direction of the relationship between the independent and dependent variables, the multiple regression equation allows us to make predictions about individuals. How do I go about doing that? Let's look at our completed equation using the information generated from the SPSS output (Figure 17.9).

Here is the general multiple regression equation with the information on the intercept and each of the independent variable partial slopes added. In other words, this equation identifies the multiple regression equation for the specific variables included in our analysis:

$$Y = 7.538 - 0.193\ (X_1) + 0.071\ (X_2) + 0.346\ (X_3)$$

The slope information for the first independent variable, for example, tells us that for each unit increase in *Age* (in other words, for each additional year of life), *Satisfaction with life* decreases by 0.193 units. You can interpret the effects of the other two independent variables the same way.

**Coefficients[a]**

| Model | | Unstandardized Coefficients | | Standardized Coefficients | | |
|---|---|---|---|---|---|---|
| | | B | Std. Error | Beta | t | Sig. |
| 1 | (Constant) | a 7.538 | 1.031 | | 7.309 | .000 |
| | age Respondent Age | X1 −.193 | .040 | −.115 | −4.829 | .000 |
| | Q14B Q14B. Copying with stress by planful problem solving | X2 .071 | .049 | .036 | 1.453 | .146 |
| | Q17 Q17.Rosenberg selt-esteem Scale | X3 .346 | .014 | .611 | 24.690 | .000 |

a. Dependent Variable: Q22 Q22. Satisfaction with life

**FIGURE 17.9** Multiple Regression Equation

Now let's discuss a hypothetical example that includes the effects of all the variables in the multiple regression equation. Let's say we have a 19-year-old who has a *Coping with stress* score of 17 (ranges from zero to 20, with higher values representing greater coping skills) and a *Self-esteem* score of 55 (ranges from 1 to 60, with higher values representing greater self-esteem). What would we predict this person's *Satisfaction with life* to be (ranges from 1 to 30, where higher values represent greater life satisfaction)? Let's plug the specific information about this person's scores on each of the independent variables into the multiple regression equation.

$$Y = 7.538 - 0.193\,(19) + 0.071\,(17) + 0.346\,(55)$$

Here are the steps in solving this equation:

$$Y = 7.538 + (-0.193 \times 19) + (0.071 \times 17) + (0.346 \times 55)$$

$$Y = 7.538 - 3.667 + 1.207 + 19.03$$

$$Y = 24.108$$

We can predict that this respondent's *Satisfaction with life* score would be 24, which is quite high (recall that the maximum score of life satisfaction is 30).

Now let's try the calculation using different numbers. Let's say we have a student who is 32 years old, has a *Coping with stress* score of 5 and a *Self-esteem* score of 20. What is the *Satisfaction with life* of this respondent? Let's solve the equation using the following steps:

$$Y = 7.538 + (-0.193 \times 32) + (0.071 \times 5) + (0.346 \times 20)$$

$$Y = 7.538 - 6.176 + 0.355 + 6.92$$

$$Y = 8.637$$

We can predict that this respondent's *Satisfaction with life* score is 9, which is quite a bit lower than the previous student's score and quite low relative to the overall scale.

As you have witnessed, multiple regression is an extremely powerful statistical tool that can be used to tell us a lot about the form, extent, and precision of a relationship between multiple independent variables on a single dependent variable. It can also help us to make predictions about individuals for whom we have no information about their measures on the dependent variable. Now that you know how to interpret the output, let's learn how to generate it.

## STEP 3 Using Computer Software

# Multiple Regression

Chapter 14 introduced the procedure for generating bivariate regression between two variables that are measured at higher (i.e. interval or ratio) levels of measurement. The steps for performing a multiple regression model are the same, only now we enter two or more independent variables. We will run a multiple regression in SPSS in order to assess the relationships between the six psychological well-being subscales: Q16A *Autonomy*; Q16B *Environmental mastery*; Q16C *Personal growth*; Q16D *Positive relations with others*; Q16E *Purpose in life*; and Q16F *Self-acceptance* with Q14A *Coping with stress by seeking social support* as our dependent variable. Remember, first, to run frequency distributions on all variables in order to assess which, if any, values need to be set to *missing*.

## Procedure

1. Select Analyze from the SPSS menu bar.
2. Go to Regression → Linear.
   a. Scroll down the list of variables on the left side to Q14A (*Coping with stress by seeking social support*) and click the arrow to move it to the Dependent box.
   b. Place the psychological well-being variables (Q16A, Q16B, Q16C, Q16D, Q16E, and Q16F) into the Independent(s) box.
   c. In the upper-right corner, select the Statistics button and check "Part and partial correlations."
   d. Select Continue.
3. Click OK.

The output viewer will display four tables, but for our purposes, we are only concerned with the Model Summary and the Coefficients tables, illustrated in Figure 17.10.

**Model Summary**

| Model | R | R Square | Adjusted R Square | Std. Error of the Estimate |
|---|---|---|---|---|
| 1 | .421a | .177 | .173 | 3.699 |

a. Predictors: (Constant), Q16F Q16F. Psychological well-being - Self-acceptance, Q16A Q16A. Psychological well-being - Autonomy, Q16E Q16E. Psychological well-being - Purpose in life, Q16D Q16D. Psychological well-being - Positive relations with others, Q16C Q16C. Psychological well-being - Personal growth, Q16B Q16B. Psychological well-being - Environmental mastery

Coefficient of determination

**FIGURE 17.10** Multiple Regression Output

There are several interpretations based on our output.

- The overall strength of our model or coefficient of multiple determination ($R^2$) is 0.177 or 17.7 per cent.
- The strongest partial correlation on the dependent variable (*Coping by seeking social support*) when all other independent variables are controlled is Q16D (*Positive relations with others*), with a partial correlation coefficient of 0.202.
- The independent variable with the greatest effect on the dependent variable (*Coping by seeking social support*) is also Q16D (*Positive relations with others*), with a standardized Beta coefficient of 0.237.
- Q16B (*Environmental mastery*) has the least effect on the dependent variable.

## STEP 4 Practice

You now have the tools for understanding the components of multiple regression outputs and generating them using computing software. To solidify your understanding you need to practise applying the statistical and software procedures.

This section provides you with opportunities to practise what you have learned about multiple regression. The first set of questions uses hand calculations. The second set uses the SPSS procedures. For each set of questions:

1. Follow the procedural steps (Set 1) or software application (Set 2).
2. Check your answers, using the Answer Key in the back of the text.
3. If your answer is incorrect, consult the Solutions section on the course website. The Solutions provide a complete step-by-step analysis of how the answers are derived.

After you have completed the next section (Interpreting the Results), return to your calculations or output and *provide complete, written interpretations of each of the statistics you have generated.*

# Set 1: Hand-Calculation Practice Questions

1. a. Write the regression equation predicting BMI from the following set of variables and coefficients:
      Hours watching TV per day: 1.37
      Number of sodas consumed per day: 1.04
      Hours of exercise per week: −0.72
      Intercept: 19.5
   b. What would be the predicted BMI of someone who watches 3 hours of TV per day, consumes 2 sodas a day, and does 3 hours of exercise per week?
   c. Which variable has the biggest effect on BMI?
   d. Interpret (put into words) what the regression coefficients in this equation tell us about the relationships between the dependent variable and each independent variable.
   e. What does the intercept tell us in this equation?

2.  a.  Write the regression equation predicting number of alcoholic drinks consumed per week from the following set of variables and coefficients:
        GPA: -0.4
        Hours of studying per week: -0.35
        Number of Twitter followers (measured in hundreds): 0.09
        Intercept: 6
    b.  Use this equation to predict the number of drinks consumed per week for the individuals in Table 17.4.

**TABLE 17.4** Data for Number of Drinks per Week Prediction

| Person | GPA | Hours Studying | Twitter Followers (hundreds) |
|---|---|---|---|
| 1 | 3.3 | 8 | 2.5 |
| 2 | 2.9 | 5 | 5 |
| 3 | 3.5 | 10 | 20 |

    c.  Interpret what this equation tells us about the relationships between the dependent variable and each independent variable.
3.  a.  Use the following set of regression coefficients and information about the intercept to create a North American model (a multiple regression equation) for predicting the number of traffic collisions per month for cities with population over 100,000.
        Population: +0.0018
        Average temperature in a given month (°C): -4.2
        Number of roundabouts: +0.35
        Intercept: 112
    b.  What is the predicted number of traffic accidents per month in the three following cities?

**TABLE 17.5** Data for Number of Accidents Prediction

| City | Population | Average Monthly Temperature (°C) | Number of Roundabouts |
|---|---|---|---|
| A | 750,000 | -30 | 8 |
| B | 1,600,000 | 15 | 18 |
| C | 500,000 | -5 | 3 |

    c.  What is your interpretation of the regression coefficients?
    d.  What information does the intercept give us?
4.  a.  Use the following set of regression coefficients and information about the intercept to create a prediction model for adult life expectancy in a post-industrial country.
        Hours of exercise per week: +0.83
        Number of servings of vegetables per day: +1.2
        Years of smoking: -0.35
        Intercept: 73
    b.  What is the predicted life expectancy for three adults with the following characteristics?

**TABLE 17.6** Data for Life Expectancy Prediction

| Person | Hours of Exercise per Week | Number of Servings of Vegetables per Day | Years of Smoking |
|--------|---------------------------|------------------------------------------|------------------|
| A | 3 | 2 | 10 |
| B | 7 | 5 | 0 |
| C | 0.5 | 0.5 | 32 |

    c.   What is your interpretation of the regression coefficients?
    d.   What information does the intercept give us?

# Set 2: SPSS Practice Questions

1.  a.   Use the Regression function to determine the simultaneous effects of *Reasons for living* (NQ24), *Negative affect* (Q3B), and *Self-esteem* (Q17) on *Satisfaction with life* (Q22).
    b.   Use the coefficients in the SPSS output to write the regression equation.
    c.   Use the regression equation to predict the level of *Life satisfaction* for the individuals in Table 17.7.

**TABLE 17.7** Data for *Life Satisfaction* Prediction

| Person | Reasons for Living | Negative Affect | Self-Esteem |
|--------|-------------------|-----------------|-------------|
| 1 | 85 | 19 | 45 |
| 2 | 175 | 9 | 55 |
| 3 | 159 | 23 | 44 |

    d.   Interpret (put into words) what the regression coefficients in this equation tell us about the relationships between the dependent variable and each independent variable.
    e.   Which independent variable has the greatest effect on *Life satisfaction*? Which has the least?
    f.   Interpret what the Beta weight for *Reasons for living* tells us about its relationship with *Life satisfaction*.
    g.   How well does this set of variables predict *Life satisfaction*?

2.  a.   Use the Regression function to determine the simultaneous effects of *Positive affect* (Q3A), *Negative affect* (Q3B), *Self-esteem* (Q17), and *Positive relations with others* (Q16D) on *Depression* (Q4).
    b.   Use the coefficients in the SPSS output to write the regression equation.
    c.   Use the regression equation to predict the *Depression* index score for the individuals in Table 17.8.
    d.   Interpret what the regression coefficients in this equation tell us about the relationships between the dependent variable and each independent variable.
    e.   Which independent variable has the greatest effect on *Depression*? Which has the least?

**TABLE 17.8** Data for *Depression* Prediction

| Person | Positive Affect | Negative Affect | Self-Esteem | Positive Relations |
|--------|-----------------|-----------------|-------------|--------------------|
| 1 | 30 | 14 | 46 | 20 |
| 2 | 23 | 19 | 54 | 15 |
| 3 | 19 | 24 | 39 | 10 |

    f.    Interpret what the Beta weight for *Self-esteem* tells us about its relationship with *Depression*.

    g.    How well does this set of variables predict *Depression*?

3.    a.    Use the Regression function to determine the simultaneous effects of *Environmental mastery* (Q16B), *Purpose in life* (Q16E), and *Reasons for living—family or friends* (Q24A) on *Strength of religious faith* (Q18).

    b.    Write out the regression equation.

    c.    Use the regression equation to predict the *Strength of religious faith* for an individual with the following scores: *Environmental mastery* = 15, *Purpose in life* = 13, and *Reasons for living—family or friends* = 22.

4.    a.    Use the Regression function to determine the simultaneous effects of *Coping with stress by seeking social support* (Q14A), *Autonomy* (Q16A), and *Personal growth* (Q16C) on *Reasons for living—one's self or one's future* (Q24B).

    b.    Write out the regression equation.

    c.    Use the regression equation to predict the *Reasons for living—one's self or one's future* score for an individual with the following scores: *Autonomy* = 10, *Personal growth* = 7, and *Coping with stress by seeking social support* = 13.

## STEP 5   Interpreting the Results

This chapter introduced a variety of tools associated with multiple regression. The discussion reviewed the tools associated with four goals that multiple regression can accomplish. Here is a review of how to interpret each of the statistical results.

# Interpreting Partial Correlation

The partial correlation statistic ($r$) provides an indicator of the relative strength of each independent variable's contribution to the dependent variable. Partial correlations are used to determine which independent variables have the strongest relationship with the dependent variable. Remember that these are *partial* correlations, which means that the identified connection between the independent and dependent variable has removed the contaminating effects of all the other independent variables. By isolating each independent variable's unique effects, it is possible to compare the relative strength of each of their contributions to the dependent variable. This can be done simply by comparing the size of each partial correlation coefficient, noting that larger sizes indicate greater contribution. If a more precise interpretation of strength is required, the partial correlation coefficients can be squared ($r^2$) and interpreted in PRE terms.

# Interpreting Partial Regression

The partial regression statistic focuses on the slopes of each independent variable in the multiple regression equation. Each unstandardized slope $(b_i)$ identifies the impact of a specific independent variable on the dependent variable, having removed the effects of the other independent (contextual) variables. Because the partial regression slopes for each independent variable are measured in different units, they are not directly comparable. To compare the relative impact of the various independent variables, the slopes must be standardized (i.e. converted into standard deviation units). This standardization process produced Beta weights. The Betas for the independent variables can be compared to determine how much impact each independent variable has on the dependent variable. The general interpretation of a Beta coefficient is as follows:

> A one-standard-deviation-unit change in $X$ [independent variable] is associated with $Z$ [Beta score] standard-deviation-unit of change in $Y$ [dependent variable].

> Example: A one standard-deviation-unit change in *Education* is associated with a half (0.50) a standard deviation unit change in *Income*.

# Interpreting Multiple Regression

The multiple regression equation describes how each of the independent variables contributes to the dependent variable. This equation is the prediction formula used to estimate dependent variable scores from independent variable conditions. The multiple regression equation identifies how the set of independent variables jointly affects the dependent variable. As such it utilizes prediction rather than interpretation. The multiple regression equation can be solved for any particular set of independent variable characteristics, and produces the best estimate of the dependent variable outcome associated with this cluster of attributes.

# Interpreting Multiple Determination

The coefficient of multiple determination $(R^2)$ characterizes how well the set of independent variables explains the dependent variable. $R^2$ is a PRE statistic whose generic interpretation is as follows:

> The set of independent variables ([IVs]) explains $X$ [$R^2$ value times 100] per cent of the total variation in $Y$ [dependent variable].

> Example: The set of independent variables (*Intelligence*, *Class attendance*, *Hours studying*) explains 53 per cent of the total variation in *Final course grade*.

The closer this value is to 100, the better our understanding of the forces contributing to the dependent variable outcomes.

# Chapter Summary

This chapter introduced the statistical techniques and tools for examining the effects of several independent variables on a dependent variable when the variables are measured at higher (interval and ratio) levels of measurement. In this chapter you learned that:

- Multiple regression techniques are an extension of simple bivariate regression and take the effects of contextual variables into account.
- Multiple regression techniques are analogous to the elaboration model ideas introduced in Chapter 16, except that they employ variables at higher levels of measurement.
- One advantage of multiple regression over the elaboration model is that statistical (rather than literal) control of contextual variables makes it easier to take multiple control variables into simultaneous consideration.
- Multiple regression procedures serve four basic purposes, including (1) identifying which independent variables have the strongest relationship with the dependent variable, (2) identifying how much impact each independent variable has on the dependent variable, (3) specifying how the set of independent variables jointly affects the dependent variable, and (4) determining how well the set of independent variables explains the dependent variable.
- Partial correlation statistics identify the relative strength of the independent variable contributors.
- Partial regression coefficients identify how much impact each independent variable has.
- Unstandardized partial regression coefficients are not directly comparable.
- Comparing partial regression coefficients requires their conversion into standard deviation units, which yields Beta weights.
- The multiple regression equation is the formula for identifying how the independent variables jointly contribute to the dependent variable.
- The multiple regression equation is used for predicting dependent variable scores for any specific combination of independent variable attributes.
- The coefficient of multiple determination $R^2$ is the statistic that describes how well a set of independent variables explains the dependent variable.
- $R^2$ has a conventional PRE interpretation.
- Comparing simple and partial correlation coefficients provides insights into what, if any, effects contextual variables are having on the original bivariate connection.

This chapter is the last one introducing statistical tools related to describing variables and their connections in an existing data set. In other words, this is the final chapter where descriptive is the answer to the first statistical selection question: Does your problem centre on a descriptive or an inferential issue? Part V turns our attention to the statistical tools and techniques related to making inferences from existing data.

# PART V
## Sampling and Inference

# 18

# Samples and Populations

## Overview

You know that the statistical maze and selection of statistical tools is organized around the answers to three key questions: Does your problem centre on a descriptive or an inferential issue? How many variables are being analyzed simultaneously? What is the level of measurement of the variables? Parts II, III, and IV (Chapters 4 through 17) introduced you to a wide variety of statistical tools based on various possible answers to these three questions. Note, however, that the differences in the variety of statistics you have learned so far centre on different answers to the second and third questions. All the statistical tools in the previous 14 chapters share the same answer to the first statistical selection question: all these statistics provided descriptive answers.

This chapter begins your exploration of a new region of the statistical maze. It introduces a variety of concepts and principles related to the use of inferential statistics. As you shall see, inferential statistics centre on the issue of generalizing the findings from a sample (descriptive statistics) to the population. In this chapter, you will learn about:

- the importance of and standards for probability samples;
- the connection between sample distributions and sampling distributions; and
- the features of sampling distributions that make inference possible.

In terms of the statistical maze, the contents of this chapter are situated here:

## Statistical Selection Checklist

1. Does your problem centre on a descriptive or an inferential issue?

    ☐   Descriptive

    ☑   Inferential

2. How many variables are being analyzed simultaneously?

    ☑   One

    ☐   Two

    ☐   Three or more

3.    What is the level of measurement of the variables?

   ☑   Nominal    ⎫

                    ⎬   for proportions

   ☑   Ordinal    ⎭

   ☑   Interval   ⎫

                    ⎬   for means

   ☑   Ratio      ⎭

This chapter begins the exploration of a new area and is largely conceptual. The chapter introduces only a few techniques but the ones it does contain are important. As always, the investigation begins with understanding the underlying ideas.

## STEP 1 Understanding the Tools

# Linking Samples to Populations

To begin, it is worth reviewing the distinction between descriptive issues and inferential ones. In practice, in most research situations it is impractical or impossible to gather all of the relevant data. When researchers collect data, they do not collect evidence from all the objects in a set. Those who want to know the attitudes of adult Canadians toward visible minorities collect data from one or two thousand respondents, not several million adults. If you have a blood chemistry test, some, not all, of your blood is extracted. In short, researchers typically collect data from a **sample**, which is a selection or subset of a population. By contrast, a **population** includes all objects in a well-defined collection. In our examples, the actual Canadian adults who were surveyed about minority groups and the drop or two of extracted blood are examples of samples. All adult Canadians and all the blood circulating in your body are populations. The process of gathering information from an entire population is called a **census**. In most research situations it is either impossible or impractical to conduct a census. It would be prohibitively expensive to interview every adult Canadian to gain a sense of prejudicial attitudes. Likewise, you should be wary of any physician who wants a blood census, rather than a blood sample.

Since the only evidence that researchers typically collect comes from a sample, the only evidence they can describe is sample evidence. All the descriptive statistical tools covered so far allow us to conduct analyses and draw conclusions on evidence from a sample. When a research study reports that 40 per cent of its sample of Canadian adults have racist attitudes, it is employing descriptive statistics. The same is true when the sample of your blood indicates high cholesterol levels. Descriptive statistics summarize various features of evidence gathered from a sample. These summaries, including measures such as means, medians, standard deviation, lambda, gamma, coefficient of determination, and the like, are statistics. In this sense, **statistics** are numerical summaries of various features of sample evidence.

In most cases, however, research is not satisfied with knowing about the features of evidence collected from samples. Researchers focusing on ethnicity are interested in the level of racism among *all* adult Canadians, not just the 1,500 or so persons they surveyed. Your physician is interested in the cholesterol level in *all* of your blood, not just the level in two drops. In other words, most research is interested in the general case, not just the specific cases from which data are gathered. Researchers are interested in the situation in a population of cases.

While statistics describe the features of variables in a sample, **parameters** describe the condition of variables in a population. While statistics describe the situation in a specific set of cases, parameters report the situation across all the cases—the general case. At this point an important practical problem becomes evident. Although researchers are interested in the general (population) situation, the only evidence they have comes from a sample of specific cases. Clearly there is a large gap between the evidence available (statistics) and the desired result (parameters). Inferential statistics are the tools researchers use to fill that gap. Inferential statistics are a set of analytical procedures for estimating population parameters from sample statistics.

Part V introduces you to a variety of inferential statistics. As expected, there are different inferential procedures depending on both the number and level of measurement of the variables involved. However, in all cases, inferential statistics are interested in the issue of generalization, of speaking about the general situation based on the information provided by specific cases. Stated in terms of the language introduced in Chapter 1, inferential questions concern the induction.

# Not All Samples Are Created Equal

Descriptive statistics summarize the evidence collected in samples. Inferential statistics take the conclusions of descriptive statistics and generalize them to the population. The analytical tools of inferential statistics, however, cannot be applied to all samples. Only samples of a specific type meet the standard for using inferential statistics. If samples don't meet the inferential requirements, then it is impossible to generalize their descriptive results. For samples that do meet the inferential requirements, it is possible to estimate population parameters from sample statistics. Given the relevance of generalization, it is important to distinguish the kinds of samples that are suitable to inferential statistics from those that are not.

In this regard, the basic distinction is between probability and non-probability samples. In **probability samples** each member of the population has a known likelihood of being included in the sample. In **non-probability samples**, members of a population are selected into the sample using some factor or factors other than probability. The distinction between these two types of samples is so important that it is worth elaborating on and illustrating the differences.

# Standards for Probability Samples

There is a variety of specific kinds of probability samples, including simple random samples, systematic samples, stratified samples, and cluster samples. The features of these various sampling types are described in any good statistical methods textbook and are not the focus here. What is important is to understand that any of these probability sampling types meets three specific scientific standards.

The first standard of all probability samples is that every member of the population has some chance of being included in the sample. Samples that do not meet this critical feature immediately fall into the non-probability type. Imagine the typical "question of the day" web poll conducted by your local television news organization. They ask questions like "Are you satisfied with the city's new garbage collection system?" or "Do you think that the mayor's recent conduct disqualifies her from office?" Viewers are then encouraged to visit the station's website and report their opinions—which then become a news item on the next day's coverage. "The majority of us are happy with our new waste management system." or "Sixty-three per cent favour the mayor's impeachment." These findings probably accurately describe how viewers who went online responded. But how confident are you that these descriptions capture the opinions of all citizens in the city? Likely, not very confident. This lack of confidence is understandable given that the sample violates the condition of all citizens having a chance of being included in the poll. What about those who were travelling outside of the country? Or those without computers or web access? Or those who never watch the specific news channel conducting the poll? Clearly, the way members of such a convenience sample are selected makes it impossible to generalize the results beyond those who participated in the poll.

The second standard that probability samples must satisfy is that the chance of any member of the population being selected into the sample must be known. Probability samples must not only be able to state that every member has some chance of being selected (the first standard) but identify what the probability of selection is. Most probability samples are designed so that all population members have an *equal chance* of being selected into the sample. However, as long as the probabilities of selection are known, it is not necessary that they be equal.

In order to calculate the likelihood of selection, researchers must have an accurate reading of who the members of a population are. Researchers determine the probability of selection by dividing the number of cases selected into the sample (sample size) by the size of the population. Think about the surveys conducted in shopping malls where interviewers approach shoppers and ask them to answer a few questions. Clearly nobody knows either who the relevant population is (mall shoppers?) or how many members of this population there are. Therefore, likelihood of selection cannot be determined, which disqualifies the results of such studies as generalizable.

Third, probability samples must meet the standard of independent selection. In this condition *independent* means that the selection of one case does not affect the likelihood of selection of some other case. Imagine you selected 100 undergraduate students at random from your university and interviewed them about their level of alienation. After completing each interview you asked them to introduce you to one of their friends so they could be interviewed, which produced a completed sample size of 200 (100 randomly selected, plus 100 of their friends). This type of snowball sample does not meet the requirements of a probability sample, since it violates the independent selection standard. The 100 friends in the sample were not independently selected; their inclusion was dependent upon their friend having been previously interviewed.

All probability samples meet these three standards, while non-probability samples violate one or more of these criteria. Meeting these standards in real-world conditions typically requires expending a lot of resources. The practical problems are illustrated in the construction of a sampling frame. A sampling frame is a list identifying all members of a population of interest. Think about the first standard for probability samples—that every member of the population has some chance of being included in the sample. Ensuring this standard requires being able to identify who the members of the relevant population are; that is, it requires having an accurate sampling frame. Creating such frames, however, is no easy matter. You might be confident of creating an accurate list of all members of your nuclear family, but how about of your university class? Are those persons who you have not seen for a few classes still members, or have they dropped out? What about the people who are officially registered but haven't bothered to attend? Or those who have quit, but failed to formally withdraw from the class? Even a quick overview of what seems like a simple question (Who are members of your class?) makes the point about the challenges of creating sampling frames. Still, since probability samples are a necessary condition for using inferential statistics, researchers expend a lot of resources making sure their samples conform to the probability rather than non-probability type.

Understanding how inferences are made from probability samples requires understanding some key concepts and principles. The remainder of this chapter is devoted to explaining the central ideas you require to make sense of the specific inferential techniques discussed in later chapters. These central ideas build upon some tools discussed in Chapter 8, which discussed the features of the normal curve. So we will begin with a brief review of the normal curve.

# A Review of the Normal Curve

Researchers are interested in identifying and accounting for differences. They ask questions like, How much inequality is there in Canada? What accounts for who becomes wealthy and who remains poor? Researchers measure differences with variables. For example, *Annual income* might be a variable used to detect inequalities, while *Level of education* could be used to explain differences in wealth. Variables are properties of objects that can differ. When researchers gather data on a variable (or set of variables) in a sample, the data they collect displays differences in the variables between the cases. For example, if you selected a probability sample of adults in your community, you would observe that different respondents have different levels of annual income. The differences among respondents' scores on a variable are evident in the presentation of frequency distributions (see Chapter 4).

When researchers analyze the results from data on a single variable collected from a sample, they are conducting univariate analysis. In doing so, they are describing the features of one variable. Univariate statistics summarize the features of the cases in the sample. Specifically, univariate statistics summarize two features of the cases. First, they can report on what is a typical value of the variable among the sample. This is what measures of central tendency (like the mode, median, and mean) do. For instance, analysis might show that the mean annual income in a sample is $73,300. Second, while the scores on a variable may be clustered around some central value, they are also spread out across the values of the variable. For example, while a typical annual income score in a sample might be $73,300, many respondents' incomes will be very different from this value. Measures of dispersion (like range, index of qualitative variation, and standard deviation) capture this second feature of univariate distributions.

For variables measured at higher levels of measurement (interval and ratio), it is common to use the mean to describe the central tendency in a distribution and the standard deviation to describe the level of dispersion.[1] In Chapter 7 you learned that differences among scores on a variable can also be displayed in graphic form.

For many variables, the graph takes on a standard shape, called a normal curve. Chapter 8 introduced you to the features of variables whose shape is normal, and it is these features that we need to review. Variables whose graph displays a normal curve share three characteristics. First, the measures of central tendency for the distribution are the same. In other words, the values of the mode, median, and mean are the same (or similar). Second, the distribution of scores is symmetrical. Symmetry means that the shape of the distribution of both sides of the centre is identical. The halves of the distribution are mirror images. Third, normal curves have a fixed relationship between the mean (measure of central tendency) and the standard

---

[1]You might find it helpful to review the calculation and interpretation of each of these measures. The mean is discussed in Chapter 5, while the standard deviation is covered in Chapter 6.

deviation (measure of dispersion). Specifically, moving a particular number of standard deviations from the mean captures a fixed number of the cases in the sample.

The fixed relationship between the mean and standard deviation of normally distributed variables allows the computation of $z$-scores (standard scores). The $z$-score translates any particular score on a variable into standard deviation units. Looking up $z$-scores in a $z$-score table allows you to determine what percentage of the sample falls between the variable's mean score and the $z$-score.

The following example illustrates these concepts by way of review. If the following illustration is unclear, please review the materials about these specific statistics covered in earlier chapters.

## Frequency Distribution

Table 18.1 contains a frequency distribution displaying the distribution of the scores on a sample of 104 respondents residing in a juvenile detention centre across the variable *Number of deviant acts in the past year*.

**TABLE 18.1**  Deviant Acts in the Past Year

| Deviant Acts | Frequency (f) | Percentage (%) | Proportion |
|---|---|---|---|
| 0 | 1 | 1.0 | 0.010 |
| 1 | 9 | 8.7 | 0.087 |
| 2 | 20 | 19.2 | 0.192 |
| 3 | 40 | 38.5 | 0.385 |
| 4 | 19 | 18.3 | 0.183 |
| 5 | 10 | 9.5 | 0.095 |
| 6 | 5 | 4.8 | 0.048 |
| Total | 104 | 100% | 1.00 |

This distribution of scores is called a **sample distribution**, since it displays the differences in scores for a single variable (*Deviant acts*) for a sample of a specific size ($N = 104$). Here are two things to note about this distribution. First, there are differences in the variable; different members of the sample committed different numbers of deviant acts. Second, as evidenced in the percentage and proportion columns, some scores on the variable are more common (probable) than others.

### Distribution Features

The distribution of *Deviant acts* scores can be summarized in terms of central tendency and dispersion. The mean number of deviant acts in this sample is 3.13. In other words, members of this sample typically committed about 3 deviant acts each in the past year. The standard deviation for this sample is 1.27, which suggests that the scores on the variable are moderately dispersed.[2]

---

[2]Again, to review how the mean and standard deviation are calculated, see Chapters 5 and 6.

### Distribution Graph

Figure 18.1 is a graph of the frequency distribution of *Deviant act* scores.

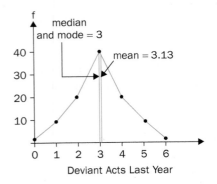

**FIGURE 18.1** Graph of *Deviant Act* Scores

Note that this graph has the shape of a normal curve. The curve is symmetrical and the mean, mode, and median all have the same value (3).

### z-Scores

Since the distribution of *Deviant acts* is normal, standard scores can be calculated for any specific respondent's reported number of deviant acts. For example, a person who reported 6 deviant acts is 2.27 standard deviations above the mean.[3] Using the *z*-score translation table tells us that 48.8 per cent of all respondents' scores fall between the mean (3.13) and this person's score (6).

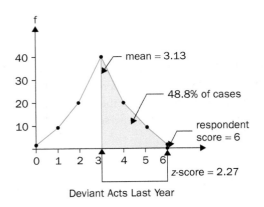

**FIGURE 18.2** *z*-Score Translation of Six Deviant Acts

---

[3] For a review of *z*-score calculations, see Chapter 8.

Note that the z-score translation of the percentage of respondents occurring between the respondent's score of 6 and the mean (3.13) occurs on the graph as a specific area under the curve. This means that if the values of all 800 juveniles residing in the detention centre were plotted on the graph, about 49 per cent of the dots (cases) would occur within the shaded area.

## From Sample Distribution to Sampling Distributions

This brief review of the features of distributions under a normal curve is important because it sets the stage for the key concepts and principles used in statistical inference. The first central concept is sampling distribution.

A researcher interested in generalizing the results draws a probability sample from a population and collects data from the cases on a set of variables. For example, imagine you draw a random sample of 1,000 adults in your city and gather data on their annual income. After the data are gathered the researcher uses descriptive statistics to summarize the findings. In our example, the researcher might find that the mean annual income in the sample is $33,599. Since a probability sampling procedure was used, the researcher is in a position to make an inference from the 1,000 adults to all adults in the city.

Before proceeding to generalize the results, imagine that the researcher conducted another study on the same population, using the same probability sampling procedure, collecting data on the same variable(s), the same way. In short, imagine the researcher replicated the original study. Because random sampling was used, data would be collected from a different subset of 1,000 respondents than were included in the original study, but the results would still be representative of the population. In the replication study, how likely is it that the mean annual income of the respondents would be exactly $33,599?

Not very likely. Why? Because probability sampling will always contain some random variation in the selected sample. Occasionally, this random variation can produce very weird results. For example, it is possible, through random selection, to select the richest 1,000 people in your community. When the mean of this probability sample is calculated, the conclusion is that a typical member of your community earns $850,000 annually! Alternatively, it is possible for a probability sample to select the poorest 1,000 adults in the community and conclude that a typical annual income is $2,600.

Now, based on other information you might have about the population, you might say that results like $850,000 and $2,600 are absurd. However, remember that the reason a researcher is conducting the study is they *don't know the results*; they don't have other information on the topic. In this situation, given that probability sampling procedures were used, the researcher has no immediate reason to think that $850,000 or $2,600 findings are any less realistic than the $33,599 sample result. You can see there is a problem.

Here is a more formal statement of the issue. Researchers who select a probability sample describe the results through summary statistics. These statistics summarize a sample distribution, the evidence on a variable from a single sample. Because there is random error in probability samples, we know that the results of any particular study may be different if the study is replicated. In other words, the statistics computed from any particular study are just one set of results from a large number of possible results: differences would occur if the study was replicated with a different probability sample. All the possible different statistical results for the sample variable, collected from different samples of the same population, using the same probability sampling technique, constitute a sampling distribution.

A **sampling distribution** is the set of different scores (statistics) that result from repeated replications of the same study using different samples. The differences in the outcomes are due to random sampling error. The three mean annual income measures ($2,600, $33,599, $850,000) are part of a sampling distribution of *all possible* mean income estimates generated from all possible probability samples of the same population. Understanding that the results of any single study come from a sampling distribution of different possible outcomes lets you appreciate that any single investigation, using the best sampling procedures, can produce very wrong results. The problem of inference, of trying to generalize from any particular sample statistic to the population parameter, centres on managing this ever-present possibility.

## Characteristics of Sampling Distributions

Any particular study that uses probability sampling procedures produces a sample distribution whose results are summarized using descriptive statistics. The cases in a *sample distribution* are *individual observations*. For example, mean annual income is computed on a sample of adults. In a *sampling distribution*, the cases are *sample statistics*.

A sampling distribution is a theoretical distribution of the different sample statistics (e.g. means) that would result from replicating the same study on the same population. Because it is a theoretical distribution, sampling distributions are based on an infinite number of study replications. Figure 18.3 illustrates a sampling distribution for our annual income study example.

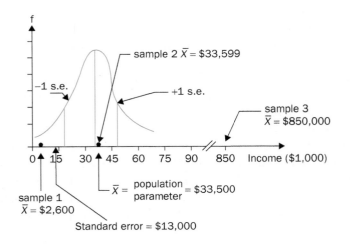

**FIGURE 18.3** Sampling Distribution of Annual Income

In Figure 18.3, the three illustrative sample results (statistics) discussed earlier are identified. Remember that this sampling distribution plots the results of all possible replications of the same study using different probability samples. Notice that, for our three particular studies, one produced exceptionally high results; one, exceptionally low results; and one, a result between these extremes.

There are some other features of sampling distributions that are important to know. First, sampling distributions are *normal curves*.[4] You can see this in the bell shape of the Figure 18.3 illustration. Like all normal curves, the mean (measure of central tendency) and standard deviation (measure of dispersion) for a sampling distribution can be calculated. Remember, however, that the observations in the sampling distribution are sample statistics (i.e. summary results from particular studies). Therefore, the mean of the sampling distribution identifies *what a typical study result would be.* In other words, on average, the most common statistical result of any probability sample is the mean of its sampling distribution. Figure 18.3 identifies the mean of the sampling distribution of annual income as $33,599. From this typical result we confirm that the extreme results are atypical and see the third sample as quite a common result.

Another critical feature of the sampling distribution is that its *mean value equals the population value*. In other words, the mean of the sampling distribution is the population parameter we are trying to estimate using inferential statistics. In our example, the real average annual income for all adults in the community is actually $33,599. The results of the three different studies identified in Figure 18.3 are approximations (estimates) of this population parameter. Notice that, using probability samples, the extreme results, although possible, are not very likely. By contrast, sample statistics that are at or near the population parameter are most likely. This feature of sampling distributions is very important to appreciate, since it is a foundation of the inferential techniques that permit the estimation of population parameters from the results of a single sample. After all, we don't know the population parameter; if we did, there would be no reason to conduct the study. Instead, we have only our single study results. The inferential question becomes one of estimating where on the sampling distribution of results our sample statistic falls.

Notice that while the mean of the sampling distribution is the most likely result of any particular study, the possible study results are wide-ranging. As you can see, another characteristic of the sampling distribution is that it has dispersion. And, since it is a normal curve, the dispersion in a sampling distribution can be measured with the standard deviation statistic. The standard deviation measure of dispersion in a sampling distribution is identified by a special name; it is called standard error. The **standard error** is a statistical measure of the amount of variation in a sampling distribution. It tells us how spread out (dispersed) the sample statistics from various studies are.

The size of the standard error affects the appearance of the sampling distribution. Since standard errors indicate the level of dispersion, a sampling distribution with a smaller standard error will have sample results that are more tightly clustered. Larger standard errors produce sampling distributions that are more widely dispersed. Figure 18.4 illustrates the effect of standard error size on distribution shape.

Notice that both examples in Figure 18.4 are normal curves. They look different because one set of results is more tightly clustered than the other.

---

[4]Sampling distributions are theoretical distributions. Since we never conduct the infinite number of studies necessary to construct a complete sampling distribution, how do we know its shape is normal? The answer lies in the Central Limit Theorem. This statistical principle states that, if statistics are calculated from samples of reasonably large size, the sampling distribution will take the shape of a normal distribution.

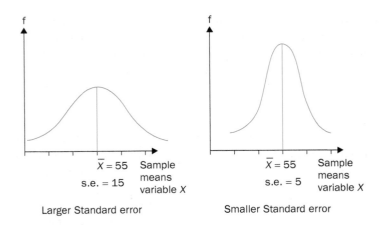

**FIGURE 18.4** Standard Errors and Sampling Distributions

The size of the standard error in a sampling distribution has important implications for inference. Appreciation of this implication stems from the word *error*. The standard error computation is really just the standard deviation calculated on a normal distribution of sample statistics. So why does it have the special label *error*? Remember that research studies are conducted to learn something new; to know about the currently unknown. Each study obtains its results from a single sample. These single sample results (descriptive statistics) come from a universe of all possible studies of the same type. Even if probability sampling techniques are used to select the sample from the population, it is possible that the sample statistics may imperfectly reflect the population parameter. The standard deviation of the sampling distribution indicates how wide the resulting sampling errors might be: hence standard *error*. In Figure 18.4, any particular study result is less likely to be a poor estimate in the sampling distribution on the left than in the one on the right. Where standard errors are larger, the findings of specific studies are more prone to error in estimating population parameters.

## A Short Summary

Several important ideas have been introduced to this point, so it is helpful to review where we are before moving on. Researchers are interested in some population of observations. They aim to talk about the general situation, characterized by population parameters. Any specific variable of interest is distributed in the population in some way. Since it is typically impractical to collect evidence from all the cases in a population, researchers use probability sampling to gather evidence from a set of cases intended to represent the population.

The sample of collected evidence is all that researchers have at their disposal. They use descriptive statistics to summarize the information on the variables of interest. These summary statistics could be univariate (e.g. mean), bivariate (e.g. gamma), or multivariate (e.g. $R^2$) measures. Researchers know that the sample used for their descriptions is only one of an infinite number of possible samples that could be drawn from the population. In other words, the sample evidence is one sample within a sampling distribution. Even using the best probability sampling techniques, there is some chance that the selected sample will be very unrepresentative. However, since the sampling distribution is a normal curve, highly unrepresentative samples are very unlikely. The opposite is, in fact, the case. It is most likely that a single sample is at, or very near, the mean of the sampling distribution. Since the mean of the sampling distribution is also the population parameter, there is a good chance that a single sample statistic is a reasonable estimate of the population parameter. What inferential statistical procedures do is provide an estimate of the likelihood of being wrong in generalizing from the sample statistic to the population parameter. Understanding how inferential statistics operate returns us to the characteristics of normal distributions.

## Sampling Distributions as Normal Curves

Sampling distributions are composed of a sample statistic (e.g. mean or proportion) plotted for every possible probability sample of a fixed size drawn from a population. Imagine a study of a probability sample of undergraduates on your campus showing that the mean number of beers consumed per week was eight. The sampling distribution of this mean is a plot of all the different mean scores that would occur if the study was replicated an infinite number of times, each time with a different probability sample drawn from the same population.

We know that the sampling distribution is a normal curve and that the mean of the sampling distribution is the population parameter of interest. For our example, the mean of the sampling distribution is the actual mean number of beers consumed per week by *all undergraduates* (not just those included in the sample). We also know that the sampling distribution's dispersion is measured by a type of standard deviation called the standard error.

Because the sampling distribution is a normal curve, we know that there is a fixed relationship between its mean (the population parameter we are trying to estimate) and its standard deviation (the standard error we can calculate). This fixed relationship tells us that about 68 per cent of all the sample statistics in the sampling distribution will fall within one standard error on either side of the mean and that about 95 per cent of the sample statistics will fall within two standard errors on either side of the mean. This situation is displayed in Figure 18.5.

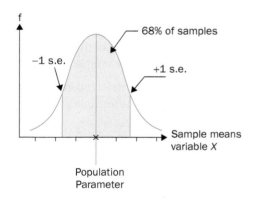

**FIGURE 18.5** Sampling Distribution as Normal Curve

Remember we know that any particular sample statistic falls somewhere on this sampling distribution of all possible statistics. For example, we are certain that the mean of eight beers per week falls somewhere on the sampling distribution in Figure 18.5. But, still, we don't know exactly where.

However, the information in the figure lets us conclude something very important. We know that 95 per cent of the observations fall within plus or minus the value of two standard errors from the mean. Therefore, we can conclude with 95 per cent confidence that our particular sample statistic (eight beers per week) is within two standard errors on either side of the mean. Remember we want to estimate what the mean of the sampling distribution is, since it is the population parameter for our particular statistic. We don't yet know what that value is, but we are getting closer to approximating it, since we have a good idea of the range in which it occurs.

The following chapters use this understanding of the sampling distribution situation to show how population parameters are estimated. These chapters will apply the concepts and principles of this chapter to make population inferences. For now, let's learn how standard errors are calculated.

## STEP 2 Learning the Calculations

Inferential procedures are based on sampling distributions. A key feature of any sampling distribution is its standard error. The standard error is the standard deviation of a sampling distribution. In other words, the standard error is a measure of how dispersed the sample statistics are in the sampling distribution.

Estimates of sampling errors are derived from evidence included in the sample (i.e. descriptive statistics). We are going to show you the steps in calculating the standard errors for two-sample statistics, means and proportions. Let's be clear on what we are doing. You have data on a single variable from a sample. It might be, for example, the mean income of persons in the sample or the proportion of single persons in the sample. The following techniques show you how to use sample information to estimate the standard error of the sampling distribution in which that particular sample statistic is located.

# Standard Error for Means

The formula for computing the standard error for means is the following:

$$\text{s.e.} = \frac{s}{\sqrt{N-1}}$$

The symbols in this equation are interpreted as follows:

- The numerator, $s$, is the standard deviation of the variable.
- The denominator is the square root of the sample size $(N)$.[5]

Imagine your sample of 1,000 adults had a mean income of $63,000 and a standard deviation of $8,000. Here are the steps in applying the formula to estimate the standard error of the sample distribution.

$$\text{s.e.} = \frac{s}{\sqrt{N-1}}$$

$$= \frac{8,000}{\sqrt{1,000-1}}$$

$$= \frac{8,000}{\sqrt{999}}$$

$$= \frac{8,000}{31.61}$$

$$= 253.08$$

---

[5] Why N-1, instead of N? Observed values will tend to fall closer, on average, to the sample mean than the population mean. Since the sample standard deviation is calculated from the deviations of scores from the sample mean, not the population mean, it will tend to underestimate the heterogeneity (the degree of dispersion) in the population (i.e. a biased estimate). Thus, given that a sample-derived estimate of the standard deviation (s) will tend to underestimate the actual standard deviation in the population ($\sigma$), we reduce the denominator by 1, which adjusts the sample estimate upward, thereby helping correct for the underestimation (i.e. an unbiased estimate).

# Standard Error for Proportions

The formula for computing the standard error for proportions is slightly different:

$$s.e. = \sqrt{\frac{P(1 - P)}{N}}$$

The symbols in this equation are interpreted as follows:

- The numerator is the proportion of persons in the sample who have a specific characteristic ($P$) and the proportion who do not have that characteristic ($Q$). Note that $P$ and $Q$ are proportions. Together they add up to 1 (i.e. the whole sample), such that: $P + Q = 1$, and thus $1 - P = Q$; and $1 - Q = P$.
- The denominator is the sample size ($N$).

Imagine your sample of 1,000 adults contains 55 per cent males (0.55). This implies that the proportion who are not males (i.e. the females) is 0.45. Using this information we can apply the formula to estimate the standard error of the sample distribution.

$$
\begin{aligned}
s.e. &= \sqrt{\frac{P(1 - P)}{N}} \\
&= \sqrt{\frac{(0.55)(0.45)}{1,000}} \\
&= \sqrt{\frac{0.2475}{1,000}} \\
&= \sqrt{0.000247} \\
&= 0.0157
\end{aligned}
$$

In short, computing the standard errors for either means or proportions is not difficult. You simply apply the sample information to the formula and generate the result.

There is one general point worth noting here. In estimating the standard errors for both means and proportions, the sample size is a very important determinant of the result. Sample size and standard error have an inverse relationship. As sample size increases, standard errors diminish. This makes sense, of course, since samples that capture a large share of the actual population are going to be more representative.

## STEP 3 Using Computer Software

# Selecting Random Samples Using SPSS

For various reasons it can be helpful to draw a random sample of cases from a data set. This can be easily accomplished by IBM SPSS statistics software ("SPSS"), using the following steps. The steps are applied to the Student Health and Well-Being Survey data set and illustrated in Figure 18.6.

# Procedure

1.  Go to Data → Select Cases → and click on the Random Sample of Cases radio button.
    a.  Check to make sure the Survey ID variable is highlighted on the left side of the window (it should be, by default).
    b.  Select Sample, which is located directly under Random sample of cases.
2.  a.  In the Select Cases: Random Sample window, make sure the "Approximately __ % of all cases" radio button is selected.
    b.  Enter the number 15, which means we are asking SPSS to select 15 per cent of the cases for our random sample.
3.  Select Continue → OK.

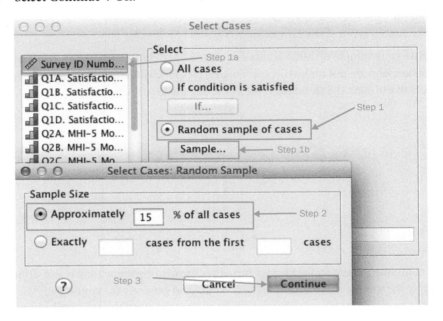

**FIGURE 18.6**  Selecting a Random Sample of Cases

Source: Reprint Courtesy of International Business Machines Corporation, © International Business Machines Corporation.

After you have OK'd the command, you should notice that along the rows in the data editor screen (Data View) there are a series of diagonal dashes for a large number of cases. These diagonal dashes represent respondents who were selected out of the random sample. The cases *without* diagonal dashes therefore represent the 15 per cent of cases who were selected to be included in the random sample.

In order to check your work, run a frequency distribution of Q22 (*Satisfaction with life*) and include the mean as your measure of central tendency (see Chapter 5 for further instruction on how this is done). Take note of the sample size and the mean for this random sample.

In order to run another random sample, you must first go back into Select Cases and check the Select All Cases radio button—then you are again working with the full sample. You can check frequencies on variables of interest to verify this. Second, to draw another sub-sample just click on the Draw Random Sample of Cases radio button again; it should still have the 15% value in the box. Click Continue →OK and you've drawn another random sample (if you want a different-sized random sample change the percentage in the box). Repeat steps 1 and 2 to draw as many random sub-samples as desired—but *do not save* your data file ("***.sav"). If you were to save the "***.sav" file, only the random sample would be saved and you would lose all other information which, obviously, we do not want to happen.

## STEP 4 Practice

You now have an understanding of the core concepts and principles underlying inferential statistics. The following illustrations provide opportunities for you to solidify your understanding of these ideas. The first set of questions uses hand calculations. The second set uses the SPSS procedures. For each set of questions:

1. Follow the procedural steps (Set 1) or software application (Set 2).
2. Check your answers, using the Answer Key in the back of the text.
3. If your answer is incorrect, consult the Solutions section on the book's website. The Solutions provide a complete step-by-step analysis of how the answers are derived.

# Set 1: Hand-Calculation Practice Questions

1. The mean high-school grade average for a probability sample of 500 undergraduate students was 78.8 per cent, with a standard deviation of 2.1 per cent. What is the standard error?

2. a. The average reported credit-card debt for a probability sample of 1,200 households was $13,577, with a standard deviation of $5,679. What is the standard error?
   b. What happens to the standard error if we increase our sample size to 1,500?
   c. What happens to the standard error if we decrease our sample size to 500?

3. In a survey designed to assess various aspects of university student quality of life, 65 per cent of student respondents ($N = 900$) indicated that they took public transit to school one or more times per week. What is the standard error?

4. a. A national opinion poll of 1,059 adult Canadians found that 59 per cent of respondents were confident in the Canadian economy. What is the standard error?
   b. What happens to the standard error if we increase our sample size to 1,500?
   c. What happens to the standard error if we decrease our sample size to 500?

5. a. A study working with a probability sample of 228 children between ages six and nine revealed that children of that age range spend, on average, 3.8 hours a day watching TV. What is the standard error when the standard deviation in the sample is 1.4 hours? 2.1 hours?
   b. How does the standard error change when the sample size changes to 600? Compute for standard deviation of 1.4 and 2.1 hours.

6.  a.  The study involving 228 children also revealed that 68 per cent of children between six and nine years old owned at least one violent video game. What is the standard error?

    b.  How does the standard error change when we change the sample size to 400?

7.  a.  Examining a probability sample of 250 university undergraduate students revealed that the average IQ was 118, with a standard deviation of 6 points. What is the standard error in this example?

    b.  How does the standard error change when the sample size is doubled? Tripled? Quadrupled?

    c.  Does the standard error change proportionately compared to the changing sample size?

8.  a.  Scientists were testing a new drug to lower blood pressure. At the end of the trial they reported results indicating the new medication successfully lowered blood pressure in 1,687 out of 2,250 people in the experimental group. Which standard error estimate is appropriate?

    b.  How large is the standard error?

# Set 2: SPSS Practice Questions

1.  Following the process for selecting random samples using SPSS outlined earlier:

    a.  Select 15 random samples from the Student Health and Well-Being Survey data set. Each of these samples should be 10 per cent of the entire sample.

    b.  Run frequencies on Q17 (*Rosenberg self-esteem scale*) for each subsample and request the mean for each (make sure missing values are set to 99 beforehand). Write down the mean score from each of the 15 subsamples.

    c.  Next, open the data set *Ch18SamplingDistExample.sav*. This data set contains five variables: *samplingdist1, samplingdist2, samplingdist3, samplingdist4,* and *samplingdist5*.

    d.  In the bottom left-hand corner are two tabs, Data View and Variable View; click Data View.

    e.  Enter the 15 mean scores from the 15 random 10 per cent samples into the *samplingdist1* column, one score each in rows 1 through 15.

    f.  Remember to save the data set after you have entered your 15 values.

    g.  Run frequencies on the variable *samplingdist1*; remember to click on the Statistics tab and then request the mean and the standard deviation.

    h.  Also click on the Charts tab, select the Histogram radio button, and check the "Normal curve on histogram" box.

The mean of these 15 sample estimates is the mean of the sampling distribution and the standard deviation of this sampling distribution is its standard error. According to the Central Limit Theorem, in a sampling distribution of an infinite number of samples, the mean of the distribution will equal the actual population parameter of interest. Our little simulated sampling distribution is small and so the mean is unlikely to be equal to the *population parameter*, which in this case is the mean *Self-esteem* score for the entire Student Health and

Well-Being Survey sample (42.71; $N = 1{,}166$). Your frequency table and histogram of your sampling distribution should resemble the one found in the Chapter 18 answers on the textbook website. Remember, it won't be exactly the same, as you will have generated 15 random samples different than the ones displayed in the answer key. Also note that, in theory, the central limit theorem applies to a distribution of infinite samples of the same size and the SPSS random sampling procedure selects an approximate proportion (e.g. 10 per cent) of the entire sample; therefore the actual size of our 15 samples varies somewhat. An additional source of variation is that some of our samples will have more or fewer *missing* responses than other samples.

2. Use the same procedures to select 15 random samples from the Student Health and Well-Being Survey, this time sizing the samples at 30 per cent of the entire sample. You can enter in the 15 mean values in the samplingdist2 column in your Sampling-DistExample data set. Once you have entered the values for this variable run a frequency for both *samplingdist1* and *samplingdist2* variables with the mean, standard deviation, and histogram with normal curve. Note the mean and standard deviation. Is there a difference between the 10 per cent sample sampling distribution and the 30 per cent sample sampling distribution? If so, why?

3. The variable *samplingdist3* is a sampling distribution comprised of one hundred 30 per cent samples. Run frequencies on the *samplingdist3* and *samplingdist2* variables requesting the mean, standard deviation, skew, and kurtosis from the Statistics tab, and a histogram with normal curve from the Charts tab. What differences do you see between the two distributions? What is the reason for these differences?

4. Following the same random sampling process, select 10 random samples of 20 per cent and find the mean *Self-acceptance* (Q16F) score for each. Enter the 10 mean scores into rows 1 through 10 of thesamplingdist4 column (the last five rows have been coded as *missing*). Find the mean and standard deviation for this new variable. Create a histogram of the new variable.

5. Repeat the steps in question 4 but this time request 10 random samples of 50 per cent of the entire population (instead of 20 per cent). Enter these mean *Self-acceptance* scores into the samplingdist5 column. Find the mean and standard deviation of this variable. Create a histogram of the new variable.

6. Which variable, *samplingdist4* or *samplingdist5*, is closer to the mean of the entire population? Which variable has the lower standard deviation? Which histogram aligns more closely to a normal curve? Explain your answers.

## STEP 5 Interpreting the Results

This chapter introduced core concepts and principles related to making inferences. The following chapters introduce you to specific inferential statistical techniques. Accordingly, interpretation of results does not apply to this chapter, although it will be important in following chapters. For now, it is important to understand the key ideas, which are summarized on the next page.

# Chapter Summary

In this chapter you learned that:

- While population parameters are of interest, collecting census evidence is impractical.
- Sample statistics include all kinds of descriptive measures.
- Inferential statistics are procedures for estimating population parameters from sample statistics.
- Inference is only possible from probability samples.
- Probability samples are ones in which all cases in the population have some known chance of independent selection.
- Accurate sampling frames are a key feature of probability sampling.
- The distribution of cases on a specific variable in a sample constitutes a sample distribution.
- Sample distributions of normal shape have similar measures of central tendency, are symmetrical, and display a fixed relationship between mean and standard deviation that is determined through z-scores.
- Any particular sample statistic comes from a distribution of possible sample results known as a sampling distribution.
- Sampling distributions are normal curves, with a mean that is the population parameter and a standard deviation called standard error.
- Sampling distributions demonstrate that any particular probability sample result is much more likely to approximate the population parameter than an unusual estimate is.

You now have the key ideas underlying statistical inference. The following chapters put these ideas into practice by showing you specific inferential techniques.

# 19

# Point Estimates, Confidence Intervals, and Confidence Levels

## Overview

The last chapter provided an overview of a new region of the statistical maze—the region dealing with inferential statistics. Inferential statistics are procedures for generalizing from sample statistics to population parameters. Inferential procedures use induction to go from the specific case (sample evidence) to the general case (population conditions).

The inferential statistics region of the statistical maze contains two basic kinds of procedures (1) point and interval estimates and (2) hypothesis testing. This chapter introduces you to point estimation and confidence interval procedures, while the following chapter covers hypothesis testing procedures. In this chapter you will learn:

- about the difference between point and interval estimates;
- about the logic of confidence intervals; and
- how to estimate population proportions and means from sample statistics.

You will gain experience with both higher and lower measurement levels, since proportions deal with variables at lower measurement levels, while means deal with variables measured at higher levels.

In terms of the statistical selection checklist defining the coordinates of the statistical maze, the contents of this chapter are situated as follows:

## Statistical Selection Checklist

1. Does your problem centre on a descriptive or an inferential issue?

    ☐ Descriptive

    ☑ Inferential

2.  How many variables are being analyzed simultaneously?

☑ One

☐ Two

☐ Three or more

3.  What is the level of measurement of each variable?

☑ Nominal
☑ Ordinal ⎤ for proportions

☑ Interval
☑ Ratio ⎤ for means

Following our model, the tour through this section of the statistical maze begins with conceptual understanding, which starts with understanding the difference between point and interval estimates.

## STEP 1 Understanding the Tools

# Point Estimates

Univariate descriptive statistics summarize information about single variables in the sample. For example, such statistics might determine that the proportion of women in a sample is 0.56, or that the mean income of sample respondents is $65,213. Remember that these are the results from one of an infinite number of possible samples of the same size that could be drawn from the population. In other words, these specific descriptions are one point on a sampling distribution of proportions and means.

Proportions and means are **unbiased estimators**, which means that, on average, they accurately reflect the proportion or mean of the variable in the population. As unbiased estimators these sample statistics will, over repeated studies, neither overestimate nor underestimate the population parameter. This is good news. The bad news is that researchers do not conduct the same study over and over. The evidence they have comes from a single study.

Proportions or means calculated on variables from a single study are not necessarily accurate reflections of the population parameter. This potential inaccuracy occurs because of random sampling error. As you learned in the last chapter, the extent of such sampling error is indicated by the size of the standard error, which measures the amount of dispersion in the sampling distribution. This fact of life has implications for point estimates.

**Point estimates** are precise descriptive numbers that generalize directly from the sample statistics to the population parameter. In the earlier example, if the researchers concluded that because the proportion of women in the sample is 0.56, the proportion of women in the population is 0.56, they would be using a point estimate. Similarly, point estimations would conclude that, because the mean income in the sample is $65,213, the mean income in the population is the same number. Point estimates are appealing inferential procedures because of their precision. To state exactly what the population parameter is provides a sense of certainty that is appealing.

Researchers use probability sampling to ensure that their sample statistics are the best estimators of the population parameters. Therefore, when the proportion or mean of some variable is calculated from sample evidence, that number is indeed the best estimate of the population parameter. But to say that the sample statistic is the best estimator of the population parameter does not say how good an estimator it is.

A football analogy might be helpful in understanding this point. It is common to see a running back charge through the offensive line and be tackled by several opposing players, resulting in a large mound of human flesh covering the ball carrier and the football hidden at the bottom of the pile. The referee blows the whistle to stop the play and then peels off players before eventually finding the ball. After the players are removed, the ball is placed at precisely some point on the field and yardsticks are brought onto the field to determine if a first down has been obtained. It is common for the ball to be just a small distance short of, or just a small distance past, the first-down marker—which evokes appropriate applause or disdain from the fans.

Placing the football on exactly one spot in this situation is like a point estimate. It brings a level of certainty that is necessary for play to proceed. But, if you think about it, how realistic is this level of precision? Is any referee, player, coach, or fan actually certain that where the ball is spotted is really where it was downed on the field? Who knows what went on in the wriggling mass of football players that affected the ball's proper location? Clearly, on the football field, precision is bought at the cost of accuracy.

The same holds true for point estimates in statistics. The sample statistic is the best estimate of the population parameter, but we all know that the sample probably contains some degree of sampling error. So to simply generalize from a sample statistic to the population parameter is risky. The procedure provides precision (i.e. a clear statement of exactly what the population parameter is) at the cost of accuracy.

# Interval Estimates

What might all stakeholders agree is a more accurate way of spotting the football when its precise location is unknown? Well, everybody knows the football is somewhere in the pile of players. It would be completely unreasonable to put it anywhere else on the field. And it is probably closer to the centre of the pile than to its edges (otherwise why would players pile on something other than the ball carrier and the football?). So it is probably reasonable for a referee to say that the football is at some specific yardage line *plus or minus one yard*. Such a ruling, of course, would create chaos for the game, but it would certainly be less absurd than the current practice.

In general, in taking measurements or making inferences, there is always a trade-off between precision and accuracy. The more precise statements are, the less accurate they are likely to be. In football, if the game rules insist on placing the football at a precise point, then these locations are less likely to be accurate. If game rules widened the range of ball location (i.e. reduced precision), accuracy would be enhanced—although the experience of players, coaches, and fans would suffer.

The "game" of social research does not require the certainty that football does, so its rules accommodate some trade-off of precision to improve the accuracy of parameter estimations. For this reason, inferential statistics rely on interval estimates. An interval estimate uses sample statistics to generate a range of scores within which the population parameter is likely to occur. Interval estimation procedures generate a confidence interval, which is a range of scores with the statistical point estimate at its centre. The wider the confidence interval (range of scores) the greater the confidence in the parameter estimate. At one extreme, if the proportion of women in the sample is 0.56 (56 per cent), a researcher can be perfectly confident that the proportion of women in the population is between 0.00 and 1.00 (zero–100 per cent). As the range of the interval narrows, confidence is reduced. The least confident estimate is the point estimate that declares the population parameter to be precisely the same as the sample statistic. The confidence interval is composed of a point estimate of the population parameter and a "fudge factor." The fudge factor is called the **margin of error**, which represents the "plus or minus factor" attached to the estimate that results in the confidence interval.

Researchers are willing to trade off some level of precision for enhanced accuracy. The results of this trade-off are interval estimates. First we shall explain the logic of interval estimation, then we shall describe how interval estimates are created in practice. In understanding both the logic and the practical application we shall use a shared example.

## Logic of Confidence Intervals

Imagine a researcher collects data from a probability sample of 1,500 Canadian adults and, through a calorie log, obtains data on the typical number of calories consumed daily. After analyzing the data, the researcher determines that the mean daily calorie consumption in her sample is 1,975 calories, and that the data has a standard deviation of 950. Using a point estimate, the researcher would simply generalize that all adult Canadians (the population), on average, consume 1,975 calories daily (the parameter). We know, however, that such a precise estimate is very unlikely to be accurate, since the chance of this one study being exactly on the mean of the sampling distribution is very small.

To increase the likelihood of making an accurate estimate, the researcher decides to employ interval estimation. Instead of stating an exact number for mean calorie consumption in the population, the researcher identifies a range within which the mean likely occurs. For instance, it is more likely that the actual population parameter falls between 1,974 and 1,976 calories, than it is precisely 1,975 calories. The researcher can be perfectly confident that the parameter falls between zero and 10,000 calories. Since increasing the range reduces the precision, the obvious issue, then, is how wide to make the range. What we know about probability samples and sampling distributions allows us to address this issue.

Since probability sampling was used, we know that the best actual estimate we have of the population parameter is 1,975 calories. Let's say the researcher chooses to make the interval estimate 50 calories wide on either side of this best estimate, concluding that the parameter falls somewhere between 1,925 and 2,025 calories. This interval estimate is graphed in Figure 19.1.

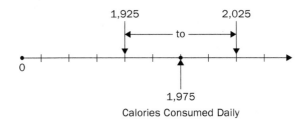

**FIGURE 19.1**  Interval Estimate of Daily Calorie Consumption

Because we know this range comes from a normal distribution, we can convert this interval into standard deviation units. In the last chapter you learned that the standard error (the standard deviation of a sampling distribution) for sample means is computed by dividing the sample standard deviation by the square root of $N - 1$:

$$s.e. = \frac{s}{\sqrt{N-1}}$$

Applied to our example, the standard error is:

$$= \frac{950}{\sqrt{1,499}}$$

$$= \frac{950}{38.71}$$

$$= 24.54$$

With this information the 100-calorie range can be translated into a standard score, using the $z$-score formula:

$$z = \frac{x - \overline{X}}{S}$$

Here is the calculation that translates the 100-calorie range into standard deviation units:

$$z = \frac{1,925 - 1,975}{24.54}$$

$$= \frac{-50}{24.54}$$

$$= -2.04$$

or

$$z = \frac{2,025 - 1,975}{24.54}$$

$$= \frac{50}{24.54}$$

$$= 2.04$$

This translation informs us that the selected estimation interval (1,925–2,025) is 2.04 standard deviation (error) units above and below the mean of 1,975 calories. This result is diagrammed in Figure 19.2.

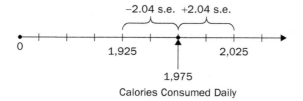

FIGURE 19.2 Standard Score Translation

Figure 19.3 places the information from Figure 19.2 on the sampling distribution, which is a conventional normal curve. Remember that the information from the researcher's sample is one result from the sampling distribution and, of course, we don't know where the sample statistic is exactly located on the sampling distribution. For illustration, Figure 19.3 places the sample results in three possible locations on the sampling distribution. In position A, the interval provides an estimate that is below the population parameter. In position B, the interval estimate is above the parameter. In position C, the sample statistic is at the population parameter. These are all possible alternatives and the researcher does not know which is actually the case.

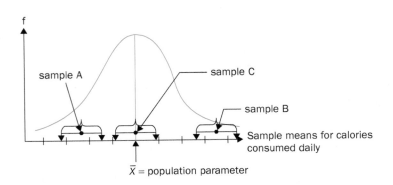

**FIGURE 19.3**  Interval Estimate Locations

As noted earlier, the researcher's interval estimate is certainly a more accurate estimate of the population parameter than a point estimate. We are still left with the question, How accurate is the interval estimate? To answer this question, look at the interval in position C on Figure 19.3. This interval clearly captures the population parameter. Now imagine the interval in position C could slide either to the left or right of its current location. If it moved just a little to the right (or left) the interval would still capture (provide an accurate estimate of) the population parameter. If it then moved a little farther right (or left), it would still provide the same accurate parameter estimate. At some point, however, a sufficiently large shift to the right (or left) would move the position C interval to a position where the range does *not* include the population parameter. At this position the interval would no longer provide an accurate parameter estimate.

At what point do such inaccurate parameter estimates occur? This question can be answered several ways. First, inaccurate parameter estimates occur when the *lower* boundary of the interval range is *above* the centre of the sampling distribution (or, alternatively, when the *upper* boundary of the interval is *below* the parameter). This first answer can be translated into standard error terms. Inaccurate estimates occur when the estimates are 2.04 standard errors above (or below) the mean (parameter). Finally, this answer can be stated in terms of the likelihood of drawing a probability sample that is more than 2.04 standard errors (deviations) above or below the mean.

This final translation is diagrammed in Figure 19.4. Notice that the shaded areas are beyond the ±2.04 standard error range. Using a *z*-score table, we can look up the area

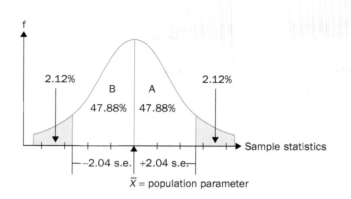

**FIGURE 19.4** Standard Score Translation

between a standard score of 2.04 and the mean. This number is 0.4788, which means that 47.88 per cent of all the samples in the sampling distribution will be between the mean and 2.04 standard units above the mean. This is identified as area A in Figure 19.4. But the same area occurs below the mean, identified as area B. So together, 95.76 per cent of all sample statistics will fall within the region of ±2.04 standard errors from the population parameter. Since the entire distribution includes 100 per cent of the samples, we can conclude that 4.24 per cent of the samples (100 − 95.76) will fall in the shaded area of Figure 19.4. The shaded areas identify the locations where the sample statistic inaccurately estimates the population parameter, since if the centre of the interval is placed beyond this location its legs do no capture the population parameter. Therefore, the researcher can expect her interval estimate to be wrong 4.24 per cent of the time. Stated positively, she can be 95.76 per cent confident that the interval range 1,925–2,025 calories per day accurately captures (estimates) the population parameter.

Here is a summary of the logic of using interval estimates:

- The sample statistic provides the best estimate of the population parameter but, as a point estimate of the parameter, this statistic is probably inaccurate.
- To increase the likelihood of an accurate parameter estimate, the researcher can create an interval estimate that estimates that the parameter occurs within a specific range of scores.
- Using the standard deviation of the sample distribution, the standard error of the sampling distribution can be calculated.
- The standard error can be used to transform the "legs" of the interval estimate into standard scores (standard deviation units).
- Using a z-score table, the standard scores can be translated into percentages of all samples that occur within and outside of the interval estimate range.
- The z-score interpretation can be translated into a statement of the probability that the interval estimate captures the population parameter.

Notice that this summary only uses descriptive information from the sample, and, although the population parameter is not precisely known, it can be estimated with a known probability as occurring within a specified range of scores. The probability of the confidence interval capturing the population parameter is called the **confidence level**.

## Research Application of Confidence Intervals and Levels

The previous section discussed the logic of how confidence intervals and levels can be used to employ sample statistics to estimate population parameters. It is important to appreciate the one point at which this discussion is unrealistic. The researcher began with a statistic that reported the mean daily consumption in the sample to be 1,975 calories. The researcher then proceeded to create a confidence interval of 1,925–2,025 calories *by adding and subtracting 50 to the sample statistic.* Why did the researcher choose 50 to create the interval? Why not 10, or 200? There is no good reason for choosing one of these numbers over another. In our example, the choice of 50 calories for the interval was *entirely arbitrary*. And, because of this arbitrariness, researchers avoid proceeding this way in actual practice.

In practice, researchers simply reverse the order of the procedures described in the previous section. In that presentation, the interval was arbitrarily established and then the probabilities of that interval capturing the population parameter were calculated. Practising researchers proceed the other way around. They first choose the level of confidence they want for their interval estimation, and then proceed to determine (calculate) the interval width associated with the chosen confidence level. The following illustration applies this approach to our previous example.

We begin with the same sample statistics as before. The sample mean is 1,975 calories and the sample standard deviation is 950. And, using the same calculations, the sampling distribution standard error is 24.54. This time, however, the researcher decides she wants to be 80 per cent confident that her confidence interval captures the population parameter. To determine the confidence interval that has the 80 per cent confidence level, here is how she proceeds.

First, let's be clear what the 80 per cent confidence level means. It means that 80 per cent of the sample statistics in the sampling distribution fall within the confidence interval. Stated negatively, the 80 per cent confidence level means that 20 per cent of the sample statistics *fall outside* the confidence interval. Figure 19.5 diagrams this situation.

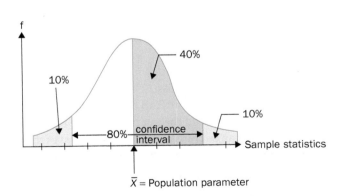

**FIGURE 19.5** Sampling Distribution 80 Per cent Confidence Level

Notice the following points about Figure 19.5. First, the figure presents a sampling distribution, whose mean is the population parameter. Applying the researcher's confidence level, 80 per cent of the samples fall within the confidence interval. The lightly shaded areas identify the 20 per cent of the samples that fall *outside* of the interval range. Since the estimation errors could just as well be too high as too low, the 20 per cent error rate has been divided in two sections. Half (10 per cent) falling in the lightly shaded right tail of the distribution, the other half in the lightly shaded left tail.

Previously we used the *z*-score table to look up a standard score and determine the percentage of observations falling within an area under the normal curve. In our current situation we know the area and can use the *z*-score table to determine the width of the confidence interval. Here's how. Remember what the values in the *z*-score table tell you: namely, the percentage of the observations (sample statistics) that fall between a specific standard score and the mean. In Figure 19.5, the area of interest is more darkly shaded[1] and includes 40 per cent of observations (samples).[2] The *z*-score table uses proportions, rather than percentages, so the number of interest in the table is the *z-score value that is closest to 0.40*. The closest table value to 0.40 is 0.3997, which is the area for a standard score of 1.28.

The standard score of 1.28 tells us that the confidence interval that is associated with an 80 per cent confidence level has legs that are 1.28 standard error (deviation) units wide on either side of the mean. In other words, for our example, the confidence interval that has an 80 per cent confidence level is 1,975 calories (the mean) ±1.28 standard errors wide. While this is the correct result, it is not intuitively appealing. What we need is the range *expressed in calories*, which requires translating the standard error units into calorie units. Converting the standard errors into calories involves multiplying the standard error (1.28) by the number of calories in each standard error (24.54). The result is 32.70 calories, which requires one final translation. The confidence interval associated with an 80 per cent confidence level is 1,975 ±32.70 calories wide, which is 1,942–2,008 calories.[3] In other words, the researcher can be 80 per cent confident that the mean population daily calorie consumption is between 1,942 and 2,008 calories.

To summarize, confidence intervals are actually established by:

- declaring the desired confidence level for the confidence interval;
- determining the area under the normal curve associated with the confidence level;
- using the *z*-table to determine the standard score that most closely approximates the identified normal curve area;
- converting the standard score value into the units of the variable being estimated; and
- adding and subtracting the size of the "legs" of the interval (margin of error) to the mean to create the confidence interval.

---

[1]We could, of course, just as easily have identified the same area to the left of the mean, since the sampling distribution is a symmetrical, normal curve.

[2]Remember the *z*-table only gives areas on one side of the curve, so the area is half the confidence level value (half of 80 per cent = 40 per cent).

[3]This range is rounded off.

## Concluding Points

You are now familiar with the ideas behind point and interval estimation. The remainder of this chapter puts these ideas into action by covering specific estimation techniques aimed at addressing particular research questions. Before turning to the steps in each of these calculation procedures, the following points deserve noting.

### Selecting Confidence Levels

For a given project, researchers could select any confidence level and determine the confidence interval associated with it. Our example selected 80 per cent confidence, but we could have used 40 per cent, 60 per cent, or any other number just as easily. Not many readers, however, would be interested in findings in which the researcher has only 15 per cent confidence (85 per cent uncertainty)! In practice, scientific norms govern the selection of confidence levels. These norms result in researchers typically using either 99 or 99 per cent confidence levels.

### Types of Distributions and Interval Estimates

The examples we provided used a $z$-table to either look up or determine standard scores. The $z$-table is used when the sample size is relatively large which, in practice, means over 100 cases. Using a $z$-table for samples with fewer than 100 cases generates poor estimates. So, when sample sizes are under 100, researchers use a **t-distribution** to estimate means. The $t$-distribution is actually a set of sampling distributions adjusted for different sample sizes. As the sample size increases, $t$-distributions generate results that are essentially similar to $z$-table results.[4] Later chapters demonstrate how $t$-distribution tables are used. For now, the important thing to recognize is that $t$-distributions are used in a similar way to $z$-tables.

Inferential interval estimates can be generated for all kinds of sample statistics. The most common ones are proportions and means, so we will focus on tools for generating interval estimates of population parameters for these two types of sample statistics. You are now ready to put these ideas into action. The following section provides the steps for calculating the confidence intervals and levels for these various inferential procedures.

## STEP 2  Learning the Calculations

The most common single variable estimates of population parameters are proportions (for variables at lower levels of measurement) and means (for variables at higher measurement levels). Here are the steps in making interval estimates for each of these sample statistics using $z$-distributions.

# Confidence Intervals for Population Proportions

Imagine a random sample of 900 students from your university responds to a survey question about whether or not they favour the legalization of marijuana. The results show that

---

[4]The results of $t$-distribution and $z$-distribution coincide when samples approach 120 cases.

60 per cent favour legalization, while 40 per cent oppose. Examining the categories of this variable (*favour, oppose*), it is clear that the variable is at the lower (nominal/ordinal) level of measurement. The sample statistic can be expressed as proportions (0.60 and 0.40) and can be generalized to the population of all students at your university. Here's the procedure to make the interval estimate for the proportion.

1.  First, you need to know the calculation formula. Here it is:

$$CI = P_s \pm z \sqrt{\frac{P(1 - P)}{N}}$$

While this formula may look a little intimidating, it is simply composed of two components. These components are identified in the formula by different colours. The blue component ($P_s$) is sample proportion you are trying to infer to the population. This value comes from your sample and, in the case of our example, is the proportion of the undergraduate sample who favour the legalization of marijuana. In this case, $P_s = 0.60$.

Notice this sample statistic is your best point estimate of the population parameter. The problem, of course, is that such a precise point estimate is likely inaccurate because of sampling error. So this point estimate needs to be broadened into an interval estimate—hence, the second component of the confidence interval estimation formula, the part of the formula in grey.

The grey component of the formula begins with the ± sign. What this tells you is that you are going to add and subtract some amount from the point estimate ($P_s$) to create the interval estimate. In other words, the confidence interval is going to be a range that indicates the population parameter falling between a lower and upper boundary.

2.  Remember that the wider the confidence intervals are, the more likely they are to be correct estimates. The width of the confidence interval is determined by the level of confidence you want to have in the results. By convention, scientific research prefers confidence levels of either 95 per cent or 99 per cent. The confidence level is built into the grey part of the confidence interval formula in terms of the z-score value.

In the formula for computing the confidence interval for proportions the z is always replaced with the value 1.96 if you want the confidence level to be 95 per cent. Alternatively, the z is replaced with a value of 2.58 if you want the confidence level to be 99 per cent. For our purposes let's agree on a 95 per cent confidence level in our estimate.

3.  The remainder of the confidence interval formula (i.e. the part in grey after the z) is the standard error estimate for proportions. Remember that the standard error is the size of the standard deviation of the sampling distribution. For population proportions, the standard error is always the square root of $0.5 \times 0.5$ divided by the sample size. In our example, the sample size ($N$) is 900.

4.  So now let's apply the entire confidence interval formula to our example and see what estimate of the population parameter it provides. We begin with the general formula:

$$CI = P_s \pm z \sqrt{\frac{P(1 - P)}{N}}$$

Now add the sample statistic proportion of students who favour legalization ($P_s = 0.60$), the 95 per cent confidence level z-value ( 1.96), and the numbers under the square root: ($0.5 \times 0.5 = 0.25$) and $N = 900$.

Here's what the completed formula looks like:

$$CI = 0.60 \pm 1.96\sqrt{\frac{0.25}{900}}$$

And now the arithmetic steps to solve the formula:

$$= 0.60 \pm 1.96(\sqrt{2.78})$$
$$= 0.60 \pm 1.96(0.017)$$
$$= 0.60 \pm 0.033$$
$$= 0.567 \text{ and } 0.633$$

The resulting confidence interval estimate is that the population parameter value falls between 0.567 and 0.633.

5.   Finally, let's interpret this calculated result in simple language. Our sample indicated that 60 per cent of our sample of 900 students favours the legalization of marijuana. The confidence interval results are in proportions, but we can return these to percentages in the following conclusion. We are 95 per cent confident that between 56.7 per cent and 63.3 per cent of all students at the university favour the legalization of marijuana.

To summarize, if your sample statistic is a proportion, then use the confidence interval formula to estimate the population parameter interval. The $P_s$ and $N$ values come from your sample. The z values and numerator under the square-root sign are fixed. So entering the values into the formula is quite simple. Then it is just a matter of completing the arithmetic.

# Confidence Intervals for Population Means

Let's imagine that in the same random sample of undergraduates at your university, the mean intelligence of students (as measured by a standardized IQ test) is 110 and that the standard deviation for this variable in the sample is 15.[5] Creating a confidence interval capturing the mean IQ of all students at your university follows a similar procedure to that used for proportions.

1.   First, we begin with the formula. Here it is, in blue and grey ink:

$$CI = \overline{X} \pm z\frac{s}{\sqrt{N-1}}$$

---

[5]Recall that the mean is a measure of central tendency; and the standard deviation, a measure of dispersion for interval/ratio variables. In this example, these statistical values are provided for you. In research situations, these values are computed from the collected evidence—using the statistical tools covered in Chapters 5 and 6.

Notice that the form of this formula is the same as the inferential formula for proportions. It begins with the sample statistic, in this case the mean (in grey). It then creates a range around this sample statistic (in blue). The range is created by multiplying two components. The $z$-score, which like the proportion formula is determined by the confidence level, is one component. The other component is the standard error of the sampling distribution, which in this case is estimated by dividing the standard deviation of the sample statistic ($s$) by the square root of the sample size ($N$) minus 1.

2.  Now let's compute the confidence interval for the 95 per cent confidence level by inserting the sample information into the formula. Here are the steps, beginning with the formula:

$$CI = \bar{X} \pm z \frac{s}{\sqrt{N - 1}}$$

Now we insert the mean IQ score (110), the $z$-value for the 95 per cent confidence level (1.96), the standard deviation value (15), and the sample size ($N = 900$):

$$CI = 110 \pm 1.96 \frac{15}{\sqrt{900 - 1}}$$

And then complete the arithmetic to solve the formula:

$$= 110 \pm 1.96 \frac{15}{\sqrt{899}}$$

$$= 110 \pm 1.96 \frac{15}{29.98}$$

$$= 110 \pm 1.96(0.50)$$

$$= 110 \pm 0.98$$

$$= 110.98 - 109.02$$

The resulting confidence interval estimate is that the population parameter value falls between 110.98 and 109.02.

3.  Finally, we interpret the confidence interval in simple language. In the sample of 900 undergraduate students, the mean (typical) intelligence score was 100. The inference indicates that we can be 95 per cent confident that the mean intelligence of *all students at the university* is between 109 and 111.[6]

To review, whether it is for means or proportions, the calculation of interval estimates uses the same steps. It takes the sample statistic and creates a range by multiplying the $z$-score of the associated confidence level by the standard error of the sampling distribution. In most research situations, IBM SPSS statistics software ("SPSS") is used to compute the confidence intervals. The following step shows you how.

---

[6]The values are rounded from the calculated results.

## STEP 3 Using Computer Software

Confidence intervals, for both proportions and means, offer a way to estimate parameters for populations using sample statistics. As you have seen, confidence intervals provide a way to inform readers of how confident they can be that the variable value in the population is within the stated confidence interval. Generating confidence intervals in SPSS is easy, but it does require you to travel to a new area of the computer program. Here is how to proceed.

# Confidence Intervals for Population Means

To illustrate how to generate confidence intervals for population means, we will use the variable Q22 *Satisfaction with life*, which is an index of a series of questions students were asked about how satisfied they were on multiple dimensions of their life. Response categories ranged from zero (completely unsatisfied with life) to 30 (completely satisfied with life). As always, before you begin your analysis, remember first to set the *no responses* to *missing*!

## Procedure

To compute the confidence interval in SPSS, follow these steps, which are portrayed in Figure 19.6:

1.  Go to Analyze → Descriptive Statistics → Explore.
2.  The Explore window should open. Scroll down on the left-hand side of the variable list to find Q22.
3.  Once highlighted, click on the arrow from the Dependent List box.
4.  a.  Click Statistics to confirm that the Descriptives–Confidence Interval for Mean is checked.
    b.  The default confidence level in SPSS is 95 per cent; however, if you want to generate results at a 99 per cent confidence level, just change it from 95 per cent to 99 per cent.
5.  Click Continue → OK.

In order to interpret your results in the SPSS Viewer window, go to the Descriptives table (see Figure 19.7). (Incidentally, you can safely ignore the stem-and-leaf plot and the boxplot graphs!) Here you will observe that the statistical mean for students' life satisfaction (Q22) is 19.31 (recall that the range was between zero and 30). Directly under the mean is the 95 per cent confidence interval for the Lower Bound calculation, which is 18.96. Below this is the

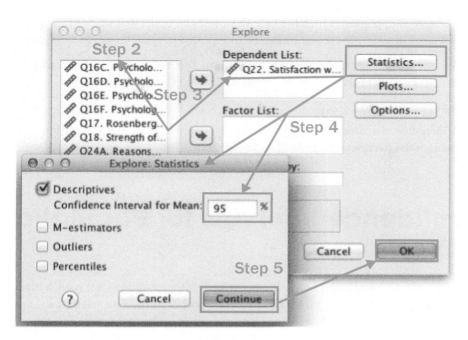

**FIGURE 19.6** Confidence Intervals for Population Means

Source: Reprint Courtesy of International Business Machines Corporation, © International Business Machines Corporation.

**Descriptives**

| | | | Statistic | Std. Error |
|---|---|---|---|---|
| Q22 Q22. Satisfaction with life | Mean | | 19.31 | .175 |
| | 95% Confidence Interval for mean | Lower Bound | 18.96 | |
| | | Upper Bound | 19.65 | |
| | 5% Trimmed Mean | | 19.53 | |
| | Median | | 20.00 | |
| | Variance | | 35.943 | |
| | Std. Deviation | | 5.995 | |
| | Minimum | | 0 | |
| | Maximum | | 30 | |
| | Range | | 30 | |
| | Interquartile Range | | 9 | |
| | Skewness | | −.485 | .071 |
| | Kurtosis | | −.158 | .143 |

**FIGURE 19.7** Confidence Interval Output

Upper Bound computation, which is 19.65. The lower and upper bounds provide the range of the confidence interval. These results can be interpreted in plain language by saying:

> We are 95 per cent confident that the mean life satisfaction of University of Manitoba students is between 18.96 and 19.65.

Or put another way:

> Ninety-five per cent of samples randomly selected from the population will contain the parameter of 19.31 (the sample mean).

# Confidence Intervals for Population Proportions

We begin with bad and good news. First, the bad news: SPSS does not have a direct way to compute confidence intervals for population proportions. The good news, however, is you can use an indirect procedure. This procedure is the same one used for computing confidence intervals for means, with two qualifications and a revision.

The first qualification is that confidence intervals can only be computed for binary variables, which are ones that include only two values or scores (e.g. *male/female* or *yes/no*). Second, in order for the interval to be calculated correctly, the values of the binary variable need to be coded as zero and 1.[7] After the interval estimation procedures are applied to the properly coded binary variable, the output requires one revision. Specifically, you need to substitute the word *proportion* for *mean* on the output (see Figure 19.8).

To illustrate the confidence intervals for proportion, let's use Q7—*In the past year, have you consumed alcohol?* By running a frequency distribution (see Chapter 4 for a review), you can see that 89.7 per cent of students answered *yes* to this question, and 10.3 per cent said *no*. These percentages can be expressed as proportions (yes—0.897; no—0.103). The confidence interval procedures can be used to generalize this sample result to the population of all undergraduate students at our university.

As noted, to generate the confidence interval for a proportion you follow the same steps described under Confidence Intervals for Population Means. However, for our example, you need to use a version of the Q7 variable whose categories have been recoded to zero and 1. The new variable you should select is NQ7, which is a new version of variable Q7 that uses zero for *no* and 1 for *yes*. Now, simply follow the previous steps, but this time select 99 per cent confidence level. The output is contained in Figure 19.8, where you see we have substituted *proportion* for *mean*.

---

[7]See Chapter 4 for a review of how to recode values of a variable.

**Descriptives**

| | | | Statistic | Std. Error |
|---|---|---|---|---|
| nq7 NQ7. In the last year, have you consumed alcohol? | ~~Mean~~ proportion | | .90 | .009 |
| | 99% Confidence Interval for ~~mean~~ proportion | Lower Bound | .87 | |
| | | Upper Bound | .92 | |
| | 5% Trimmed Mean | | .94 | |
| | Median | | 1.00 | |
| | Variance | | .093 | |
| | Std. Deviation | | .304 | |
| | Minimum | | 0 | |
| | Maximum | | 1 | |
| | Range | | 1 | |
| | Interquartile Range | | 0 | |
| | Skewness | | −2.614 | .069 |
| | Kurtosis | | 4.842 | .139 |

**FIGURE 19.8** Output for Confidence Interval for Population Proportions

Recall that the confidence interval provides a range of values in which the population parameter is likely to occur. Notice in the output that the lower and upper bounds are expressed as proportions. The results in the output are interpreted as follows:

> The sample estimate for the proportion of participants who have consumed alcohol in the last year is 0.90. A 99 per cent confidence interval calculated from the sample reveals that the population proportion is between 0.87 and 0.92.

In other words, you can be 99 per cent confident that between 87 per cent and 92 per cent of all undergraduate students have consumed alcohol in the last year.

## STEP 4 Practice

You now have the tools for understanding how confidence intervals and confidence levels are generated for different kinds of research questions. To solidify your understanding you need to practise applying these statistical procedures.

This section provides opportunities to practise what you have learned about interval estimates. The questions focus on hand calculations. For each question:

1. Follow the procedural steps to compute the results.
2. Check your answers, using the Answers section in the back of the text.
3. If your answer is incorrect, consult the Solutions section on the book's website. The Solutions provide a complete step-by-step analysis of how the answers are derived.

After you have completed the next section (Interpreting the Results), return to your calculations or output and *provide complete, written interpretations of each of the statistics you have generated.*

# Set 1: Hand-Calculation Practice Questions

1.  In Chapter 18 you calculated the standard error for a sample of 500 undergraduates, where the mean high-school grade average was 78.8 per cent. Calculate the 95 per cent confidence interval for this estimate. What are the upper and lower limits of this interval? Write a clear statement interpreting this confidence interval.

2.  In Chapter 18 you calculated the standard error for a sample of 1,200 households where the average household debt was $13,577. Calculate the 99 per cent confidence interval for this estimate. How wide is this interval? Interpret this confidence interval in a clear written statement.

3.  a.  In Chapter 18 you calculated the standard error for a sample of 900 university students in which 65 per cent of students reported taking public transit to school at least once a week. Calculate the 95 per cent confidence interval for this estimate. What are the upper and lower limits of this interval? Write a clear statement interpreting this confidence interval.

    b.  In this same survey, 44 per cent of students reported being happy with the variety of food offerings provided by food vendors on campus. Calculate the 99 per cent confidence interval for this estimate. How wide is this interval? Write a clear statement interpreting this confidence interval.

    c.  In this same survey, 8 per cent of respondents indicated that they were gay, lesbian, bisexual, or transgendered. Of these 72 respondents, 35 per cent indicated they had experienced some form of harassment on campus. Calculate the 95 per cent confidence interval for this statistic. What are the upper and lower limits of this interval? Write a clear statement interpreting this confidence interval.

    d.  Also in this survey, 9 per cent of respondents indicated they were international students. Of these 81 international students, 67 per cent reported being satisfied with their university experience in Canada. Calculate the 99 per cent confidence interval for this estimate. How wide is this interval? Write a clear statement interpreting this confidence interval.

    e.  The average monthly rent paid by international students in this survey was $685, with a standard deviation of $99. Calculate the 95 per cent confidence interval for this estimate. What are the upper and lower limits of this interval? Write a clear statement interpreting this confidence interval.

    f.  The mean number of months that international students had been attending university in Canada was 8, with a standard deviation of 3. Calculate the 99 per cent confidence interval for this estimate. How wide is this interval? Write a clear statement interpreting this confidence interval.

4.  In the latest pre-election opinion poll, 21 per cent of people expressed their support for the Green party. Construct a 95 per cent confidence interval, state its limits, and formulate a verbal statement about the limits of the interval when:

    a.  the sample size is 500
    b.  the sample size is 1,000 (size doubled)
    c.  the sample size is 1,500 (size tripled)

    d.   the sample size is 2,000 (size quadrupled)

    e.   Compare the margin of error over the different sample sizes. Is the change in the confidence interval directly proportionate to the change in sample size?

    f.   Considering that costs for carrying out an opinion poll are almost directly proportionate to its sample size, do the gains in the confidence interval precision justify the extra costs once the sample size reaches, say, 1,000?

5.    A representative sample of 215 individuals who took the Medical College Admission Test (MCAT) in the previous year scored, on average, 23.1 points, with a standard deviation of 7.8.

    a.   Compute a 95 per cent and 99 per cent confidence interval.

    b.   What are its limits?

    c.   Formulate a verbal statement about what the confidence interval reveals about the population of individuals who completed the MCAT in the past year.

    d.   How would your 95 per cent confidence interval change if your random sample had only 75 individuals?

6.    In Chapter 18 we mentioned a study involving a probability sample of 228 children between ages six and nine which revealed that children in this age range spend, on average, 3.8 hours a day watching TV.

    a.   What is the 95 per cent confidence interval if the standard deviation is 1.4 hours? State its limits and write a statement about the population of children between the ages of six and nine and TV watching.

    b.   What is the 99 per cent confidence interval if the standard deviation is 2.1 hours? State its limits and write a statement about what it says about the television viewing habits of children between the ages of six and nine.

7.    As mentioned in the previous chapter, the study of 228 children between ages six and nine also revealed that 68 per cent of children owned at least one violent video game.

    a.   Compute the 99.9 per cent confidence interval for this estimate.

    b.   Write down the limits of the interval and write a statement regarding your estimate of the population parameter.

    c.   How does the 99.9 per cent confidence interval change when the sample size decreases to only 35 children? State the limits of the interval and write a verbal statement regarding your estimate of the population parameter.

8.    In Chapter 18, the mean IQ of a probability sample of 250 university undergraduate students was 118, with a standard deviation of 6 points.

    a.   Compute a 99 per cent confidence interval and state its limits.

    b.   Write a verbal statement about the average IQ value of all undergraduate students in that university.

    c.   How would the confidence interval change if we decreased the sample size from 250 to 41? Put your answer into words with respect to the change in the margin of error.

# Set 2: SPSS Practice Questions

1.    Compute and interpret the 95 per cent confidence interval for the mean *Self-esteem* score (Q17) in the Student Health and Well-Being Survey.

2.    Compute and interpret the 95 per cent confidence interval for mean *Depression* level (Q4).

3.  Compute and interpret the 99 per cent confidence interval for average *Strength of religious faith* (Q18).

4.  Compute and interpret the 95 per cent confidence interval for the proportion of respondents that report regularly attending religious services (NQ35). (Remember, to generate a confidence interval for proportions with SPSS, variables must be dichotomous; see Chapter 4 for instructions on recoding).

5.  Compute and interpret the 99 per cent confidence interval for the proportion of respondents that were born in Canada (Q27 recoded as dichotomous nQ27).

6.  Compute and interpret the 99 per cent confidence interval for the proportion of respondents that report having volunteered in the past 6 months (Q19 recoded as dichotomous nQ19).

7.  Compute and interpret the 95 per cent confidence interval for mean *Feelings of positive affect* (Q3A).

8.  Compute and interpret the 99 per cent confidence interval for mean *Personal growth* (Q16C).

9.  Compute and interpret the 99 per cent confidence interval for mean *Reasons for living—one's self or one's future* (Q24B).

10. Recode Q25 as 0 = *male* and 1 = *female* (nQ25). Then compute and interpret the 99 per cent confidence interval for the proportion of respondents that are female.

11. Recode Q9 as 0 = *no* and 1 = *yes* (nQ9). Then compute and interpret the 99 per cent confidence interval for the proportion of respondents that in the past year have consumed soft drugs.

12. Recode Q13B as 0 = *no* and 1 = *yes* (nQ13B). Then compute and interpret the 95 per cent confidence interval for the proportion of respondents that in the past six months have had problems with school.

# STEP 5  Interpreting the Results

As demonstrated in the first section, all interval estimation procedures share a similar logic. The procedures produce four pieces of information. First, they identify a *point estimate*. The point estimate is the sample statistic (e.g. proportion, mean) that is the best estimate of the population parameter for the variable of interest. Second, they identify a *margin of error*. The margin of error is a "plus or minus" factor identifying a likely distance by which the point estimate is mistaken. Third, they state a *confidence interval*. The confidence interval is a range of scores within which the population parameter likely occurs. Confidence intervals are generated by adding and subtracting the value of the margin of error to the point estimate. Fourth, interval estimates produce a *confidence level*, which is a statement of the likelihood that the confidence interval accurately estimates the population parameter.

Even though the specifics of the interval estimation techniques differ, their results are interpreted in similar ways. Two forms of interval interpretive statements are common (Szafran, 2012:284). The first of these statements takes the following form:

> The sample result is _____ [sample statistic identifying the variable(s) of interest] with a margin of error of _____ [margin of error value] and a level of confidence of _____ [confidence level expressed as a percentage].

An example of this kind of interpretation is:

> The sample result shows that 76 per cent favour the current laws on abortion, with a margin of error of ±3.8 per cent and a confidence level of 95 per cent.

This is the most common kind of interpretive statement, since it provides optimal uncertainty reduction by emphasizing the point estimate and then adding some qualifiers.

The second kind of interpretive statement suggests more uncertainty in the estimate, since it emphasizes the confidence levels and intervals and leaves aside the point estimate. The general form of this interpretation is as follows:

> The results indicate we can be _____ [confidence level expressed as a percentage] confident that the actual _____ [statistic and variable under consideration] value lies between _____ [confidence interval].

Applied to the earlier example, this statement would read:

> The results indicate we can be 95 per cent confident that the actual percentage of the population who favour current abortion laws is between 72.2 and 79.8 per cent.

# Chapter Summary

This chapter introduced you to the concepts, principles, and techniques for using statistics from a sample to generate estimates of population parameters. In this chapter you learned that:

- Point estimates are precise numbers that generalize directly from the sample statistic to the population parameter.
- Means and proportions provide unbiased estimates of parameters.
- Precision and accuracy of estimates are inversely related; as one increases, the other decreases.
- Interval estimates provide less precise but more accurate estimates of parameters than point estimates do.
- Interval estimates provide a range of values in which the population parameter is likely to occur.
- Interval estimates are created by adding and subtracting a margin of error to the point estimate.
- Margin of error is the "fudge factor" attached to a point estimate.
- Interval estimates are associated with specific confidence levels.
- Confidence levels express the likelihood that the confidence interval estimate captures the population parameter.
- Confidence intervals and confidence levels are produced from connecting sample statistics to the features of sampling distributions with known characteristics.
- Interval estimation procedures rely on the features of either z-distributions (for larger sample sizes) or t-distributions (for smaller sample sizes).
- Scientific norms dictate that confidence levels of either 95 per cent or 99 per cent are required for sufficient confidence in research results.

- Interval estimation procedures are commonly used for generalizing the proportions or means from single variables.
- Interval estimation procedures can also be used for generalizing differences between proportions or means in different groups.

This chapter began by noting that inferential techniques are of two types. This chapter covered one set of procedures; namely, those used for generating point and interval estimates. The other set of inferential techniques relates to hypothesis testing. The discussion of hypothesis testing procedures begins in the next chapter.

# 20

# Hypothesis Testing

## Overview

The last chapter introduced you to a new region of the statistical maze by discussing the procedures for making point and interval estimates. These procedures constitute one of the two principal categories of inferential statistics. This chapter introduces you to the second category of inferential statistics, hypothesis testing. In this chapter you will learn about:

- null and research hypotheses;
- the logic of hypothesis testing; and
- the calculation and interpretation of chi-square.

At the end of this chapter your conceptual understanding will be greater and your statistical toolkit bigger.

In terms of the statistical selection checklist defining the coordinates of the statistical maze, the contents of this chapter are situated as follows:

## Statistical Selection Checklist

1. Does your problem centre on a descriptive or an inferential issue?
   - ❏ Descriptive
   - ☑ Inferential
2. How many variables are being analyzed simultaneously?
   - ❏ One
   - ☑ Two
   - ❏ Three or more
3. What is the level of measurement of each variable?
   - ☑ Nominal
   - ☑ Ordinal
   - ❏ Interval
   - ❏ Ratio

As always, let's begin with introducing you to key concepts and principles.

## STEP 1 Understanding the Tools

# Hypothesis Testing versus Interval Estimation

Let's begin by clarifying how these two categories of inferential statistics are similar and different. The clear similarity is that both interval estimation and hypothesis testing are inferential procedures; they aim to make generalizations about the situation in the population based on evidence from a probability sample. The procedures, however, are quite different in their approach.

Interval estimation procedures begin with *no prior idea of the value of the population parameter*. These procedures use the sample statistic as the best estimate of the parameter and then determine a range of values (confidence interval) in which the parameter is likely to occur (confidence level). **Hypothesis testing** proceeds differently; it begins with *an assumption (i.e. an educated guess) about what the value of the population parameter is*. Hypothesis testing then proceeds to use an odd form of logic to determine whether or not the predicted population parameter is credible.

There is a wide variety of specific inferential procedures that use the hypothesis testing approach. They all, however, share a similar language and logic. The first part of this section introduces you to the general approach of hypothesis testing. Then we apply this approach in one set of inferential tools called chi-square.

## Kinds of Hypotheses

Chapter 1 stated that the method of science aims to make statements about reality and that the method is based on formulating and testing hypotheses. Hypotheses are tentative statements about the nature of reality; as you probably learned in grade school, hypotheses are "educated guesses." In other words, researchers have some informed (the *educated* part) expectation (the *guess* part) about the nature of some aspect of reality. In hypothesis testing, the hypotheses refer to the population parameters.

The logic of hypothesis testing employs two kinds of hypothesis. The key hypothesis is called a null hypothesis. The **null hypothesis** is a statement about the population parameter that is assumed to be true. The null hypothesis is symbolized as $H_o$. As the name suggests, the null hypothesis often takes the form of stating that no relationship exists between the variables of interest in the population. So, for example, a null hypothesis might state:

> There is no difference between Aboriginal and non-Aboriginal rates of incarceration.

In hypothesis testing, the null hypothesis is contrasted against a research hypothesis.

As you know, researchers prepare before conducting their investigations. They review the existing literature, make preliminary observations, and think hard about the implications of a theory. This preliminary work positions the researcher to devise an informed expectation about what the research study will reveal. The informed expectation that occurs prior to examining the evidence is called a **research hypothesis**, symbolized as $H_1$. Sometimes a research hypotheses is called the **alternative hypothesis** or **substantive hypothesis**. The research hypothesis is derived from the preliminary research activity, and in that sense it is substantive (as opposed to wild speculation), and is an alternative to the null hypothesis. So, for example, based on a researcher's preliminary understanding of the social conditions of Aboriginal people in Canada, her research hypothesis might be:

> Aboriginals will have higher rates of incarceration than non-Aboriginals in Canada.

In short, the null hypothesis makes an assumption about conditions in the population, while the research hypothesis makes a prediction about population parameters based on existing understanding. Shortly, you shall see how these competing hypotheses (null and research) are used in hypothesis testing logic. Before that, however, several points related to these hypotheses need to be emphasized.

First, the null hypothesis is at the *centre of hypothesis testing*, which probably seems a little odd. After all, before collecting evidence, a researcher expends a lot of energy trying to establish her best educated guess about the situation in some field of interest. This effort culminates in the research hypothesis. But, despite this effort, the research hypothesis is not the centre of attention in hypothesis testing. The null hypothesis is. The research hypothesis acts as a fallback idea (alternative hypothesis) if the null hypothesis (the centre of attention) is not supported. We shall see why this is the case shortly.

The null hypothesis is always the *reverse of the research hypothesis*. If a research hypothesis states that there is a particular kind of relationship between the independent and dependent variables, the null hypothesis will state that something other than the expected relationship exists between the variables (i.e. that the expected relationship is not correct). If the research hypothesis states:

> First-year students have a GPA of under 2.5.

The null hypothesis states:

> The GPA of first-year students is equal to or greater than 2.5.

The important point to note here is that the research and null hypotheses must contradict each other. They cannot both be true. Next you will see that the logic of hypothesis testing hinges on the fact that the research and null hypotheses are contradictory. When this is the case, when you reject one of these hypotheses, the other is supported.

## Basic Logic of Hypothesis Testing

Hypothesis testing begins by assuming the null hypothesis is true, which is why it is the centre of attention. Since the research hypothesis is the opposite of the null hypothesis, if the evidence supports the null hypothesis, then the research hypothesis is contradicted. Alternatively, if the null hypothesis is rejected, the research hypothesis gains credibility.

The evidence for rejecting either the null or research hypothesis comes, of course, from the sample. And, as you know, the results from any particular probability sample are prone to sampling error. Therefore, although the sample results may support one hypothesis or the other, you cannot be certain that these sample results reflect the population situation. It is the job of hypothesis testing to help you decide whether or not it is reasonable to generalize your sample findings about the null hypothesis (either acceptance or rejection) to the population.

Because of sampling error, when you generalize findings about the null hypothesis, there is a risk of being wrong. Researchers identify two types of inferential error, tediously labelled Type I and Type II error. Type I error occurs when the null hypothesis is rejected when it is actually true. Type II error occurs when a false null hypothesis is accepted. Table 20.1 summarizes these inferential error conditions.

**TABLE 20.1**  Type I and Type II Error

| Research Decision Based on Sample | Actual Condition in Population | |
|---|---|---|
| | **Null Hypothesis Is True** | **Null Hypothesis Is False** |
| Reject null hypothesis | Type I error | correct decision |
| Accept null hypothesis | correct decision | Type II error |

Let's elaborate on each of these possible errors. First, remember that these are possible errors in inferential statistics, that is, in generalizing from observations in the sample to what occurs in the population. Second, notice that the focus is on generalizing conclusions about the null hypothesis, since this is the centre of attention in hypothesis testing. Third, let's be clear about what each of the errors means for research conclusions. When a true null hypothesis is rejected, Type I error occurs. In other words, in the Type I situation, the researcher concludes that the research hypothesis is true when it is actually false. In general, the character of Type I error is to *claim that something exists (e.g. a relationship between variables, or a difference between groups) when, in fact, it does not.* The conclusions following from Type I errors are false positives.

By contrast, Type II error involves accepting the null hypothesis when it is actually false. In general, the character of Type II error is to *claim that something does not exist when, in fact, it does.* Type II errors generate conclusions that are false negatives.

Both Type I and Type II errors involve inferential mistakes, causing the researcher to incorrectly generalize from her sample results to the population situation. Related to these errors, three things are important to note. First, the researcher knows which type of error is in play in specific research situations. In Table 20.1, the researcher's evidence is listed on the left side of the table. In specific research situations, the researcher knows whether the descriptive results are either rejecting or not rejecting the null hypothesis. If the null hypothesis is rejected (top row), the researcher is either correct (the null hypothesis is false) or incorrect (Type I error). Alternatively, if the researcher does not reject the null hypothesis (lower row), she is either correct (null hypothesis is true) or incorrect (Type II error).

Second, researchers do not treat Type I and Type II errors equivalently. Chapter 1 emphasized that the credibility of the method of science as a way of knowing rests on its superior ability to make claims about what is real. Consequently, scientists are very guarded in making claims about what is real; they are conservative in their claims about what exists. Like all methods of knowing, science is imperfect and mistakes are going to be made. Given the conservatism of the method, scientists try to err on the side of being cautious. They would rather claim something does not exist (when it does), then claim something is real (when it is not). In other words, scientists are more guarded against making Type I errors than Type II errors. The focus on Type I error is expressed in a study's alpha level. **Alpha level (α)** reports the likelihood of rejecting a null hypothesis that is true. Alpha levels vary between 0.00 and 1.00, with conventional criteria being either 0.05 or 0.01. The former criterion means that there is a 5 per cent chance of committing a Type I error; the latter, a 1 per cent chance. In related terms, the alpha level expresses the level of statistical significance of a study.

Third, Type I and Type II errors are inversely related to each other; decreasing the risk of one increases the risk of the other. An analogy is helpful in understanding this point. Imagine a person charged with a crime and brought to trial. In reality, the individual is either guilty or not guilty of the charge. The claim of *guilty* is analogous to a research hypothesis, since this is the prosecutor's best educated guess. *Not guilty* is analogous to a null hypothesis, since the claim is that the charged person is no different than the general population (i.e. is innocent). Using this analogy, the information in Table 20.1 can be recast as Table 20.2.

**TABLE 20.2** Type I and Type II Error Analogy

| Verdict | Reality | |
| --- | --- | --- |
| | **Not Guilty** | **Guilty** |
| Guilty | Type I error | correct decision |
| Not guilty | correct decision | Type II error |

Analogous to the researcher's situation, the jury does not know what really happened; all they can do is make their best judgment based on the evidence at hand. In two situations they are going to make the correct decision (inference). In one, the jury will find the person is guilty when, in fact, they are. Alternatively, the jury may find the person is not guilty when this is actually the case.

By contrast, the jury can make two kinds of errors. They can find a person guilty when they are not, or they may proclaim a guilty person is innocent. The first error is analogous to Type I error, since the null hypothesis (*not guilty*) is being rejected when it is actually true. The second situation is analogous to Type II error, where the null hypothesis is accepted as true when it is not. In the first error type, an innocent person is found guilty; in the second error type, a guilty person is found innocent. Our judicial system assumes the first error (convicting an innocent person) is more egregious than committing the second type (not convicting a guilty person). Since the judicial system takes extensive measures to reduce the chances of convicting an innocent person (e.g. proof beyond a reasonable doubt), it increases the chances of letting guilty persons go free. Of course, both judicial and scientific institutions seek to make correct decisions. But they also recognize their imperfection and seek to guide mistakes in a particular (conservative) direction.

## Steps in Hypothesis Testing

As you shall learn, there are a variety of specific hypothesis testing procedures. They all, however, share the logic of forwarding a null hypothesis and then trying to reject it. The various procedures are also organized around the following set of steps.

1. *State the null and research hypotheses*: The research hypothesis is the best educated guess about the situation; the null hypothesis is its opposite. It is essential that these hypotheses be formulated so that, if evidence supports one, it cannot support the other.

2. *Examine the descriptive evidence*: Remember that hypothesis testing is an inferential procedure; it is concerned with generalization from the sample to the population. Sample statistics, however, are the only actual evidence a researcher has for decisions. So the researcher must consult the evidence and see whether the sample findings (statistics) support either the null hypothesis or the research hypothesis. This investigation can only yield two possible results. The first possibility is that the actual evidence supports the null hypothesis. When this is the case, then the researcher concludes she cannot reject the null hypothesis.

   The alternative possibility is that the sample evidence supports the research hypothesis. When this occurs, the researcher does *not* conclude that the research hypothesis is correct, since these results may be the result of sampling error. But at least the results are encouraging enough to proceed to the next step.

3. *Conduct an appropriate inferential test*: In this and the following chapters, you will learn about specific inferential tests and their applications. However, the outcome of all these tests is the same: *They report the probability of obtaining your sample results if the null hypothesis is actually true.* In other words, inferential tests report the likelihood of the sample statistics occurring because of sampling error when the population parameter is the null hypothesis. In inferential statistics, this probability is called the significance level of the findings.

   Significance level is the probability of rejecting the null hypothesis when it is true. In other words, the significance level (also called alpha) is simply the probability of making a Type I error.

4. *Make a decision about the null hypothesis*: Remember that the method of science is a conservative procedure interested in minimizing Type I error. The norms of science indicate that Type I error should not be greater than 5 per cent. For that reason, researchers typically set the alpha level at either 0.05 or 0.01. The 0.05 indicates that 95 per cent of the time the results will not be due to sampling error, while the 0.01 level indicates the results are probable 99 per cent of the time.

   Using either the 0.05 or 0.01 significance level as their cut-off value for decision making, a researcher draws a conclusion about whether or not to reject the null hypothesis. If the significance level is greater than the established alpha value, the null hypothesis is not rejected. This means that, despite what the sample results show, the researcher is unwilling to accept their credibility. Alternatively, if the significance level is at or smaller than the established alpha level, the researcher rejects the null hypothesis. Rejecting the null hypothesis at this particular level of certainty provides support for the research hypothesis, by indicating that obtaining such a result by chance or fluke alone (i.e. sampling error) is highly improbable.

5. *Draw a conclusion*: Hypothesis testing begins with competing hypotheses—the null hypothesis and research hypothesis. The null hypothesis is the focus of attention in inferential tests. If the researcher determines not to reject the null hypothesis, what is the conclusion? Interestingly, the conclusion is neither that the null hypothesis is accepted nor that the research hypothesis is accepted. Rather, the conclusion is indeterminate. Not rejecting the null hypothesis means that further research is necessary for clarifying the research question under consideration.

Alternatively, if the null hypothesis is rejected, then the conclusion is that the research hypothesis is supported. When hypothesis testing leads to the conclusion that the research hypothesis is supported, the findings are called statistically significant. Statistical significance means that the descriptive findings reported are not likely due to chance (sampling error).

Two points about statistical significance needs emphasizing. First, statistical significance is not the same as substantive significance. Too often the term *statistical significance* is wrongly interpreted as "*importance.*" The declaration of statistical significance is a conclusion based on inferential statistics and, therefore, it only speaks to the likelihood that the reported findings are a fluke. Statistical significance says nothing about the substantive significance of the findings. Substantive significance refers to how important the descriptive findings are. It is quite possible for substantially insignificant findings to be statistically significant, which simply means that trivial findings in a research study may likely reflect trivial results in the population.

Second, statistically significant findings are not certain; they are infallible. The declaration of statistical significance occurs at either the 0.05 or 0.01 significance level, which means that researchers are only 95 or 99 per cent confident in the results. There is always a chance of error.

## Chi-Square

All hypothesis testing inferential procedures begin with competing hypotheses and then follow the set of procedures just outlined for determining whether the findings warrant rejecting the null hypothesis at a specified significance level. The third step in the hypothesis testing procedure is to conduct an appropriate inferential test. There are several kinds of inferential tests; chi-square (symbolized as $\chi^2$) is a common one and will let you see the inferential testing steps in action.

Conducting an appropriate inferential test means using a procedure designed for the research situation under consideration. Chi-square is an appropriate inferential test under the following conditions:

- The research hypothesis specifies a relationship between at least two variables. In other words, the relationship is bivariate or multivariate.
- The variables are at lower (nominal or ordinal) levels of measurement.

Taken together, these two conditions indicate the chi-square is an appropriate inferential test of relationships expressed in tables. Chapters 10 and 13 introduced you to descriptive statistical procedures for reporting on relationships expressed in tables. Chi-square provides a test of whether these descriptive results are statistically significant. Here are the basic ideas for conducting a chi-square test.

### *State the Null and Research Hypotheses*

The research hypothesis is the expected relationship between the independent and dependent variables. For example, for the variables *Sex* and *Attitudes toward capital punishment*, the research hypothesis might be:

Men and women will have different attitudes toward capital punishment.

The null hypothesis is the opposite of the research hypothesis, so it would state:

There are no differences between men and women in their attitudes toward capital punishment.

### Examine the Descriptive Evidence

The descriptive evidence is that collected by, or provided to, the researcher. Imagine the data set for our example yielded the following table of evidence related to the research question.

**TABLE 20.3** Observed Frequencies

| Attitudes toward Capital Punishment | Sex | | |
| --- | --- | --- | --- |
| | Men | Women | Total |
| Favour | 150 | 70 | 220 |
| Not favour | 60 | 180 | 240 |
| Total | 210 | 250 | 460 |

An analysis of this evidence reveals the following. First, the form of the relationship is that men have more favourable attitudes toward capital punishment than women. Second, the strength of this relationship, as indicated by lambda, is 0.41. In other words, knowledge of a respondent's sex decreases the error in predicting attitudes toward capital punishment by 40 per cent.[1] In short, the evidence from this sample clearly favours the research hypothesis. It appears as if there are real sex differences in capital punishment attitudes.

### Conduct an Appropriate Inferential Test

The finding that there are actual sex differences in capital punishment attitudes is a result that describes the evidence in the sample. Lambda is a sample statistic. It is possible, of course, that the 460 persons in this sample are atypical of the population. Inferential statistics allow us to test this likelihood. As noted previously, chi-square is an appropriate inferential test for tabular evidence. The next section will provide you with detailed steps for conducting a chi-square test, but the basic ideas are the following.

The chi-square test begins by creating a table of expected frequencies. This table of expected frequencies is compared to the table of observed frequencies. The observed frequencies table is the evidence that the research has analyzed (observed). In our example, lambda was calculated on the data in Table 20.3. The expected frequencies table is one that shows what the evidence would look like *if the null hypothesis was true*. In our example, an expected frequencies table would report how the 460 cases would be displayed if there was no relationship between *Sex* and *Attitudes toward capital punishment*. The next section will show you how to create a table of expected frequencies, but the expected frequencies for our example are presented in Table 20.4.

The next step in chi-square involves computing how different the table of expected frequencies is from the table of observed frequencies. These computation steps are detailed in the next section. This comparison determines how different the actual evidence (observed frequencies) is from evidence that would favour the null hypothesis (expected frequencies). The computed result of this comparison is the chi-square value. For our example, the chi-square result is 88.3.

---

[1]For a review of the computation and interpretation of form and strength in tables, see chapters 10 and 13.

**TABLE 20.4** Expected Frequencies

| Attitudes toward Capital Punishment | Sex | | |
| --- | --- | --- | --- |
| | Men | Women | Total |
| Favour | 100.43 | 119.57 | 220 |
| Not favour | 109.57 | 130.43 | 240 |
| Total | 210 | 250 | 460 |

### Make a Decision about the Null Hypothesis

Remember that inferential tests centre on the null hypothesis and its potential rejection. Making a decision about the null hypothesis involves comparing the computed chi-square value to a value in a chi-square table. The chi-square table provides a list of critical values. A **critical value** is the minimum size that the inferential test value has to be in order to reject the null hypothesis. There are different critical values for different levels of significance. So, for example, the critical value for the 0.01 significance level will be larger than that for the 0.05 significance level. This is because the test for being 99 per cent confident in your results is more demanding than the test of being 95 per cent confident. If the computed chi-square value is *greater than* the relevant critical value, then the researcher rejects the null hypothesis at the selected confidence level. In our example, the critical values of chi-square are 3.841 for the 0.05 significance level and 6.635 for the 0.01 significance level. Our computed chi-square value is 88.3.

### Draw a Conclusion

Hypothesis testing begins with competing hypotheses—the research and the null hypothesis. The goal of such testing is to conclude which is more credible in the population. For our example, comparing the computed value of chi-square 88.3 to the critical values (3.841 or 6.635) lets us draw a clear conclusion. Since the computed value is much higher than the critical value, we can be at least 99 per cent confident in rejecting the null hypothesis that there are no sex differences in attitudes toward capital punishment. This rejection of the null hypothesis lends support for the research hypothesis that there are sex differences in capital punishment attitudes. Stated in other words, the substantive findings of form and strength are statistically significant at the 0.01 level.

This introduction shows how a particular inferential test (chi-square) follows the standard template used by all inferential tests. In our example the chi-square components were provided. The next section shows you how the components of chi-square procedures are generated.

## STEP 2 Learning the Calculations

# Calculating Chi-Square

You are now familiar with the general concepts underlying the computation and use of the chi-square inferential test. Let's work through a detailed example that shows you the execution of the steps in chi-square computation. Remember that chi-square is an inferential test appropriate for testing the bivariate relationship between variables at low (nominal/ordinal) levels of measurement. In other words, the descriptive sample evidence being generalized to the population through chi-square testing is in the form of tables.

The example we will use to illustrate the chi-square computation steps involves the connection between *Motivation for marriage* (the independent variable) and *Marital happiness* (the dependent variable). Evidence was gathered from 40 married individuals (the sample size, *N*). Each person was asked to rate whether their principal reason for marrying was *romantic love* or *practicality*. They were also asked to reported whether their current level of *Marital happiness* was *low* or *high*. The evidence connecting their responses to these two variables is displayed in Table 20.5.

**TABLE 20.5**  Marriage Motivation and Marital Happiness

| Marital Happiness | Marriage Motivation | | |
| --- | --- | --- | --- |
| | Romantic Love | Practicality | Total |
| Low | 6 | 22 | 30 |
| High | 10 | 2 | 10 |
| Total | 16 | 24 | 40 |

The researcher collecting the evidence was a romantic who hypothesized that persons motivated by romantic love experience higher levels of marital happiness than those motivated by other reasons. Here are the steps in testing this hypothesis using chi-square.

## 1. State the Null and Research Hypotheses

The research hypothesis is the researcher's informed expectation about the relationship between the variables under consideration. In this case, the *research hypothesis* is:

> Persons motivated by romantic love will experience higher levels of marital happiness than those motivated by practicality.

The null hypothesis is the reverse of the research hypothesis; these competing hypotheses must contradict each other. Therefore, the *null hypothesis* is:

> Persons motivated by romantic love will experience marital happiness levels equal to or lower than those motivated by practicality.

A couple of points to note about these hypotheses. First, the evidence will support one of these hypotheses or the other, not both or neither. Second, the research hypothesis specifies the direction of the expected difference.[2] This means that a *one-tailed test* is in order.

---

[2] A non-directional hypothesis would simply state that those who marry for romantic reasons will experience different marital happiness levels than those who marry for practical reasons—without specifying the kind of difference expected.

## 2. Examine the Descriptive Evidence

Chi-square is an inferential test; it informs us whether the sample evidence can reasonably be generalized to the population. In this case, the sample evidence is a description of whether, among the 40 respondents in this study, those motivated by romantic love experience higher marital happiness than those marrying for practical reasons. So, before conducting the chi-square test, it is helpful to know what the actual relationship is in the table—since this is the connection we aim to generalize from the 40 respondents to the population.

The form and strength of the relationship in Table 20.5 are as follows.[3] Regarding form, the evidence shows that those who marry for practical reasons experience lower levels of marital happiness than those whose marriage is motivated by romantic love. The strength of the relationship, as measured by lambda, is 0.20. This means that the errors in predicting marital happiness are reduced by 20 per cent when the reason for marriage is known. In short, there is a relationship between *Marriage motivation* and *Happiness* among the 40 persons included in the sample, and this relationship supports the research hypothesis.

The question becomes, Can this reported relationship be reasonably generalized from the sample to the population of married persons?

## 3. Conducting the Chi-Square Test

Chi-square is the appropriate inferential test for evidence like that contained in Table 20.5. We will demonstrate the application of chi-square to this table by identifying each procedure and then performing it. Each application will be added to a table of evidence that eventually produces the chi-square statistic.

a. List the observed frequencies: The observed frequencies are the joint frequencies included in the actual evidence. In other words, the observed frequencies are the numbers *inside the table*. In a chi-square calculation, the observed frequencies are identified as $f_o$. Here they are:

**TABLE 20.6** Observed Frequencies

| Observed Frequencies |
| --- |
| 6 |
| 10 |
| 22 |
| 2 |

Notice that the column of observed frequencies includes each of the four joint frequencies listed inside Table 20.5. Also note, the marginal frequencies (i.e. those included on the *outside* of Table 20.5) are not included in the chi-square calculation, since they do not represent the connection between the variables.

b. List the expected frequencies: The observed frequencies are the actual evidence of the relationship between the variables. Expected frequencies are hypothetical evidence, not actual evidence. Expected frequencies are the numbers that we would expect to

[3] See chapters 10 and 13 for a review of how these conclusions were generated.

find in the table, *if there was no relationship between the variables*. In other words, expected frequencies identify what the evidence would look like *if the null hypothesis was supported*. Expected frequencies are identified as $f_e$.

There is an expected frequency for each of the cells inside the table. Expected frequencies are generated through application of the following two-step rule: (1) for a particular cell, multiply the *row frequency* by the *column frequency*; (2) divide this product by the *sample size* ($N$). For example, the upper-left cell is 6, which is the observed frequency. The expected frequency for this cell is computed by applying the two-step rule as follows:

$$(1)\ 30 \times 16 = 480$$
$$(2)\ 480 \div 40 = 12$$

So, for the upper-left cell, the expected frequency is 12. Repeating this procedure for the other three cells (observed frequencies) yields the following column of evidence:

**TABLE 20.7**  Observed and Expected Frequencies

| Observed Frequencies $f_o$ | Expected Frequencies $f_e$ |
|:---:|:---:|
| 6 | 12 |
| 10 | 4 |
| 22 | 18 |
| 2 | 6 |

c.  Compute the difference between the observed frequencies and expected frequencies: The key idea of chi-square is to examine how different the actual evidence (i.e. the observed frequencies) is from what the evidence would look like if the null hypothesis (i.e. expected frequencies) was supported. For this reason each expected frequency value is subtracted from the associated observed frequency. So we add another column to the table, as follows:

**TABLE 20.8**  Difference between Observed and Expected Frequencies

| Observed Frequencies $f_e$ | Expected Frequencies $f_o$ | Difference between $f_o$ and $f_e$ $(f_o - f_e)$ |
|:---:|:---:|:---:|
| 6 | 12 | − 6 |
| 10 | 4 | + 6 |
| 22 | 18 | + 4 |
| 2 | 6 | − 4 |

d.  Square the differences between $f_o$ and $f_e$: In order to know how different the actual evidence is from the null hypothesis, you need to compute the total difference across all cells. As you see from the previous step, this is currently a problem since there are as many negative values as positives ones. Without adjustment, the total difference between observed and expected frequencies would always be zero, this is not instructive. This issue is solved by squaring each value from the previous step. Squaring either positive or negative values results in positive values. We end up with a column list of the squared differences between observed and expected frequencies, as follows:

**TABLE 20.9** Squared Differences

| Observed Frequencies $f_o$ | Expected Frequencies $f_e$ | Difference between $f_o$ and $f_e$ $(f_o - f_e)$ | Squared Differences $(f_o - f_e)^2$ |
|---|---|---|---|
| 6 | 12 | − 6 | 36 |
| 10 | 4 | + 6 | 36 |
| 22 | 18 | + 4 | 16 |
| 2 | 6 | − 4 | 16 |

e. Divide each of the squared differences [$(f_o - f_e)^2$] by its expected frequency ($f_e$): this step standardizes each of the differences. So, for the first squared difference (36), standardized adjustment would divide this number by 12, resulting in 3. The new column in our table provides the computation for all of the squared differences:

**TABLE 20.10** Standardized Squared Differences

| Observed Frequencies $f_o$ | Expected Frequencies $f_e$ | Difference between $f_o$ and $f_e$ $(f_o - f_e)$ | Squared Differences $(f_o - f_e)^2$ | Standardized Squared Differences $(f_o - f_e)^2 \div f_e$ |
|---|---|---|---|---|
| 6 | 12 | − 6 | 36 | 3 |
| 10 | 4 | + 6 | 36 | 9 |
| 22 | 18 | + 4 | 16 | 0.89 |
| 2 | 6 | − 4 | 16 | 2.67 |

f. Total the standardized squared differences: Here is the complete formula for chi-square.

$$\chi^2(\text{obtained}) = \sum \frac{(f_o - f_e)^2}{f_e}$$

Notice that you have completed all the calculation operations except one, which is related to the summation sign ($\Sigma$). Remember this sign tells you to add up the results of the operations following the sign. In other words, total the last column in the table. This result will be chi-square. Let's perform the operation.

g. Determine the critical chi-square value: The chi-square value is a sophisticated calculation of how different the actual evidence is from what the evidence would be like if the null hypothesis was true. Interpreting the computed chi-square result requires comparing this value to a critical value in a chi-square table. The table of critical chi-square values is found in Appendix A. In order to use this table a few points need clarification.

**TABLE 20.11**  Chi-Square

| Observed Frequencies $f_o$ | Expected Frequencies $f_e$ | Difference between $f_o$ and $f_e$ ($f_o - f_e$) | Squared Differences $(f_o - f_e)^2$ | Standardized Squared Differences $(f_o - f_e)^2 \div f_e$ |
|---|---|---|---|---|
| 6 | 12 | − 6 | 36 | 3 |
| 10 | 4 | + 6 | 36 | 9 |
| 22 | 18 | + 4 | 16 | 0.89 |
| 2 | 6 | − 4 | 16 | 2.67 |
| | | | Chi-square $\chi^2 =$ | 15.56 |

First, the table contains critical values of chi-square. Critical values identify the cut-off point in the sampling distribution determining whether the null hypothesis is rejected or accepted. For a null hypothesis to be rejected, *the computed value needs to be as large as or larger than the critical value.* In short, rejecting the null hypothesis requires the computed chi-square be *at least as large* as the table value. Our computed value is 15.56. Notice that the chi-square table includes lots of columns and rows. You need to understand the columns and rows in order to select the correct critical value.

The columns in the table let you select a level of significance (alpha). By convention, the usual levels are either 0.05 or 0.01. Let's choose an alpha of 0.05. This means that we are willing to accept a 5 per cent chance that our generalization from sample results to the population are wrong.

The rows in the table are identified as *df*, which stands for degrees of freedom. **Degrees of freedom** is a complicated idea referring to the number of pieces of information that can vary independently; it tells you how much data was used to calculate a particular statistic. For practical purposes, all you need be concerned about is which df (i.e. which row) to use in the chi-square table. For any chi-square table, the degrees of freedom (df) is calculated by the following computation: (rows − 1) (columns − 1). In other words, look at the original table (Table 20.5), subtract 1 from the number of rows *inside the table*, subtract 1 from the number of columns *inside the table*, and multiply. Table 20.5 includes 2 rows and 2 columns, so:

$$df = (2 - 1)(2 - 1) = 1 \times 1 = 1$$

Our example includes one degree of freedom.

You can now use the critical values of the chi-square table (Appendix A) to determine the critical value for our example. What critical value inside the chi-square table is at the intersection of the 0.05 level of significance column and the degrees of freedom (*df = 1*) row? The answer is 3.841.

## 4. Make a Decision about the Null Hypothesis

To recap, you have computed the chi-square value for the table (steps 3a to 3f). Determine the appropriate critical value for the table (3g), which tells you how large the chi-square needs to be in order to reject the null hypothesis. In our example, the chi-square value is 15.56 and the critical value is 3.841. Since the computed value is much larger than the critical value, we are confident in rejecting the null hypothesis.

## 5. Draw a Conclusion

Our guiding hypothesis stated that:

> Persons motivated by romantic love will experience higher levels of marital happiness than those motivated by practicality.

The alternative null hypothesis stated that:

> Persons motivated by romantic love will experience marital happiness levels equal to or lower than those motivated by practicality.

Our evidence indicates we can reject the null hypothesis at the 0.05 alpha level. Therefore, our conclusion is that we are 95 per cent confident that the descriptive findings (step 2) can be generalized to the population. In short, the results are statistically significant.

When described in detail, the steps in computing and interpreting chi-square seem drawn out. The actual formula (see step 3f) is straightforward and, with some practice, you will find the computation and interpretation routine. It is even easier, however, to have the computer conduct the calculations.

## STEP 3 Using Computer Software

# Running a Chi-Square Test

This section illustrates how to run a chi-square test of significance in IBM SPSS statistics software ("SPSS"), which is one of many different types of inferential statistics.

## Procedure

1. Select Analyze from the menu bar.
2. Select Descriptive Statistics → Crosstabs.
   a. Place the independent variable Q25 (*What is your sex*) into the Column box.
   b. Place the dependent variable Q9 (*In the past year, have you consumed soft drugs*) into the Row box.
3. Click Cells → Column percentages → Continue.
4. Click Statistics → Chi-square → Continue → OK.

This procedure creates a table called Chi-Square Tests in the Output window (See Figure 20.1).

To interpret your results, you only need to pay attention to the first row of the Chi-Square Tests table (the Pearson Chi-Square row). In this row, the Value column provides the calculated chi-square value (e.g. value 4.52), which is a measure of how different the observed and expected frequency tables are. The second column, df, represents the number of degrees of

**Chi-Square Tests**

| | Value | df | Asymp. Sig. (2-sided) | Exact Sig. (2-sided) | Exact Sig. (1-sided) |
|---|---|---|---|---|---|
| Pearson Chi-Square | 4.520[a] | 1 | .034 | | |
| Continuity Correction[b] | 4.261 | 1 | .039 | | |
| Likelihood Ration | 4.495 | 1 | .034 | | |
| Fisher's Exact Test | | | | .036 | .020 |
| Linear-by-Linear Association | 4.516 | 1 | .034 | | |
| N of Valid Cases | 1223 | | | | |

a. 0 cells (0.0%) have expected count less than 5. The minimum expected count is 161.80.
b. Computed only for a 2x2 table

**FIGURE 20.1**  Chi-Square Output

freedom.[4] The third column in this row ("Asymp. Sig. 2-sided") provides you with the statistical significance value (0.034). Note that because an actual significance value is given, you do not have to look up the critical value in a table. You simply need to concentrate on an acceptable level of significance ($p < 0.05$, $p < 0.01$, $p < 0.001$). In the above example, a significance value of .034 is greater than both the $p < .01$ and $p < 0.001$ levels, but is less than the $p < 0.05$ level. As such, we can interpret our results as follows: With a chi-square value of 4.52, we reject the null hypothesis (at the $p < 0.05$ level with 2 degrees of freedom) that there is no relationship between respondent sex and consumption of soft drugs in the past 6 months. With p =0.034, the probability of making a Type I error (false rejection error) in this example is 3.4 per cent.

## STEP 4 Practice

You now have the general model of how hypothesis testing inferential techniques work and how they can be applied to evidence in tables through chi-square. To solidify your understanding you need to practise applying the statistical and software procedures.

This section provides you with opportunities to practise what you have learned about hypothesis testing ideas and chi-square application. The first set of questions encourages conceptual understanding and uses hand calculations. The second set uses the SPSS procedures. For each set of questions:

1. Follow the procedural steps (Set 1) or software application (Set 2).
2. Check your answers, using the Answer Key in the back of the text.
3. If your answer is incorrect, consult the Solutions section on the book's website. The Solutions provide a complete step-by-step analysis of how the answers are derived.

After you have completed the next section (Interpreting the Results), return to your calculations or output and *provide complete, written interpretations of each of the statistics you have generated.*

---

[4]Hint: By following the above procedures, SPSS has calculated the degrees of freedom and the overall chi-square formulas you have learned in this chapter. If you want SPSS to also calculate the expected frequencies, after you have inserted your independent and dependent variables click Cells → Expected frequencies → Continue.

# Set 1: Conceptual Understanding and Hand-Calculation Practice Questions

1. State the null hypothesis for each of the following research hypotheses:
   a. There is a positive relationship between daily caloric intake and BMI.
   b. There is a negative relationship between female literacy rates and birth rates in the developing world.
   c. Americans are more likely to own a gun than are Canadians.
   d. The monetary return on years of education obtained is less for visible minority Canadians than for Caucasians.
   e. The life expectancy for Aboriginal Canadians is less than for other Canadians.
   f. Wealth inequality is rising in Canada.
   g. University students consume more alcohol per capita than the general population.
   h. Public transit usage rates increase as gas prices increase.
   i. Divorce rates are higher among dual-earner marriages than single-earner marriages.
   j. Home sales decrease as interest rates increase.
   k. There is a significant difference between males and females in the mean number of Facebook friends.
   l. There is an association between early childhood nutrition and school readiness at age five.

2. State the research hypothesis for each of the following null hypotheses:
   a. There is no relationship between parental educational attainment and the educational attainment of their children.
   b. There is no relationship between one's socio-economic position and one's health status.
   c. There is no relationship between sex and personal income.
   d. Affluent people are no more likely to enjoy more disability-free years than poor people.
   e. Women are no more likely to perpetrate violence against their own children than are men.
   f. There is no difference between Canadians and Americans in their attitude toward beavers.
   g. There is no relationship between hours spent partying per week and one's GPA.
   h. Married people are no more likely to experience optimal mental well-being than are single people.
   i. Divorced older women are no more likely to live in poverty than are their male counterparts.
   j. Couples with small children are no more likely to experience higher relationship strain than are couples with older children.
   k. There is no difference between men and women in their mathematical skills.
   l. There is no relationship between the number of hours you spend studying this textbook and your performance in your research methods course.

3.  Using Table 20.12, answer the following questions.

**TABLE 20.12**  Nationality and Gun Ownership

|  | Nationality | |
| --- | --- | --- |
| **Own Gun** | **Canadian** | **American** |
| Yes | 75 | 135 |
| No | 153 | 97 |

a.  What are the expected cell frequencies?
b.  What is the obtained chi-square value?
c.  How many degrees of freedom are there?
d.  Does the obtained chi-square exceed the critical value?
e.  Interpret what these findings mean in terms of the null and research hypotheses.

**TABLE 20.13**  Residence and Political Affiliation

|  | Residence | | |
| --- | --- | --- | --- |
| **Political Affiliation** | **Rural** | **Suburban** | **Urban** |
| NDP | 22 | 35 | 39 |
| Liberal | 17 | 39 | 47 |
| Conservative | 42 | 32 | 24 |

4.  Using Table 20.13, answer the following questions.
    a.  Calculate chi-square and compare it to the relevant critical value.
    b.  Interpret what your results mean in terms of the null and research hypotheses.

5.  In Chapter 16 we used the elaboration model to further examine the relationship between an individual's marital status and their mental well-being. After controlling for sex, however, we found that the original relationship holds for men, but not for women. Use Table 20.14 to test if this relationship in the male subsample is statistically significant at the 0.05 alpha level.

**TABLE 20.14**  Marital Status and Mental Well-Being

|  | Marital Status | |
| --- | --- | --- |
| **Mental Well-Being** | **Married** | **Single** |
| Good | 6 (75%) | 3 (37.5%) |
| Poor | 2 (25%) | 5 (62.5%) |

a.  Compute the expected frequencies.
b.  Set up a computational table and calculate the value of chi-square (obtained).
c.  Find out the number of degrees of freedom for this table and find the critical chi-square value.
d.  Write a conclusion about your results.

6. Use Table 20.15 to repeat the test of significance in the previous exercise for the relationship between the marital status of men and their mental well-being, only this time with a larger sample.

   Note: The proportions, as you can see from the table percentages in parentheses, remain the same. The only thing that has changed is the size of the sample.

   **TABLE 20.15** Marital Status and Mental Well-Being—Larger Sample

   | Mental Well-Being | Marital Status | |
   |---|---|---|
   | | Married | Single |
   | Good | 24 (75%) | 12 (37.5%) |
   | Poor | 8 (25%) | 20 (62.5%) |

   a. Compute the expected frequencies.
   b. Set up a computational table and calculate the value of chi-square (obtained).
   c. Find out the number of degrees of freedom for this table and find the critical chi-square value.
   d. Write a conclusion about your results.

7. In Chapter 10 we used the percentage difference technique to determine whether there was a relationship between a country's health spending per capita and its infant mortality rate. Use Table 20.16 below to test if the relationship is statistically significant at the 99 per cent confidence level.

   **TABLE 20.16** Infant Mortality and Health-Care Spending per Capita

   | Infant Mortality Rate | Health-Care Spending per Capita | | |
   |---|---|---|---|
   | | Low | Medium | High |
   | Low | 2 | 10 | 16 |
   | Medium | 13 | 18 | 8 |
   | High | 26 | 6 | 1 |

   a. Before you proceed to the significance testing itself, state your research hypothesis. (Hint: you will need to first compute table percentages and inspect the conditional distribution of the dependent variable, in order to be able to comment on the direction of the relationship. You can refresh your memory of this technique in Chapter 10.)
   b. Compute the expected frequencies.
   c. Set up a computational table and calculate the value of chi-square (obtained).
   d. Calculate the number of degrees of freedom for this table and find the critical chi-square value.
   e. Write a conclusion about your results.

8. In Chapter 10 we introduced the concept of health gradient, exemplified by a hypothetical example of a relationship between a person's social class and their self-rated health. Is the relationship between the two variables statistically significant at a 99.9 per cent level of significance according to the data in Table 20.17?

**TABLE 20.17** Self-Rated Health and Social Class

| Self-Rated Health | Social Class | | |
| --- | --- | --- | --- |
| | Low | Medium | High |
| Poor | 28 | 18 | 7 |
| Fair | 54 | 32 | 15 |
| Good | 32 | 45 | 28 |
| Excellent | 13 | 21 | 21 |

a. State the research hypothesis first. (**Hint:** Compute table frequencies and inspect conditional distributions of the DV to find the direction of the relationship between the two variables. Revisit Chapter 10 for more help.)
b. Compute the expected frequencies.
c. Set up a computational table and calculate the value of chi-square (obtained).
d. Calculate the number of degrees of freedom for this table and find the critical chi-square value.
e. Write a conclusion about your results.

9. Is there an association between level of educational attainment and maternal education? Using Table 20.18, answer the following questions.

**TABLE 20.18** Educational Attainment and Maternal Education

| Education | Mother's Education | | | |
| --- | --- | --- | --- | --- |
| | < High School | High School | Undergraduate | Graduate/ Professional |
| < High school | 14 | 12 | 9 | 5 |
| High school | 25 | 19 | 21 | 22 |
| Undergraduate | 15 | 15 | 24 | 27 |
| Graduate / Professional | 8 | 11 | 16 | 19 |

a. Calculate chi-square and compare it to the relevant critical value ($\alpha = 0.05$) in Appendix A.
b. Interpret what your results mean in terms of the null and research hypotheses.

# Set 2: SPSS Practice Questions

Use chi-square to assess the relationship between the following pairs of variables from the Student Health and Well-Being Survey. For each table also request the relevant PRE statistic (lambda or gamma) from the Statistics tab and column percentages from the Cells tab.

1. Independent variable Q25 (*Sex*); dependent variable Q19 (*Volunteering*).
   a. State the research and null hypotheses.
   b. Interpret what the obtained results mean in terms of the null and research hypotheses.

2. Independent variable Q2 (*Depressive symptoms*); dependent variable Q24 (*Reasons for living*)
   a. State the research and null hypotheses.
   b. Interpret what the obtained results mean in terms of the null and research hypotheses.

3. Independent variable nQ1D (*Satisfaction with mental health*); dependent variable Q13B (*Problems with school*)
   a. State the research and null hypotheses.
   b. Interpret what the obtained results mean in terms of the null and research hypotheses.

4. Independent variable nQ30 (*Social class*); dependent variable nNQ35 (*Frequency of religious service attendance*)
   a. State the research and null hypotheses.
   b. Interpret what the obtained results mean in terms of the null and research hypotheses.

5. State the research and null hypotheses for each of the following pairs of variables. Then compute the appropriate PRE measure of association and the chi-square test of statistical significance. Interpret the results.
   a. Independent variable (Q23C) *How many times in your life were you called names or insulted*; dependent variable (nQ17) *Self-esteem* (recoded in chapter 4 as *low*, *medium*, and *high*)
   b. Independent variable (Q23A) *How many times in your life were you treated with less courtesy than other people*; dependent variable (Q21F) *Wishing you had more people in your life that enjoyed the same things*
   c. Independent variable (Q9) *Have you consumed drugs in the past year*; dependent variable (Q13D) *Have you been excessively partying in the past 6 months*
   d. Independent variable (Q11) *Did your parents have a history of mental illness while you were growing up* (recode this variable as nQ11 with 0 = *no* and 1 = *yes*); dependent variable (Q19) *Have you volunteered in the past six months*

## STEP 5 Interpreting the Results

# Interpretations of Hypothesis Testing

In general, the interpretation of hypothesis testing results needs to speak to the following points. First, the interpretation needs to state whether or not the null hypothesis was rejected, including the significance level. Then it needs to state the implication of this acceptance or rejection for the research hypothesis. Finally, the interpretation should combine the inferential test results with the substantive findings to provide a statement of confidence.

Here are the general forms of the interpretations of hypothesis testing results for the alternative null hypothesis results:

Null Hypothesis Rejected: The results indicate that the null hypothesis of _____ [null hypothesis] is rejected at the _____ [either 0.05 or 0.01] level of significance.

This result supports the hypothesis that _____ [research hypothesis]. These statistically significant results suggest that we are _____ [confidence level] per cent confident that _____ [substantive findings of the study].

Null Hypothesis Not Rejected: The results indicated that the null hypothesis of _____ [null hypothesis] cannot be rejected at the _____ [either 0.05 or 0.01] level of significance. This result indicates that further research is necessary to confirm statistically significant results.

Step 1 provided an example of chi-square findings related to the proposed relationship between *Sex* and Attitudes Toward Capital Punishment. Using the template, here is how these results are interpreted:

The results indicate that the null hypothesis of no sex differences in attitudes toward capital punishment is rejected at the 0.01 level of significance. This result supports the hypothesis that there are sex differences in capital punishment attitudes. These statistically significant results suggest that we are 99 per cent confident that men favour capital punishment more than women and that knowing a respondent's sex will reduce prediction errors in capital punishment attitudes by 41 per cent.

## A Word of Caution

Although chi-square is a powerful inferential tool for determining statistical significance, there are conditions that limit its application. These conditions relate to conditional frequencies. Conditional frequencies are the numbers (frequency counts) inside a table. Chi-square becomes unreliable when either any cell has an expected frequency of less than 1.00 or when more than 20 per cent of the cells have expected frequencies of less than 5.00. So, if either of these conditions apply, chi-square should not be used to test statistical significance.

The practical problem then arises: What should be done if this is the case? The practical solution is to consider reclassifying the attributes of one or both variables. The reason these conditional problems arise is that there are too many cells relative to the sample size. A researcher cannot change sample size after the fact. But she can consider collapsing some of the categories of the variable, which will result in fewer cells in the table. This intervention often results in conditional frequency conditions that are appropriate for chi-square application.

# Chapter Summary

This chapter introduced you to the basic ideas behind hypothesis testing inferential procedures. It also introduced you to one specific hypothesis testing test, chi-square. In this chapter you learned that:

- Hypothesis testing inferential procedures proceed from an assumption about the value of the population parameter.
- Hypothesis testing procedures employ both null and research hypotheses.
- Null and research hypotheses are contradictory.
- The null hypothesis is the centre of attention of hypothesis testing; the aim is to decide if the null hypothesis can be rejected.
- Hypothesis testing runs the risk of inferential mistakes called Type I and Type II errors.
- Type I error is to claim that something exists when, in fact, it does not; Type II error is to claim that something does not exist when, in fact, it does.
- Type I and II errors are reciprocal; decreasing one increases the risk of the other.
- The method of science is conservative in its reality claims and, therefore, places a higher priority on reducing Type I rather than Type II errors.
- All hypothesis testing inferential tests share a common logic, which results in accepting or rejecting the null hypothesis at a specified significance level.
- A declaration of statistical significance says nothing about the substantive significance of the findings.
- Chi-square is a common inferential test of relationship between variables measured at lower levels of measurement.
- Chi-square computations involve comparing the observed frequencies in a table to those expected if the null hypothesis is true.
- Comparing the computed chi-square value to a critical value determines the statistical significance of the results.
- Chi-square is problematic if conditional frequencies are deficient.

This chapter provided a general overview of hypothesis testing inferential tests and gave you the specifics of one such test, chi-square. The next chapter introduces several other inferential tests that employ hypothesis testing logic.

# 21

# Various Significance Tests

## Overview

You have entered the final region of the statistical maze. Just a few more steps and your journey will be complete! The last chapter introduced you to some central concepts and principles, as well as the logic, of the hypothesis testing form of inferential statistics. In addition, the chapter illustrated hypothesis testing procedures by application to chi-square, one specific inferential test. This chapter introduces a few additional inferential concepts and then shows you several additional inferential tests. In this chapter you will learn about:

- one- and two-tailed inferential tests;
- $t$-distributions and $t$-tests, and $z$-distributions and tests;
- inferential tests for means and proportions using one sample; and
- inferential tools applicable to testing differences between means and proportions in two samples.

With these tools in place, you are positioned to understand how all manner of descriptive statistics can be tested for their generalizability.

In terms of the statistical selection checklist defining the coordinates of the statistical maze, the contents of this chapter are situated as follows:

## Statistical Selection Checklist

1. Does your problem centre on a descriptive or an inferential issue?

   ❏ Descriptive

   ☑ Inferential

2. How many variables are being analyzed simultaneously?

   ☑ One

   ☑ Two

   ❏ Three or more

3. What is the level of measurement of each variable?

☑ Nominal  
}  for proportions  
☑ Ordinal

☑ Interval  
}  for means  
☑ Ratio

Before we get to the inferential tests themselves, we need to return to the starting point for all inferential tests—the construction of the null and research hypotheses.

## STEP 1 Understanding the Tools

# Null and Research Hypotheses Revisited

The logic of hypothesis testing revolves around the null hypothesis. We need to keep several things about the null hypothesis clearly in mind. First, the null hypothesis is a statement about the value of some variable(s) *in the population*, not the sample. It is a statement about some population parameter, not a sample statistic. After all, the researcher already knows the situation in the sample, since she has analyzed the data and generated descriptive statistics. Second, the null hypothesis is a statement that the researcher *assumes to be true*. It is important to recognize that this claim is different from the condition that the researcher *expects to be true*. What the researcher expects to be true (on the basis of existing literature, prior research, theoretical deduction, or whatever) is the research hypothesis. Which leads us to the third point. The null hypothesis and the research hypothesis are the *inverse of one another*. In other words, evidence that does not support one of these hypotheses necessarily supports the other.

One clear way of identifying a null hypothesis is by noting that it claims that a variable (or variables) *equals some specific value*. A null hypothesis can add something to the equality specification, but it always includes that feature. Here are some illustrations of null hypotheses:

1. The average (mean) weekly beer consumption among undergraduates is four or fewer beers.
2. At least 80 per cent of undergraduates had a date in the last month.
3. There are no differences in the average income of men and women in Canada.
4. For the same offence, the likelihood of Aboriginals being arrested by police is equal to, or less than, the likelihood of non-Aboriginals being arrested.

Perhaps it is easier to see the equality specification if we express each of these null hypotheses in arithmetic terms:

1. undergraduate weekly beer consumption $\leq 4$
2. proportion of undergraduates who dated last month $\geq 0.80$
3. sex differences in income $= 0$
4. difference between proportion of Aboriginals and non-Aboriginals arrested $\leq 0$

Notice that each of these null hypotheses includes an equal sign (=). In several examples, the less than (<) or greater than (>) sign was added to the equal sign.

The research hypothesis is the inverse of the null hypothesis. It is the expectation that contradicts the claim assumed to be true in the null hypothesis. So, for our examples, here are the competing research hypotheses:

1.  On average, undergraduates consume more than four beers per week.
2.  Less than 80 per cent of undergraduates had a date last month.
3.  There are sex differences in average income.
4.  Aboriginals are more likely than non-Aboriginals to be arrested for the same offence.

Notice that in all cases the null and research hypotheses are competing claims. If one is supported by the evidence, the other is not. If undergraduates consume more than four beers, they cannot consume just four or fewer beers. If under 80 per cent of undergraduates had a date, 80 or more per cent cannot have had one. If there are sex differences in income, then the sexes do not have equal incomes. If Aboriginals are more likely than non-Aboriginals to be arrested, then they cannot be less likely to be arrested.

Just as the arithmetic expression of a null hypothesis contains an equal sign, the arithmetic expression of a research hypothesis is going to use one of the greater-than sign, the less-than sign, or the not-equal sign ($\neq$). This is evident in the following arithmetic expressions of the illustrative research hypotheses:

1.  undergraduate beer consumption > 4
2.  proportion of undergraduates who had dates last month < 0.80
3.  male incomes $\neq$ female incomes
4.  likelihood of Aboriginals arrests > likelihood of non-Aboriginal arrests

## Null Hypotheses and Sampling Distributions

Inferential statistics test the generalizability of the sample findings reported in descriptive statistics. Inferential testing is necessary because any probability sample is prone to sampling error and may produce descriptive findings that are unrepresentative of the population situation. Stated in inferential testing language, the particular sample a researcher has gathered is one sample from a sampling distribution that contains a range of possible results. The inferential statistics question involves estimating where the researcher's sample falls on the sampling distribution. Is it likely a typical sample? Or is it more likely an atypical one?

You will recall that sampling distributions are normal distributions and, like all normal curves, they have a centre and dispersion around that centre. Remember that hypothesis testing using inferential procedures begins with a null hypothesis, which is a claim about the population parameter that the researcher *assumes to be true*. This null hypothesis claim is transferable to the sampling distribution. The null hypothesis is an assumption about *where the centre of the sampling distribution is located*. As previously noted, the null hypothesis always contains an equal sign. The value of the equal sign in the null hypothesis is its claim about where the centre of the sampling distribution is located. Figure 21.1 illustrates this set-up for our four illustrations.

The location of each marked position in Figure 21.1 identifies where the sample statistic would fall on the sampling distribution *if the null hypothesis was true*. The researcher, of

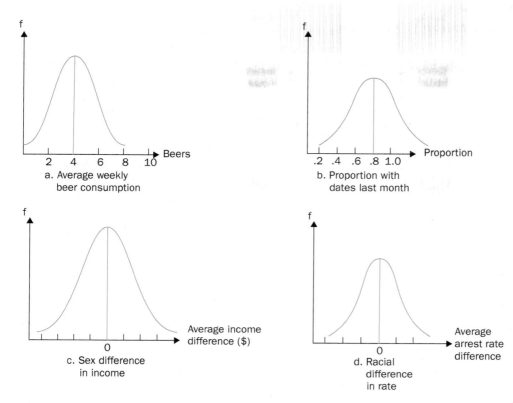

**FIGURE 21.1** Illustrative Sampling Distributions

course, does not anticipate the null hypothesis is correct; instead, the researcher's best estimate of the population parameter is contained in the research hypothesis. In order to reject the null hypothesis, and in doing so provide support for the research hypothesis, the sample results need to *fall far from the centre* of these respective sampling distributions. The farther the actual results fall from the centre, the less likely it is that the null hypothesis is credible and the more likely it is that the research hypothesis is probable. Applying this idea to the illustrations, the research hypothesis becomes *more likely* when:

1. the average weekly undergraduate beer consumption is *much higher than* four beers
2. the proportion of undergraduates who had a date last month is *much less than 80 per cent*
3. *the difference is larger* between male and female incomes
4. Aboriginals are *much more likely* than non-Aboriginals to be arrested

Look again at the sampling distributions in Figure 21.1. The farther the *actual sample result (statistic)* is away from the centre and shaded area on the distribution the *more likely* it is that the descriptive statistics can be safely generalized to the population. Why? If the results were not generalizable, they would fall at or near the centre of the sampling distribution; in other words, the study results would be similar to the null hypothesis expectation. If the study results are far away from the null hypothesis expectation, it is very unlikely that these results will occur when the null hypothesis is true—which means that the alternative (research) hypothesis is credible.

### An Illustration

Rejecting a null hypothesis provides support for the research hypothesis. Inferential tests provide support for the research hypothesis at a particular level of confidence. By convention, research standards require confidence levels of 95 per cent or 99 per cent. These levels mean that the researcher is 95 (or 99) per cent confident in rejecting the null hypothesis. Notice that researchers are never completely certain that their findings are not due to sampling error, since they can never eliminate this possibility. However, they can express very high degrees of confidence in generalizing their sample findings to the population.

The ever-present risk in making inferences is expressed as the alpha level, which is the probability of making a Type I error—rejecting the null hypothesis when it is true. The alpha value is the level of "unconfidence" or inferential risk. If the level of confidence is set at 95 per cent, then there is a 5 per cent risk of incorrect generalization; at 99 per cent confidence, the risk level is 1 per cent. The risk level can be visualized on the sampling distribution. It is the percentage of cases (samples) that are beyond the point of the sample result (statistic) on the distribution. To see this clearly, let's diagram some potential results from our illustrative hypothesis about sex differences in income.

Here are the research and null hypotheses for this example:

$H_1$: There are sex differences in average income.

$H_0$: There are no sex differences in average income.

Remember that the equal sign in the null hypothesis identifies the centre of the sampling distribution. In this case, the null hypothesis states that the sex differences equal zero (*males' average income – females' average income = zero*). So the value at the centre of the sampling distribution is zero, as illustrated in Figure 21.2.

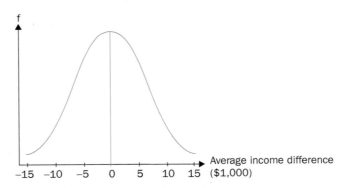

**FIGURE 21.2** Sample Distribution for Sex Differences in Income

If our sample result showed no sex differences in income, then the sample result occurs at the centre of this sampling distribution. This location would confirm that the sample result is what is expected if the null hypothesis is true.[1] The greater the sex differences, the farther away the sample result (statistic) will be from the zero point at the centre of the sampling distribution. Table 21.1 shows four possible results from four different studies:

---

[1]Remember, the sampling distribution displays what is expected *if the null hypothesis is true.*

**TABLE 21.1** Sex Differences in Income

| Study Number | Male Average Income | Female Average Income | Sex Difference in Average Income |
|---|---|---|---|
| 1 | $75,000 | $55,000 | $20,000 |
| 2 | $47,000 | $45,000 | $2,000 |
| 3 | $55,000 | $75,000 | −$20,000 |
| 4 | $45,000 | $47,000 | −$2,000 |

Figure 21.3 plots these various study results on the sampling distribution.

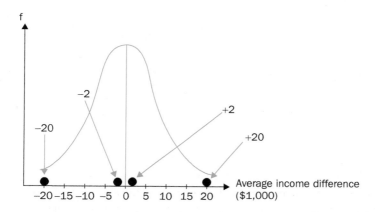

**FIGURE 21.3** Various Study Results of Sex Differences in Average Income

Figure 21.3 shows that the results from studies 1 and 3 are farther away from the centre of the sampling distribution than are the results of studies 2 and 4. This means that the study 1 and 3 results are more likely to be statistically significant than are the results of studies 2 and 4. Why? Because these results occur much less frequently when the null hypothesis is true.

## One-Tailed and Two-Tailed Tests

There is an additional issue you need to understand in order to apply specific hypothesis testing procedures, which concerns the distinction between one-tailed and two-tailed tests. This distinction relates to the alpha level of the inferential test.

To review, hypothesis testing uses a sampling distribution with the predicted null hypothesis value at its centre. The farther the descriptive sample results are from this centre, the more likely the results are *not like* the null hypothesis. In other words, the farther the test results are from the centre of the sampling distribution, the more confident the researcher is in *rejecting* the null hypothesis. The null hypothesis is rejected when the sample results occur in the *tails of the sampling distribution*. The researcher's confidence level in rejecting the null hypothesis is associated with the percentage of the sampling distribution cases that fall in the tails (alpha level). The 0.05 significance level states that the sample result is so far from the centre of the sampling distribution that only 5 per cent of the cases occur in that area. Similarly, the 0.01 significance level reports that only one per cent of the samples fall in the region of the sample result when the null hypothesis is true.

We know that those 5 per cent or 1 per cent of the cases fall in the tails of the distribution. Exactly where they are is a function of the how the null hypothesis is framed. Let's examine two null hypotheses considered earlier. Here they are in both statement and arithmetic form:

Example 1 $H_o$:

There are no differences in the average income of men and women in Canada.

Sex differences in average income = 0

Example 2 $H_o$:

For the same offence, Aboriginals are as likely as or less likely than non-Aboriginals to be arrested by police.

Difference between proportion of Aboriginals and non-Aboriginals arrested ≤ 0

Now let's determine the location on the sampling distribution where the actual sample evidence would have to occur in order to reject each null hypothesis. We know that the results will have to be in the tails of the distribution (i.e. far from the centre).

Consider the first null hypothesis, which states that there are no sex differences in average income. The zero (no sex differences) will be at the centre of the sampling distribution. Noting that sex differences in average income result from subtracting females' from males' income (*male income − female income = income difference*), what kind of results would refute this null hypothesis? One situation occurs when males' incomes are much higher than females' incomes (i.e. *big income − small income = large difference*). The other situation is where males' incomes are much lower than females' income (i.e. *small income − big income = large difference*). Figure 21.4 shows where the conditions that refute the null hypothesis are located on the sampling distribution.

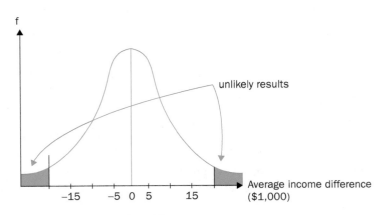

**FIGURE 21.4** Two-Tailed Test Condition

This sampling distribution shows that there are two kinds of "not zero" income difference that allow for refutation of the null hypothesis. If the males' average incomes are much larger than those of females, then the large differences occur in the positive tail of the distribution. If the males' average incomes are much smaller than those of females, then the large differences occur in the negative tail of the distribution.

This type of inferential test is a two-tailed test. A **two-tailed test** occurs when results that appear in either tail of the distribution are sufficient to reject the null hypothesis. The designation as a two-tailed test does not affect the significance level, which stays at either 0.05 or 0.01. It does, however, affect the distribution of these 5 per cent or 1 per cent of the cases. Since the 5 per cent (or 1 per cent) rejection area can fall in two regions (i.e. either tail), each of these regions include only 2.5 per cent (or 0.5 per cent) of the cases. In two-tailed tests the cases in the alpha level are split in half in order to cover the two-tailed rejection regions. In Figure 21.4, each shaded region would include 2.5 per cent of the cases (using a 0.05 alpha) or 0.5 per cent of the cases (using a 0.01 alpha).

Now let's look at the second illustrative null hypothesis which, in arithmetic form, asserts that the difference between the proportion of Aboriginals and non-Aboriginals arrested is ≤ 0. Like the previous example, the centre of the null hypothesis sampling distribution is zero. And, as in the earlier example, the difference between Aboriginal and non-Aboriginal arrests is computed by subtracting the latter from the former (*Aboriginal arrests – non-Aboriginal arrests = arrest difference*). Finally, as in the earlier illustration, actual sample differences in arrests that are far from the centre of the sampling distribution indicate that the null hypothesis is rejected. Figure 21.5 illustrates the sampling distribution for an inferential test of this null hypothesis.

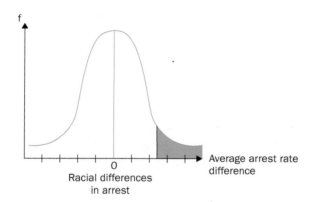

**FIGURE 21.5** One-Tailed Test Condition

Using an alpha of 0.05, the sample results need to be so distinctive that only 5 per cent of the cases would be different to that extent in order to reject the null hypothesis. However, for this hypothesis test, where are these 5 per cent of sample results located on the sampling distribution? The answer is indicated by the shaded area in Figure 21.5. The 5 per cent of cases that lead to rejection of the null hypothesis all occur on the *right tail* of the sampling distribution. This circumstance is called a **one-tailed test** because evidence for rejecting the null hypothesis only occurs on one-end (tail) of the distribution.

One-tailed tests result from directional hypotheses. A directional hypothesis specifies *how* the sample results are supposed to differ from the null expectation. The research hypothesis in our example does not just indicate that Aboriginal arrests are different from non-Aboriginal arrests; it states *how* Aboriginal rates will differ (they will be *greater than* non-Aboriginal rates). This restricts the evidence that would disconfirm the null hypothesis to one tail of the sampling distribution—the positive side since Aboriginal minus non-Aboriginal rates have to be positively different from each other. If Aboriginal arrest rates were substantially lower than non-Aboriginal rates, they would be different, but not different in a way that would lead to a rejection of the null hypothesis. This kind of difference would occur on the left tail of the sampling distribution and, as indicated on Figure 21.5, would not fall into the shaded 1 or 5 per cent rejection area.

### Selecting Appropriate Inferential Tests

Chapter 20 reported that inferential hypothesis testing follows a five-step procedure, including:

1. stating the null and research hypotheses
2. examining the descriptive evidence
3. conducting an appropriate inferential test
4. making a decision about the null hypothesis
5. drawing a conclusion

You know how to construct null and research hypotheses (step 1). If the sample evidence supports the null hypothesis in step 2, then you move directly to steps 4 and 5, concluding that the null hypothesis cannot be rejected and that further research is necessary to clarify the situation. If, however, the sample evidence counters the null hypothesis, the next move is to step 3, which requires selecting an appropriate inferential test to determine the generalizability of the descriptive results.

Following these steps, Chapter 20 introduced one hypothesis testing statistic, chi-square. Chi-square is an appropriate procedure for testing relationships between variables at lower levels of measurement (i.e. in tables). Chi-square is only one kind of inferential test; other tests include z-tests, t-tests, and F tests. Like chi-square, these different tests are appropriate under different conditions. Since they are the most common, we will focus on z-tests and t-tests.

As you have seen, all inferential hypothesis tests compare the actual descriptive sample results to a sampling distribution. The sampling distribution displays all the possible sample results obtainable if the null hypothesis is true. As the actual sample result becomes sufficiently different (far away from) the centre of the sampling distribution, the test indicates that the null hypothesis is not credible. In this way, unlikely results provide support for the research hypothesis. All inferential hypothesis tests follow this model.

Hypothesis tests differ, however, in the sampling distributions they use. Without going into technical detail, here is the basic situation. When the sample size is sufficiently large, the appropriate inferential statistic is a *z-test*. As a general rule, sufficiently large samples are those with 120 or more cases. Chapter 8 introduced z-distributions, and you are familiar with the idea that z-scores are used in conjunction with normal distributions. The reason z-scores are connected with sample sizes of 120 or more is that samples smaller than this rarely yield normal distributions.

When sample sizes are smaller than 120, adjustments have to be made to the sampling distribution in order to estimate the correct location of the descriptive statistic on the sampling distribution. When sample sizes are smaller, the appropriate sampling distributions are *t*-distributions, which use *t-tests*. Sampling distributions used for *t*-tests are adjusted for sample size. The sample size adjustments to *t*-distributions employ degrees of freedom, a calculation that helps you use the *t*-table correctly.

For now, the important point to appreciate is that z-tests are used when sample sizes are 120 or larger, and that t-tests are used for smaller sample sizes. Because z-distributions and *t*-distributions are just different kinds of sampling distributions, the logic of inferential hypothesis testing using these statistics does not change. In both cases, the distance of the sample result from the centre of the sampling distribution is calculated and a decision is made whether this distance is sufficiently large to justify rejecting the null hypothesis.

To review, a first consideration in selecting an appropriate inferential hypothesis test is sample size; use *z*-tests for larger tests and *t*-tests for smaller ones. A second consideration in appropriate selection is whether the research issue involves one sample or two. The distinction between one- and two-sample tests can sometimes be confusing, so let's be clear about what the difference is. The simplest clarification is to remind yourself of the three questions guiding the selection of appropriate statistical tools.[2] The second of these questions is whether the research issue involves one variable (univariate analysis), two variables (bivariate analysis), or three or more variables (multivariate analysis). The distinction between one-sample and two-sample inferential tests hinges on the answer to this question. One-sample tests are univariate tests; they test whether the value of a *single variable* in the sample can be generalized to the population. Two-sample tests are bivariate tests; they test if a difference between *two sample statistics is significant*.

For current purposes, one final consideration related to selecting appropriate inferential hypothesis tests needs mentioning. This relates to the issue of levels of measurement. As you know, different statistical tests are used for variables measured at different levels. For variables at lower levels (nominal and ordinal) it is common for inferential tests to be conducted on *proportions*. For variables at higher levels (interval and ratio) it is common for inferential tests to be conducted on *means*.

Taken together, these considerations for selecting appropriate inferential hypothesis tests yield the following possibilities:

- one-sample tests of population means
- one-sample tests of population proportions
- two-sample tests of population means
- two-sample tests of population proportions

Each of these possibilities can be conducted on smaller samples or larger ones, which means each could employ either *t*-tests or *z*-tests. In the next section we will show you how calculations are made for some of these eight possibilities. Before getting to these details, however, it is worth emphasizing that all these tests share a common form.

### Shared Form of Hypothesis Tests

Dealing with one- rather than two-sample tests affects the specific calculations in an inferential test. Likewise, whether the sample statistic is a proportion or a mean also affects the computations required. Similarly, the calculations for smaller samples are somewhat different than for tests that use larger samples. The next section introduces you to these specific calculations. It is important to recognize, however, that the different inferential test formulas share a common core. And better yet, you are already familiar with that core logic.

---

[2] A reminder: These questions include: (1) Does the research question involve a descriptive or inferential issue? (2) How many variables are being considered simultaneously? (i.e. univariate, bivariate, multivariate)? (3) What are the levels of measurement of each variable?

Chapter 8 introduced you to normal curves, which have a fixed relationship between their centre and the percentage of cases that fall within specific distances from that centre. The length of the specific distances from the centre of normal distributions is measured in standard deviations. This fixed relationship between the centre of normal distributions and their standard deviation allows for the location of any specific score to be measured in standard deviation units, called standard scores. Here is how any standard score is calculated:

$$\text{standard score} = \frac{\text{specific score} - \text{centre score}}{\text{standard deviation}}$$

After a standard score is computed, its value can be looked up in a table that reports what percentage of cases are above and below the location of the standard score.

All hypothesis tests are rooted in exactly this procedure, with the following translations.

- The normal distribution used is a sampling distribution; it is a distribution of how all possible sample statistics would be distributed if the null hypothesis was true.
- Specific score is the value of the descriptive sample statistic. For example, it could be a mean or proportion, or differences between means or proportions.
- The value at the centre of the distribution (centre score) is the value specified in the null hypothesis.
- The standard distribution is the standard error of the sampling distribution.

Figure 21.6 portrays the hypothesis testing set-up.

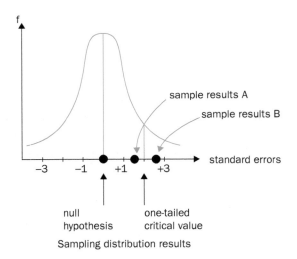

**FIGURE 21.6** Hypothesis Testing Set-Up

Remember that hypothesis testing gains support for the research hypothesis by rejecting the null hypothesis. In Figure 21.6 the null hypothesis value is at the centre of the sampling distribution. In order for the null hypothesis to be rejected, the sample statistic has to be far away from the centre of the distribution. How far away the sample statistic has to be from the centre is set by the alpha value. The specific point on the sampling distribution at which the null hypothesis can be rejected is called the critical value. The critical value is the point or points[3] on the sampling distribution beyond which the null hypothesis is rejected. In Figure 21.6, the critical value is identified for a one-tailed test. Notice that sample result A would not lead to rejection of the null hypothesis, while sample result B would lead to it.

This appreciation leads to one final point in understanding hypothesis testing. The hypothesis testing computation (which uses some variation on the standard score calculation procedure) is always compared to the value in some specific table score. The table score identifies the critical value. Therefore, in hypothesis testing, researchers are always looking to see how their calculated score compares to the table value. If the calculated score is as large as or larger than the table value, the null hypothesis is rejected at the chosen significance level.

You now have all the concepts and principles used in hypothesis testing. The next section shows you how these ideas are put into practice by taking you through the calculations of several different hypothesis tests. Before turning to these procedures it is worth noting that all kinds of descriptive sample results can be tested to determine if they can be generalized to the population. In Chapter 14, for example, you learned how to take sample data and determine the intercept ($a$) and slope ($b$) of the best-fit regression equation:

$$(Y = a + bX)$$

You also learned how to measure how well this regression line fit the data by calculating the correlation coefficient ($r$) and measuring the strength of the overall relationship with the coefficient of determination ($R^2$). All of these statistics *describe* the situation in the sample data. And all of them can also be subject to inferential tests that determine whether these results are generalizable.

## STEP 2 Learning the Calculations

# Hypothesis Testing

As noted, hypothesis testing for different sample statistics (e.g. means and proportions) for one- and two-sample situations with small and large sample sizes yields a wide variety of different tests. We will illustrate this variety by showing you how to calculate the following four tests:

- For the one-sample case:
  - *t*-test for means
  - *z*-test for proportions
- For the two-sample case:
  - *t*-test for means
  - *z*-test for proportions

Remember, one-sample tests are univariate tests; they test whether the value of a *single variable* in the sample can be generalized to the population. Specifically, when a mean or a proportion is calculated for a variable in a sample, one-sample tests can be used to infer whether this sample statistic can reasonably be generalized to the population. Here is how you proceed in each case.

---

[3]The critical value is a single point in a one-tailed test; it is described by two points in a two-tailed test.

# One-Sample *t*-Test for Means

### Scenario

A professor read in an academic article that Canadian undergraduate students have a grade-point average of 2.67. She wonders whether the 45 students in her class are different than the national average. To do so she looks up the GPAs of her students and determines their mean GPA is 2.81, with a standard deviation of 1.35.

### Null Hypothesis and Research Hypothesis

The null hypothesis ($H_o$) states that the professor's students are no different than the national average in terms of GPA. The research hypothesis states that the professor's students are different than the national average.

### Test Characteristics

The sample size is small ($N = 45$), so a *t*-test is appropriate. Since the research hypothesis only states that the professor's students are different than the national average (and does not specify *how* they are different), a two-tailed test is appropriate. The professor selects a conventional 95 per cent confidence level, so the alpha is 0.05.

### Computing the *t*-Statistic

Here is the *t*-test formula for means:

$$t(\text{obtained}) = \frac{\overline{X} - \mu}{s / \sqrt{N - 1}}$$

The following legend indicates what each of the symbols in the formula means:

- $\overline{X}$ is the mean of the variable in the sample; in our example it is the mean GPA of the professor's students (2.81).
- $\mu$ is the mean of the variable in the population; in our example it is the mean GPA of all Canadian undergraduate students (2.67).
- $s$ is the standard deviation of the variable in the sample; in our example it is 1.35.
- $N$ is the sample size; in our example, 45.

Next we put these values in the appropriate place in the formula and solve it, using the following steps:

$$
\begin{aligned}
t(\text{obtained}) &= \frac{\overline{X} - \mu}{s / \sqrt{N - 1}} \\[2mm]
&= \frac{2.81 - 2.67}{1.35 \div \sqrt{45 - 1}} \\[2mm]
&= \frac{0.14}{1.35 \div \sqrt{44}} \\[2mm]
&= \frac{0.14}{1.35 \div 6.63} \\[2mm]
&= \frac{0.14}{0.20} \\[2mm]
&= 0.70
\end{aligned}
$$

The obtained *t*-test value for our example is 0.70.

### Selecting the Critical Value

Hypothesis tests always compare the obtained test value to a critical value that identifies the cut-off value for rejecting the null hypothesis. Like chi-square, the critical values of $t$-tests are included in a table like the one included in Appendix B. As before, selecting the appropriate critical value for the $t$-statistic is a matter of deciding whether the test is one-tailed or two-tailed, what the alpha level is, and the number of degrees of freedom ($df$). In our example, we know we are using a two-tailed test with an 0.05 alpha level. That only leaves degrees of freedom. For sample means:

$$df = N - 1$$

In our example, with a sample size of 45:

$$df = 45 - 1 = 44$$

Now you can look up the critical $t$-value in the table. Find the number in the table (Appendix B) that uses an alpha of 0.05, a two-tailed test, and $df$ of 44.[4] In our case the critical value from the table is 2.021.

### Make a Decision

The critical value of 2.021 means that the obtained $t$-value must be at least that large for the null hypothesis to be rejected. Our obtained $t$-test value is 0.70, which is a lot smaller than 2.021. So, we do not have sufficient grounds for rejecting the null hypothesis. Therefore, we must conclude that the professor's students had GPAs that were not significantly different than the national average.

## One-Sample z-Test for Proportions

### Scenario

A newspaper article reported that a national study found that 60 per cent of Canadians ate chocolate at least once a week. The other 40 per cent ate chocolate less than once a week. When 225 parents of students at a local elementary school were asked about their weekly chocolate consumption, 70 per cent reported consuming chocolate at least once weekly. Are the parents different from the national sample?

### Null Hypothesis and Research Hypothesis

The null hypothesis ($H_o$) states that the parents are no different than the national average in terms of chocolate consumption. The research hypothesis ($H_1$) states that the parents' chocolate consumption is different than the national average.

### Test Characteristics

The sample size is large enough ($N = 225$) that a $z$-test is appropriate. Since the research hypothesis only states that the parents are different than the national average (and does not specify *how* they are different), a two-tailed test is appropriate. Let's be cautious and use a 99 per cent confidence level, so the alpha is 0.01.

---

[4]If the exact $df$ value is not reported in the table, select the closest $df$ value.

### Computing the z-Statistic

Here is the $z$-test formula for proportions:

$$z(\text{obtained}) = \frac{P_s - P_u}{\sqrt{P_u(1 - P_u)/N}}$$

The following legend indicates what each of the symbols in the formula means:

- $P_s$ is the proportion for the variable in the sample; in our example the proportion of parents who ate chocolate at least weekly was 0.70 (70 per cent).
- $P_u$ is the proportion for the variable in the population of interest; in our example 60 percents of Canadians (0.60) ate chocolate at least once a week.
- $N$ is the sample size; in our example this is the 225 parents of elementary school children.

Next we put these values in the appropriate place in the formula and solve it, using the following steps:

$$
\begin{aligned}
z(\text{obtained}) &= \frac{P_s - P_u}{\sqrt{P_u(1 - P_u)/N}} \\
&= \frac{0.70 - 0.60}{\sqrt{0.60(1 - 0.60) \div 225}} \\
&= \frac{0.10}{\sqrt{0.60(1 - 0.60) \div 225}} \\
&= \frac{0.10}{\sqrt{0.60(0.40) \div 225}} \\
&= \frac{0.10}{\sqrt{0.24 \div 225}} \\
&= \frac{0.10}{0.033} \\
&= 3.03
\end{aligned}
$$

The obtained $z$-test value for our example is 3.03.

### Selecting the Critical Value

As always, the computed $z$-test value needs to be compared to a critical value. The critical values for the $z$-test are found in Areas under the Normal Curve table (Appendix C) you are familiar with from previous tests. Selecting the appropriate critical value from this table uses an inverted procedure. In this case you find the alpha level in the body of the table and then determine what $z$-score (the critical value) is associated with it. In our case the alpha level is 0.01, so we are interested in finding the $z$-score that has 1 per cent (0.01) of the cases beyond it. If you look in the Appendix C table, you will see this value is 2.58. This is the appropriate critical value.

### Make a Decision

The critical value for this $z$-test is 2.58. Our obtained $z$-score value is 3.03. Since the computed value is larger than the critical value we are confident in rejecting the null hypothesis at the selected confidence level. In this case we are 99 per cent confident that the parents' chocolate consumption is significantly different (higher) than the national average.

As you have seen, the one-sample tests compare the sample statistic (mean or proportion) to some population parameter and ask whether the sample evidence is significantly different from the population. Two-sample tests take a different approach. Two-sample tests are bivariate tests; they test if a difference between *two-sample statistics is statistically significant*. Two-sample tests are similar in form to one-sample tests, as you will see in the following demonstrations.

# Two-Sample *t*-Test for Means

### Scenario

A professor had a sense that students enrolled in morning classes performed better than students in afternoon classes. To test her idea, she examined the most recent test grades of a sample of morning students and a separate sample of students in afternoon courses. When she computed the mean and standard deviation for each sample, she noticed some differences. Table 21.2 shows the information about central tendency and dispersion for each section.

Are the differences in test performances between morning and afternoon students statistically significant?

**TABLE 21.2** Central Tendency and Dispersion for Morning and Afternoon Classes

|  | Mean | Standard Deviation | N |
|---|---|---|---|
| Morning section | 67% | 6 | 33 |
| Afternoon section | 63% | 8 | 37 |

### Null Hypothesis and Research Hypothesis

The null hypothesis ($H_o$) states that the morning and afternoon students are no different in terms of their most recent test performance. The research hypothesis states that the morning students perform better than afternoon students.

### Test Characteristics

The sample sizes ($N_1 = 33$; $N_2 = 37$) are small, so a *t*-test is appropriate. Since the research hypothesis states that one group will perform better than the other (a directional hypothesis), a one-tailed test is appropriate. Following convention, an alpha of 0.05 is used.

### Computing the *t*-Statistic

Here is the *t*-test formula for differences in means for small samples:

$$t(\text{obtained}) = \frac{\overline{X}_1 - \overline{X}_2}{\sigma_{\overline{X}_1 - \overline{X}_2}}$$

The following legend indicates what each of the symbols in the formula means:

- The $\bar{X}_1$ and $\bar{X}_2$ are the means of the two samples; in our example $\bar{X}_1 = 67$ and $\bar{X}_2 = 63$.
- The denominator $\sigma_{\bar{X}_1 - \bar{X}_2}$ is an estimate of the standard deviation of the sampling distribution for differences between means.

Before we can solve this $t$-test formula we need to know how to calculate the denominator, which involves a separate bit of arithmetic. Here is the formula for the pooled estimate of the sampling standard deviation (the denominator):

$$\sigma_{\bar{X}_1 \bar{X}_2} = \sqrt{\frac{N_1 S_1^2 + N_2 S_2^2}{N_1 + N_2 - 2}} \sqrt{\frac{N_1 + N_2}{N_1 N_2}}$$

The symbols in this formula are translated as follows:

- $N_1$ is the sample size of the first group, and $N_2$ the sample size of the second group; in our example $N_1 = 33$ and $N_2 = 37$.
- $S_1^2$ and $S_2^2$ are the variances for the first and second groups; in our example $S_1^2 = 36$ and $S_2^2 = 64$.

Computing the $t$-test statistic is a two-step procedure. First, we solve the equation for the denominator, and then we proceed to solve the complete $t$-test equation. Here are the steps for the first step.

$$= \sqrt{\frac{33(36) + 37(64)}{33 + 37 - 2}} \sqrt{\frac{33 + 37}{33(37)}}$$

$$= \sqrt{\frac{1,188 + 2,368}{68}} \sqrt{\frac{70}{1,221}}$$

$$= \sqrt{\frac{3,556}{68}} \sqrt{0.057}$$

$$= \sqrt{52.29}(\sqrt{0.057})$$

$$= 7.23(0.239)$$

$$= 1.73$$

Whew! The heavy lifting is done. Now all we need to do is complete the $t$-test formula, using the following steps:

$$t(\text{obtained}) = \frac{\bar{X}_1 - \bar{X}_2}{\sigma_{\bar{X}_1 - \bar{X}_2}}$$

$$= \frac{(67 - 63)}{1.73}$$

$$= \frac{4}{1.73}$$

$$= 2.31$$

The obtained $t$-test value for our example is 2.31.

### Selecting the Critical Value

The remainder of the *t*-test procedure is essentially similar to the procedure for single sample tests. We must compare the obtained test value to a critical value that identifies the cut-off value for rejecting the null hypothesis. The critical values of *t*-tests are included in a table like the one in Appendix B. As before, selecting the appropriate critical value for the *t*-statistic is a matter of deciding whether the test is one-tailed or two-tailed, what the alpha level is, and the number of degrees of freedom (*df*). In our example, we know we are using a one-tailed test with a 0.05 alpha level. For two-sample tests, the degrees of freedom is computed as:

$$df = N_1 + N_2 - 2$$

So, applied to our example, the calculation is:

$$df = 33 + 37 - 2 = 68$$

Now we look up the critical *t*-value, which means finding the number in the table (Appendix B) that uses an alpha of 0.05, a one-tailed test, and *df* of 68. In our case the critical value from the table is 1.671.

### Make a Decision

The critical value of 1.671 means that the obtained *t*-value must be that at least that large in order for the null hypothesis to be rejected. Our obtained *t*-test value is 2.31, which is larger than 1.671. We therefore conclude that we have sufficient grounds for rejecting the null hypothesis of no difference between morning and afternoon students' performance. In this study, morning students do display statistically significant higher test performances than afternoon students.

## Two-Sample *z*-Test for Proportions

### Scenario

A professor suspects that there are sex differences in students' motivation to do well in university. To test her idea, she draws random samples of males and females from her university and, based on their responses to a series of questions, rates their motivation as *high* or *low*. The samples include 575 males and 544 females and show that the proportion of each sex in the *high* motivation category is *Males* = 0.51 and *Females* = 0.48. Is this difference statistically significant?

### Null Hypothesis and Research Hypothesis

The null hypothesis ($H_o$) states that there is no difference in the proportions of males and females with high motivation. The research hypothesis ($H_1$) states that there are statistically significant sex differences in motivation.

### Test Characteristics

Our samples include over 500 cases, so a *z*-test is appropriate. Since the research hypothesis only specifies that sex differences in motivation exist, the hypothesis is non-directional and therefore implies a two-tailed test. Let's set the alpha level at 0.05. Using this information, we can now compute the test statistic.

### Computing the z-Statistic

Here is the two-sample z-test formula for proportions:

$$z\,(\text{obtained}) = \frac{P_{S1} - P_{S2}}{\sigma_{P_1 - P_2}}$$

The following legend indicates what each of the symbols in the formula means:

- $P_{S1}$ and $P_{S2}$ are the proportions for the variable under consideration in the sample; in our example it is the proportion in each sample in the *high* motivation category—males, 0.51; and females, 0.48.
- The denominator is the standard deviation of the sampling distribution for difference between proportions.

We have the information for the numerator for this test; obtaining the denominator is slightly more challenging and requires the following two interim steps.

First, we need to obtain an estimate of the population proportion for the variable $(P_u)$. This is estimated from the samples sizes and sample proportions using the following formula:

$$P_u = \frac{N_1 P_{S1} + N_2 P_{S2}}{N_1 + N_2}$$

Let's solve this formula using the following steps:

$$P_u = \frac{575(0.51) + 544(0.48)}{575 + 544}$$

$$= \frac{293.25 + 261.12}{1,119}$$

$$= \frac{554.37}{1,119}$$

$$= 0.495$$

Next, using this estimate of the population proportion, we can estimate the value of the standard deviation of the sampling distribution for difference between proportions—which is the denominator of the original equation. Here's the formula:

$$\sigma_{P_1 - P_2} = \sqrt{P_u(1 - P_u)}\,\sqrt{\frac{N_1 + N_2}{N_1 N_2}}$$

And here's its application to our current problem:

$$= \sqrt{0.4954(0.5046)}\,\sqrt{\frac{575 + 544}{575(544)}}$$

$$= \sqrt{0.2499}\,\sqrt{\frac{1,119}{312,800}}$$

$$= \sqrt{0.2499}\,\sqrt{0.0036}$$

$$= 0.4999(0.06)$$

$$= 0.03$$

Okay, now we can return to solving the original $z$-test formula:

$$z \, (\text{obtained}) = \frac{P_{S1} - P_{S2}}{\sigma_{P_1 - P_2}}$$

$$= \frac{(0.51 - 0.48)}{0.03}$$

$$= \frac{0.03}{0.03}$$

$$= 1.0$$

Whew! It took a few steps but the obtained $z$-test value for the difference in proportions between the two samples is 1.0.

### Selecting the Critical Value

This obtained $z$-test value needs to be compared to a critical value, which is found in the Areas under the Normal Curve table (Appendix C) you are familiar with from previous tests. As before, selecting the appropriate critical value from this table uses an inverted procedure. In this case you find the alpha level in the body of the table and then determine what $z$-score (the critical value) is associated with it. In our case the alpha level is 0.05, so we are interested in finding the $z$-score that has 5 per cent (0.05) of the cases beyond it. If you look in the Appendix C table, you will see this value is 1.96. This is the appropriate critical value.

### Make a Decision

The critical value for this $z$-test is 1.96. Our obtained $z$-score value is 1.0. Since the computed value is much smaller than the critical value, we are not confident in rejecting the null hypothesis at the selected confidence level. The evidence indicates that there are no statistically significant differences in motivation between undergraduates males and females.

As is typical of more sophisticated procedures, you appreciate that computing results by hand is challenging. As the next section shows, using IBM® SPSS® Statistics software ("SPSS") makes this task considerably easier.

## STEP 3 Using Computer Software

Means and proportions are two-sample statistics commonly used for making inferences. And, as you have seen, these statistics can be used in both one-sample and two-sample situations. Rather than take you through a variety of possibilities, this section illustrates two SPSS procedures. First we show you the one-sample test using means, then the two-sample test using proportions.

# One-Sample Tests of Population Means

One sample $t$-tests are useful in determining if the mean of a sample is different from a particular value. In our example, we will determine if the mean age of the participants in our data set is greater than 18. Why 18? According to a report on Canadian undergraduate students who are in their first year of studies, their average age is 18.[5] We can use a one-sample test of population means to determine if the average age of participants in our sample is greater than 18, the national average.

In this example, our $H_0$ would be:

The survey participants' average age is not greater than 18.

We also know that a one-tailed test is needed, because the $H_0$ uses a greater-than symbol. Further, because the age variable in our data set is a continuous measure, either a *t*-test or a *z*-test would be appropriate to use. Because our sample is well over 100 respondents, you have been taught to use *z*-test. However, in SPSS determining *z*-test scores is a difficult endeavour. As such, we will use the *t*-test option, which does not pose a problem for our large sample, because *t*-distributions approximate *z*-distributions as sample size increases.

## Procedure

1. Select Analyze from the menu.
2. Select Compare Means → One-Sample T-Test.
   a. Move the *Age* variable into the Test Variable(s) box.
   b. In the Test Value box enter 18, which is the value we want to compare.
3. Click OK.

The third column of the first section of output (see Figure 21.7) tells us that the mean age in our sample is 20.17. At first glance, we see that students from our sample are, on average, over our cutting point of 18; however, what we still need to determine is whether the difference is statistically significant.

The second part of the output aids us in determining whether the difference is statistically significant or not. The second column represents the *t*-test value, which in our example is 21.380. The third column is the degrees of freedom (in our example, it is quite large at 1,227).

One-Sample Statistics

|  | N | Mean | Std. Deviation | Std.Error Mean |
|---|---|---|---|---|
| Age respondent age | 1228 | 20.17 | 3.564 | .102 |

One-Sample Test

|  | Test value = 18 | | | | | |
|---|---|---|---|---|---|---|
|  | t | df | Sig- (2-tailed) | Mean Difference | 955 Confidence Interval of the Difference | |
|  |  |  |  |  | Lower | Upper |
| Age respondent age | 21.380 | 1227 | .000 | 2.174 | 1.97 | 2.37 |

**FIGURE 21.7** One-Sample Mean Test Output

---

[5]Student Services (University of Manitoba) (2007). Undergraduate experience at Canadian universities: First-year university students. *University Affairs*, 17, 1, retrieved on 14 March 2013 at: http://www.umanitoba.ca/student/media/Research_Report_V19_no1.pdf

The fourth column gives the two-tailed significance score, which can be confusing since our hypothesis is based on a one-tailed test. In SPSS, there is no option for a one-tailed test. However, if you look in any *t*-test table of critical values, you can determine the critical value for a one-tailed test. By referring to the *t*-test critical value table, we see that the critical *t* with 1,227 degrees of freedom at $p = 0.05$ for a one-tailed test is 1.645 (tables generally do not display critical values over 120 or 200, but often at the bottom have an infinity symbol ($\infty$) to indicate critical values for larger sample sizes).

Because the *t*-value calculated in SPSS is much greater than 1.645 (recall it is 21.38), we reject the $H_0$. That is, students from the Student Health and Well-Being Survey data set are significantly older than 18. While statistically we reject the $H_0$, substantively there could be some reasonable explanations accounting for such a difference. For example, some of the participants were in a second-year course (that had an intro course as a prerequisite). In addition, although the vast majority of the survey respondents were enrolled in Introduction to Sociology (a first-year course), we are merely assuming that these participants are in fact first-year students. One interpretation of the statistical results is that perhaps we are incorrect to make such an assumption!

# Two-Sample Tests of Population Proportions

Unlike one-sample tests, two-sample tests represent bivariate calculations. Recall as well that population means tests are designed for interval or ratio variables. In SPSS the procedure used to satisfy both requirements is the Independent Samples T-Test.

For example, is there a statistically significant difference between the two samples found in Q25 (*Respondent sex*) and Q14B (*Coping with stress by planful problem solving*), which is a ratio variable? In this example, the $H_0$ is:

> There is no difference/relationship between sex and coping with stress by planful problem solving.

Because the $H_0$ does not specify an order, we should use a two-tailed test.

## Procedure

1. Select Analyze from the menu.
2. Select Compare Means → Independent Samples T-Test (Figure 21.8).
   a. Move the dependent variable Q14B (*Coping*) into the Test Variable(s) box.
   b. Move the independent variable Q25 (*Respondent sex*) into the Grouping Variable box.
   c. Click Define Groups to specify the values for the two groups that you want to compare. In our case, we want to compare females and males, so we enter 1 under Group 1 and 2 under Group 2.
3. Click Continue → OK.

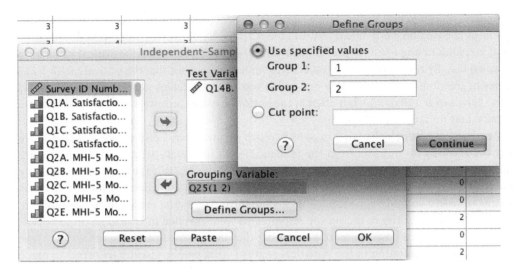

**FIGURE 21.8** Means Comparison Procedure

Source: Reprint Courtesy of International Business Machines Corporation, © International Business Machines Corporation.

The output appears in two parts: Group Statistics (i.e. descriptive statistics) and Independent Samples Test (i.e. inferential statistics). The descriptive statistics table presents the univariate results for the two independent samples (female and male).

As illustrated in Figure 21.9, looking in the Mean column, you will observe that females have, on average, a higher score (12.33), meaning they tend to use coping with stress by planful problem solving more than males (11.98).

**Group Statistics**

| | Q25 Q25. What is your sex | N | Mean | Std. Deviation | Std. Error Mean |
|---|---|---|---|---|---|
| Q14B Q14B. Coping with stress by planful problem solving | 1 Female | 768 | 12.33 | 2.984 | .108 |
| | 2 Male | 459 | 11.98 | 3.280 | .153 |

**Independent Samples Test**

| | | Levene's Test for Equality of Variances | | t-test | | | | |
|---|---|---|---|---|---|---|---|---|
| | | F | Sig. | t | df | Sig. (2-tailed) | Di |
| Q14B Q14B. Coping with stress by planful problem solving | Equal variances assumed | 1.953 | .163 | 1.938 | 1225 | .053 | |
| | Equal variances not assumed | | | 1.893 | 8920.847 | .059 | |

**FIGURE 21.9** Mean Comparison Output

But, is this difference statistically significant? Or is it probably a result of sampling error? To answer these questions, we need to refer to the second part of the output: Independent Samples Test. Before we can decide whether or not to reject the $H_0$, we need to find whether an important assumption of the two-sample $t$-test has been met, which is that the variability of each group is approximately equal. If this assumption is not achieved, a special form of the *t*-test result in SPSS should be used. At this point, take a deep breath. For your purposes, all you need to do is look at the columns labelled "Levene's Test for Equality of Variances." In the second column "Sig." there is a *p*-value of 0.163, which is greater than all scientifically acceptable alpha or *p*-levels. For our purposes this means that we use the "Equal variances assumed" row when interpreting the *t*-test results.[6] If the "Sig." level was less than or equal to the *p*-level 0.05, then we would use the bottom row labelled "Equal variances not assumed."

In the "equal variances assumed" row, under the column "t," our *t*-score is 1.938 and the two-tailed significance score, in the column labelled "Sig. (2-tailed)", is 0.053, which is greater than an alpha or *p*-level of 0.05, meaning that the test fails and we do not reject the $H_0$. Put another way, there is no significant difference between sex and coping with stress by using planful problem solving; or, on average males and females engage in this coping strategy at similar rates.

## STEP 4 Practice

You now have the tools for understanding a variety of inferential tests. To solidify your understanding you need to practise applying the statistical and software procedures.

This section provides you with opportunities to practise what you have learned about scatterplot analysis. The first set of questions uses hand calculations. The second set uses the SPSS procedures. For each set of questions:

1.  Follow the procedural steps (Set 1) or software application (Set 2).
2.  Check your answers, using the Answer Key in the back of the text.
3.  If your answer is incorrect, consult the Solutions section on the book's website. The Solutions provide a complete step-by-step analysis of how the answers are derived.

After you have completed the next section (Interpreting the Results), return to your calculations or output and *provide complete, written interpretations of each of the statistics you have generated.*

# Set 1: Hand-Calculation Practice Questions

1.  A random sample of students from different high schools in an urban school division completed a survey regarding their use of social media. The mean number of Facebook friends across all high schools in the Central School Division was 170. In comparison, the average number of Facebook friends among Major Smith High School students ($N = 34$) was 159 ($s = 26$). Use the one-sample *t*-test for means with a small sample ($\alpha = 0.05$) to test whether the average number of Facebook friends for Major Smith students differs from the rest of the high school population in the Central School Division.
    a.  State the relevant hypotheses.
    b.  What is the obtained *t*-test value?
    c.  What is the critical *t*-test value?

---

[6]More specifically, a significant *p*-value of the Levene's test means you can reject the $H_0$ that the variability of the two groups is equal, implying that the variances are unequal. In the output found in Figure 21.9, we do not reject the $H_0$, which means that we are assuming that the variances are equal.

d.  Interpret the results of this test.

2.  In a national survey 42 per cent of Canadian high school students indicated that they logged on to Facebook daily, while only 35 per cent of sampled Central School Division students ($N = 1,150$) did. Use the one-sample $t$-test for proportions with a large sample to assess whether this difference is significant ($\alpha = 0.01$).

a.  State the relevant hypotheses.

b.  What is the obtained $t$-test value?

c.  What is the critical $t$-test value?

d.  Interpret the results of this test.

3.  A small study on social determinants of health found that shift workers ($N = 45$) reported an average of 4.7 sick days per year ($s = 1.1$) compared to 3.9 ($s = 0.9$) for those who worked straight days ($N = 49$). The study's authors expect this difference to be consistent with previous research that indicates shift workers tend to have more health problems. Use the two-sample $t$-test of difference between means for a small sample to test this ($\alpha = 0.01$).

a.  State the relevant hypotheses.

b.  Calculate the standard deviation of the difference between means.

c.  What is the obtained $t$-test value?

d.  What is the critical $t$-test value?

e.  Interpret the results of this test.

4.  One of the questions on a survey of second- and third-year undergraduate students' consumer habits asked whether or not buying fair-trade products was important to them. Sixty-eight per cent of sociology majors ($N = 315$) and 64 per cent of psychology majors ($N = 370$) indicated that it was. Test if this difference is significant using the two-sample test of difference between proportions for a large sample ($\alpha = 0.01$).

a.  State the relevant hypotheses.

b.  Calculate the estimate of the population proportion.

c.  Calculate the standard deviation of the sampling distribution for the difference between proportions.

d.  What is the obtained $z$-test value?

e.  What is the critical $z$-test value?

f.  Interpret the results of this test.

5.  Given the data in Tables 21.3, 21.4, and 21.5, state the null and research hypotheses, calculate the appropriate test of statistical significance at the given alpha level, and interpret the result:

a.  for Table 21.3, alpha = 0.05, one-tailed

**TABLE 21.3** Motor Vehicle Data

| | Motor Vehicles | |
| --- | --- | --- |
| | Rural | Urban |
| Average number per household | 2.4 | 1.9 |
| Standard deviation | 0.33 | 0.37 |
| Sample size | 45 | 42 |

b.  for Table 21.4, alpha = 0.05, two-tailed

c.  for Table 21.4, alpha = 0.05, one-tailed

d.  for Table 21.5, alpha = 0.01, one-tailed

**TABLE 21.4**  University Degree Data

| | Have a University Degree | |
| --- | --- | --- |
| | Female | Male |
| Percentage | 62 | 55 |
| Sample size | 305 | 250 |

**TABLE 21.5**  Approval Data

| | Approve of Stephen Harper's Performance as Prime Minister | |
| --- | --- | --- |
| | Alberta | Canada |
| Percentage | 59 | 52 |
| Sample size | 255 | 2,500 |

6.  According to the Association of American Medical Colleges (AAMC), the average Medical College Admission Test (MCAT) score in 2013 was 25.3.[7] A private tutoring company, however, claims that their methods of preparation guarantee "significantly better" results. To support their claim, they show records indicating that in 2013 a sample of 41 of their students achieved, on average, an MCAT score of 29.1, with a standard deviation of 7.6. Can we conclude that the preparation course truly had a significant effect on bettering the students' performance (i.e. is the difference between means statistically significant?), or is there a high probability that the "significantly better" results are due to chance alone?
    a.  State the relevant hypotheses.
    b.  Perform a one-tailed significance test, with 95 per cent significance level.
    c.  Is the difference statistically significant? Interpret the results of your test.
7.  Social psychologists theorized that seeing unrealistically beautiful women in music videos and magazines would make men perceive "real" women as less attractive. The researchers selected two representative samples, each consisting of 45 undergraduate men, and randomly assigned them to one of the two conditions. First (control) group watched neutral images and videos of nature and architecture for 30 minutes, while the second (experimental) group viewed 30 minutes of music videos and pictures of models from magazines. Afterwards both groups were asked to rate, on a scale from 1 to 10 (10 was the highest score), 15 pictures of "real" women. While the average score of the first (control) group was 7.8, with a standard deviation of 1.4, the second (experimental) group gave the "real" women a mean score of 5.5, with a standard deviation of 1.8. Are the two groups significantly different?
    a.  State the relevant hypotheses.
    b.  Perform a two-tailed significance test, with 95 per cent significance level.
    c.  Is the difference between the two groups statistically significant? Interpret the results of your test.
8.  Researchers were testing a new "light" medication for treating people diagnosed with insomnia. The medication was proven to produce no physical addiction, but does it cure insomnia? The researchers used two randomly selected samples of 1,500 insomniac patients each, and randomly assigned them to one of the two following conditions.

---

[7]Data from AAMC, 2013. Data retrieved on 29 December 2013 from: https://www.aamc.org/students/download/361080/data/combined13.pdf.pdf

The first (control) group received a placebo, while the second (experimental) group received the actual medication. After one month of therapy, 36 per cent of the patients in the first (control) group reported significant improvement, while in the second (experimental) group there was improvement in 39 per cent of patients. Can the researchers conclude that the new medication is effective at curing insomnia?

   a. State the relevant hypotheses.

   b. Perform a one-tailed significance test, at alpha of 0.05.

   c. Is the difference statistically significant? Interpret the results of your test.

   d. Comment on the results of your significance testing considering the notion of both statistical and substantive significance.

9. Researchers found that 23 per cent of children between three and five years old have an anxious–ambivalent attachment to their mother. They theorized, however, that this number would be much higher in case of children of mothers suffering from a bipolar affective disorder. They examined a sample of 105 children of bipolar mothers and found that 44 per cent of them had an anxious–ambivalent attachment to their mother. Is this difference statistically significant?

   a. State the relevant hypotheses.

   b. Perform a one-tailed significance test at the 99 per cent significance level.

   c. Is the difference statistically significant? Interpret the results of your test.

10. A group of social-science researchers decided to examine a claim frequently made by bio-tech corporations: that there is a strong consensus within the scientific community about safety of genetically engineered foods and crops. They asked 81 nutritional and agricultural scientists whether they agree with the following statement: "There is a widespread scientific consensus about safety of genetically engineered foods and crops." The scientists were divided into two categories—bio-tech (n=43) or other (n=38)—depending on the source of their funding: 83 per cent of biotech-funded scientists said yes there was consensus, while only 21 per cent of scientists funded from other sources agreed with this statement.

   a. State the relevant hypotheses. (**Hint**: putting the data into a bivariate table may help you.)

   b. Perform a two-tailed significance test at alpha of 0.05.

   c. Is the difference statistically significant? Interpret the results of your test.

11. National data indicated that the average waiting time for a magnetic resonance imaging (MRI) scan in Country $X$ was 21 days. Researchers theorized, however, that the average waiting time would be shorter for people from the upper socio-economic class despite the country's fully socialized public health-care system. The assumption was that in absence of a private health sector, people from higher social circles draw upon influential contacts from their social networks to gain considerable privileges within the existing health-care system. The researchers found that the average waiting time for an MRI scan for a sample of 70 people from the upper class was 14.5 days, with a standard deviation of 5 days. Can the researchers safely conclude that the difference is statistically significant?

   a. State the relevant hypotheses.

   b. Perform a test of statistical significance at the 95 per cent confidence level. Decide whether you want to use a one- or a two-tailed test but provide a sound rationale for your choice.

   c. Is the difference statistically significant? Interpret the results of your test.

12. A researcher wondered whether the sociology or the psychology honours undergraduate program at his university was more competitive. He recorded data about cumulative GPA of a sample of graduates from each program from the past five years and got the results shown in Table 21.6.

**TABLE 21.6** GPA in Psychology and Sociology Departments

| | Department | |
| --- | --- | --- |
| | **Psychology** | **Sociology** |
| Average cumulative GPA at graduation | 4.15 | 4.09 |
| Standard deviation | 0.09 | 0.08 |
| Sample size | 42 | 28 |

    a. State the relevant hypotheses.
    b. Perform a two-tailed significance test at the 95 per cent confidence level.
    c. Is the difference statistically significant? Interpret the results of your test.

# Set 2: SPSS Practice Questions

1. Use the one-sample $t$-test procedure to assess whether the mean Student Health and Well-Being Survey CES-D *Depression* scale score (Q4) differs significantly from the mean score in a reference population (9.25) used to validate the scale originally.[8]

2. Use the one-sample $t$-test procedure to assess whether the mean Student Health and Well-Being Survey *Rosenberg self-esteem scale* (Q17) score differs from the mean score in the general Canadian population (30.22) as measured in a previous study.[9]

3. Use the Independent-samples $t$-test procedure to assess whether the mean level of *Satisfaction with life* (Q22) differs significantly between respondents born in Canada and those not (Q27).

4. Use the Independent-samples $t$-test procedure to assess whether the mean *Self-esteem* score (Q17) differs significantly between male and female respondents (Q25).

5. Use the Independent-samples $t$-test procedure to assess whether the mean *Depression* score (Q4) significantly differs between male and female respondents (Q25).

6. Use the independent-samples $t$-test procedure to assess whether the mean *Level of religious faith* (Q18) differs significantly between respondents who report having used soft drugs in the past year and those who have not (Q9).

7. A national poll by Forum Research commissioned by the National Post in 2012 found that 5 per cent of Canadians identify as LBGTQ. Use the one-sample $t$-test procedure to assess whether the proportion of Student Health and Well-Being Survey respondents identifying as *LBGTQ* (Q31) is different than the proportion in the national survey. Remember to recode Q31 into a dichotomous variable as 1 = *LBGTQ*.

---

[8]*Radloff (1977).*
[9]*Schmitt and Allik (2005).*

8. Use the independent-samples *t*-test procedure to assess whether there is a difference in the proportion of males compared to females (Q25) who report having volunteered in the last six months (nQ19).

9. Using the one-sample *t*-test procedure, determine whether the mean *Feelings of positive affect* (Q3A) is significantly different from a hypothetical population mean of 10.

10. Using the one-sample *t*-test procedure, determine whether the mean *Coping with stress by planful behavior* (Q14B) is significantly different from a hypothetical population mean of 17.

11. Using the independent-samples *t*-test procedure, assess whether the mean *Feelings of negative affect* (Q3B) differ significantly for people who have consumed alcohol in the past year (Q7) compared with those that have not.

12. Using the independent-samples *t*-test procedure, assess whether the mean *Coping with stress by seeking social support* (Q14A) differs significantly for people who have consumed drugs in the past year (Q9) compared with those that have not.

13. Using the independent-samples *t*-test procedure, assess whether the mean *Coping with stress by avoidant behavior* (Q14C) differs significantly between people who have been excessively partying in the past 6 months (Q13D) and those who have not.

14. Using the independent-samples *t*-test procedure, assess whether the mean *Coping with stress by planful problem solving* (Q14B) differs significantly for people who have moved to a new home in the past 6 months (Q13A) and those who have not.

15. Using the one-sample *t*-test procedure, assess whether the proportion of Student Health and Well-Being Survey participants who have volunteered in the past 6 months (nQ19) is significantly different from a hypothetical population proportion of 20 per cent.

16. Using the independent-samples *t*-test procedure, assess whether the proportion of respondents who report having consumed alcohol in the past year (Q7) differs between those born in Canada and those not born in Canada (Q27).

## STEP 5 Interpreting the Results

Interpreting the results of any hypothesis testing procedure is straightforward. Appreciating the interpretation, however, requires a review of the context. Using descriptive statistical tools, a researcher reports on the findings from her sample. If the results do not support the research hypothesis, that is pretty much the end of the story. Alternatively, if the sample results support the research hypothesis, the question of generalizability enters the picture. The specific research results are findings from a single sample which, because of sampling error, have some risk of being unrepresentative of the population.[10] Inferential testing checks to see how likely it is that the sample results are due to sampling error (i.e. are unrepresentative).

Inferential tests centre on the null hypothesis, which is a statement that contradicts the research hypothesis. The null hypothesis value is located at the centre of a sampling distribution. A location is identified on the sampling distribution where it is highly unlikely that the sample results could have occurred if the null hypothesis was true. This location, associated

---

[10]Crudely stated, the operation of sampling error means that the study results are a fluke.

with either a 0.05 or a 0.01 significance level, is the critical value of the specific hypothesis test. The computed value of the test statistic is compared to the critical value and a decision is made. At this point, two possibilities emerge:

- If the computed value is *as large as or larger than* the critical value, the result is interpreted as a *rejection* of the null hypothesis.
- If the computed value is *smaller than* the critical value, the null hypothesis is accepted.

Remember that acceptance or rejection is of a *specific null hypothesis,* and this acceptance or rejection has implications for a *specific research hypothesis.* Therefore, your interpretation of the hypothesis testing results must refer to these specific hypotheses. It must also refer to the level of significance of the test.

Here are the general forms for interpreting hypothesis testing results:

Null Hypothesis Rejected: The results indicate that the null hypothesis of _____ [specific null hypothesis] is rejected at the _____ [either 0.05 or 0.01] level of significance. This result supports the hypothesis that _____ [specific research hypothesis]. These statistically significant results suggest that we are _____ [confidence level] per cent confident that _____ [substantive findings of the study].

Example: The null hypothesis that there is no difference in the motivation of males and females is rejected at the 0.05 level of significance. This result supports the hypothesis that there are statistically significant sex differences in motivation. These statistically significant results suggest we are 95 per cent confident that there are sex differences in motivation.

Null Hypothesis Not Rejected: The results indicate that the null hypothesis of _____ [specific null hypothesis] cannot be rejected at the _____ [either 0.05 or 0.01] level of significance. This result indicates that further research is necessary to confirm statistically significant results.[11]

Example: The results indicate that the null hypothesis of no differences in the motivation of males and females cannot be rejected at the 0.05 level of significance. This results indicates that further research is necessary to confirm statistically significant results.

---

[11] It is important to remember that the conclusion of "not rejecting the null hypothesis" is *not* the same as concluding that the null hypothesis is correct. A finding of statistical *in*significance tells us that the results are probably due to sampling error—so we don't really know what is going on. Hence the recommendation for further research to clarify the situation.

# Chapter Summary

This chapter expanded your understanding of hypothesis testing inferential procedures. The chapter introduced several new concepts and you learned a set of specific hypothesis tests and their interpretations. In summary, this chapter taught you that:

- The null and research hypotheses are competing assertions about some aspect of reality.
- The null hypothesis includes the assertion of some specific value of a variable or relationship between variables.
- The specific value of the null hypothesis is located at the centre of the sampling distribution used in hypothesis testing.
- Rejecting the null hypothesis becomes more likely as the sample statistic is located farther from the centre of the sampling distribution.
- A two-tailed hypothesis test only states that the research finding is different from the null hypothesis. A one-tailed test specifies the direction of difference from the null.
- In a two-tailed test the alpha is split between the two tails of the sampling distribution; in a one-tailed test the alpha proportion is concentrated in one-tail of the distribution.
- On larger sample sizes, $z$-tests are used; $t$-tests, on smaller ones.
- A $t$-test adjusts the sampling distribution for degrees of freedom.
- One-sample tests are univariate tests that examine whether the value of a single variable is generalizable to the population.
- Two-sample tests are bivariate tests that examine whether the observed difference between two variables can be inferred to the population.
- Both $z$-tests and $t$-tests, as well as one- and two-sample tests, can be conducted on both one- and two-sample tests of means or proportions.
- All inferential hypothesis tests end up comparing a calculated value to a critical value.
- In all inferential hypothesis tests, the null hypothesis is rejected if the calculated value is as large as or larger than the critical value.

This chapter concludes our coverage of inferential statistics. You now have a full complement of both descriptive and inferential tools in your social statistics toolbox. Looking back over the chapters, you should be impressed by the wide range of techniques you have mastered. As we have repeatedly stressed, different statistical tools are designed for different kinds of research issues. In all cases, selecting the appropriate tool is related to answering (1) whether the issue is a descriptive or an inferential one, (2) how many variables are being analyzed simultaneously, and (3) the level of measurement of the variables involved.

# Epilogue

## A Few Final Words

Your tour through the maze of social statistics is complete. Congratulations on your efforts working your way through! We are pleased to have been your tour guides. Now that you have been through the maze and know the grid pattern, there is no reason why you cannot return with confidence any time you like.

Whenever you return to the maze, just remember the three central questions that act as the compass that provides orientation:

- Does your problem centre on a descriptive or an inferential issue?
- How many variables are being analyzed simultaneously?
- What is the level of measurement of each variable?

As you have seen, different combinations of answers to these three questions place you in different locations in the statistical maze. In each location you will find the appropriate statistical tools. If it has been a while since you visited any specific region of the maze, you can always review any of the five steps to refresh your understanding of the statistical tools found there, including their use, calculation, and interpretation.

As noted in the Preface, there are many reasons for gaining statistical competence. Perhaps you have learned that (at least sometimes) solving statistical problems can be fun! Alternatively, statistical knowledge can be profitable—in attaining better grades in this course, in obtaining paid employment, in helping you understand the world.

The world is a complicated place and there are many ways to approach understanding it. No single way of knowing is perfect, which is why it is useful to have multiple methods at your command. The quantitative research tradition, of which social statistics is a part, is one important way of gaining insight into questions of interest. If you look at government reports, non-government assessments, private sector planning documents, or any of the varied forms of mass media, you will see they often contain statistical findings. The contents of this book provide you the tools for, at least, being an informed consumer of this information.

Competence in social statistics is both uncommon and important. If you attend carefully to reports of statistical results, rather than just skip over them, you may be amazed at the number of times basic errors appear. Since a lot of public opinion and policy is based on such statistical findings, it is important that you can make your own assessments of such reports about reality. With the tools you learned in this book, you do not have to simply believe whatever you are told.

Following the cartoon reprinted in the Preface, the statistical tools at your command locate you "three standard deviations above the norm"! We trust you will be able to put your understanding of these tools to good use. If you want to share your stories about either the frustrations or glories of using statistical tools, we are happy to hear them. Feel free to send us a note at Lance.Roberts@umanitoba.ca.

LWR

JDE

TP

LA

# Appendix A

## Chi-Square

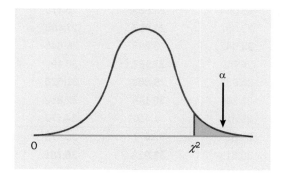

| df | Critical values of chi-square Level of significance for two-tailed test | | | | |
|---|---|---|---|---|---|
| | 0.1 | 0.05 | 0.025 | 0.01 | 0.001 |
| 1 | 2.706 | 3.841 | 5.024 | 6.635 | 10.828 |
| 2 | 4.605 | 5.991 | 7.378 | 9.210 | 13.816 |
| 3 | 6.251 | 7.815 | 9.348 | 11.345 | 16.266 |
| 4 | 7.779 | 9.488 | 11.143 | 13.277 | 18.467 |
| 5 | 9.236 | 11.070 | 12.833 | 15.086 | 20.515 |
| 6 | 10.645 | 12.592 | 14.449 | 16.812 | 22.458 |
| 7 | 12.017 | 14.067 | 16.013 | 18.475 | 24.322 |
| 8 | 13.362 | 15.507 | 17.535 | 20.090 | 26.125 |
| 9 | 14.684 | 16.919 | 19.023 | 21.666 | 27.877 |
| 10 | 15.987 | 18.307 | 20.483 | 23.209 | 29.588 |
| 11 | 17.275 | 19.675 | 21.920 | 24.725 | 31.264 |
| 12 | 18.549 | 21.026 | 23.337 | 26.217 | 32.910 |
| 13 | 19.812 | 22.362 | 24.736 | 27.688 | 34.528 |
| 14 | 21.064 | 23.685 | 26.119 | 29.141 | 36.123 |
| 15 | 22.307 | 24.996 | 27.488 | 30.578 | 37.697 |
| 16 | 23.542 | 26.296 | 28.845 | 32.000 | 39.252 |
| 17 | 24.769 | 27.587 | 30.191 | 33.409 | 40.790 |
| 18 | 25.989 | 28.869 | 31.526 | 34.805 | 42.312 |
| 19 | 27.204 | 30.144 | 32.852 | 36.191 | 43.820 |
| 20 | 28.412 | 31.410 | 34.170 | 37.566 | 45.315 |
| 21 | 29.615 | 32.671 | 35.479 | 38.932 | 46.797 |
| 22 | 30.813 | 33.924 | 36.781 | 40.289 | 48.268 |
| 23 | 32.007 | 35.172 | 38.076 | 41.638 | 49.728 |
| 24 | 33.196 | 36.415 | 39.364 | 42.980 | 51.179 |
| 25 | 34.382 | 37.652 | 40.646 | 44.314 | 52.620 |
| 26 | 35.563 | 38.885 | 41.923 | 45.642 | 54.052 |
| 27 | 36.741 | 40.113 | 43.195 | 46.963 | 55.476 |
| 28 | 37.916 | 41.337 | 44.461 | 48.278 | 56.892 |
| 29 | 39.087 | 42.557 | 45.722 | 49.588 | 58.301 |
| 30 | 40.256 | 43.773 | 46.979 | 50.892 | 59.703 |
| 31 | 41.422 | 44.985 | 48.232 | 52.191 | 61.098 |
| 32 | 42.585 | 46.194 | 49.480 | 53.486 | 62.487 |
| 33 | 43.745 | 47.400 | 50.725 | 54.776 | 63.870 |
| 34 | 44.903 | 48.602 | 51.966 | 56.061 | 65.247 |
| 35 | 46.059 | 49.802 | 53.203 | 57.342 | 66.619 |
| 36 | 47.212 | 50.998 | 54.437 | 58.619 | 67.985 |
| 37 | 48.363 | 52.192 | 55.668 | 59.893 | 69.347 |
| 38 | 49.513 | 53.384 | 56.896 | 61.162 | 70.703 |

| df | Critical values of chi-square Level of significance for two-tailed test | | | | |
|---|---|---|---|---|---|
| | 0.1 | 0.05 | 0.025 | 0.01 | 0.001 |
| 39 | 50.660 | 54.572 | 58.120 | 62.428 | 72.055 |
| 40 | 51.805 | 55.758 | 59.342 | 63.691 | 73.402 |
| 41 | 52.949 | 56.942 | 60.561 | 64.950 | 74.745 |
| 42 | 54.090 | 58.124 | 61.777 | 66.206 | 76.084 |
| 43 | 55.230 | 59.304 | 62.990 | 67.459 | 77.419 |
| 44 | 56.369 | 60.481 | 64.201 | 68.710 | 78.750 |
| 45 | 57.505 | 61.656 | 65.410 | 69.957 | 80.077 |
| 46 | 58.641 | 62.830 | 66.617 | 71.201 | 81.400 |
| 47 | 59.774 | 64.001 | 67.821 | 72.443 | 82.720 |
| 48 | 60.907 | 65.171 | 69.023 | 73.683 | 84.037 |
| 49 | 62.038 | 66.339 | 70.222 | 74.919 | 85.351 |
| 50 | 63.167 | 67.505 | 71.420 | 76.154 | 86.661 |
| 51 | 64.295 | 68.669 | 72.616 | 77.386 | 87.968 |
| 52 | 65.422 | 69.832 | 73.810 | 78.616 | 89.272 |
| 53 | 66.548 | 70.993 | 75.002 | 79.843 | 90.573 |
| 54 | 67.673 | 72.153 | 76.192 | 81.069 | 91.872 |
| 55 | 68.796 | 73.311 | 77.380 | 82.292 | 93.168 |
| 56 | 69.919 | 74.468 | 78.567 | 83.513 | 94.461 |
| 57 | 71.040 | 75.624 | 79.752 | 84.733 | 95.751 |
| 58 | 72.160 | 76.778 | 80.936 | 85.950 | 97.039 |
| 59 | 73.279 | 77.931 | 82.117 | 87.166 | 98.324 |
| 60 | 74.397 | 79.082 | 83.298 | 88.379 | 99.607 |
| 61 | 75.514 | 80.232 | 84.476 | 89.591 | 100.888 |
| 62 | 76.630 | 81.381 | 85.654 | 90.802 | 102.166 |
| 63 | 77.745 | 82.529 | 86.830 | 92.010 | 103.442 |
| 64 | 78.860 | 83.675 | 88.004 | 93.217 | 104.716 |
| 65 | 79.973 | 84.821 | 89.177 | 94.422 | 105.988 |
| 66 | 81.085 | 85.965 | 90.349 | 95.626 | 107.258 |
| 67 | 82.197 | 87.108 | 91.519 | 96.828 | 108.526 |
| 68 | 83.308 | 88.250 | 92.689 | 98.028 | 109.791 |
| 69 | 84.418 | 89.391 | 93.856 | 99.228 | 111.055 |
| 70 | 85.527 | 90.531 | 95.023 | 100.425 | 112.317 |
| 71 | 86.635 | 91.670 | 96.189 | 101.621 | 113.577 |
| 72 | 87.743 | 92.808 | 97.353 | 102.816 | 114.835 |
| 73 | 88.850 | 93.945 | 98.516 | 104.010 | 116.092 |
| 74 | 89.956 | 95.081 | 99.678 | 105.202 | 117.346 |
| 75 | 91.061 | 96.217 | 100.839 | 106.393 | 118.599 |
| 76 | 92.166 | 97.351 | 101.999 | 107.583 | 119.850 |

(*continued*)

| df | 0.1 | 0.05 | 0.025 | 0.01 | 0.001 |
|---|---|---|---|---|---|
| | Critical values of chi-square Level of significance for two-tailed test | | | | |
| 77 | 93.270 | 98.484 | 103.158 | 108.771 | 121.100 |
| 78 | 94.374 | 99.617 | 104.316 | 109.958 | 122.348 |
| 79 | 95.476 | 100.749 | 105.473 | 111.144 | 123.594 |
| 80 | 96.578 | 101.879 | 106.629 | 112.329 | 124.839 |
| 81 | 97.680 | 103.010 | 107.783 | 113.512 | 126.083 |
| 82 | 98.780 | 104.139 | 108.937 | 114.695 | 127.324 |
| 83 | 99.880 | 105.267 | 110.090 | 115.876 | 128.565 |
| 84 | 100.980 | 106.395 | 111.242 | 117.057 | 129.804 |
| 85 | 102.079 | 107.522 | 112.393 | 118.236 | 131.041 |
| 86 | 103.177 | 108.648 | 113.544 | 119.414 | 132.277 |
| 87 | 104.275 | 109.773 | 114.693 | 120.591 | 133.512 |
| 88 | 105.372 | 110.898 | 115.841 | 121.767 | 134.746 |
| 89 | 106.469 | 112.022 | 116.989 | 122.942 | 135.978 |
| 90 | 107.565 | 113.145 | 118.136 | 124.116 | 137.208 |
| 91 | 108.661 | 114.268 | 119.282 | 125.289 | 138.438 |
| 92 | 109.756 | 115.390 | 120.427 | 126.462 | 139.666 |
| 93 | 110.850 | 116.511 | 121.571 | 127.633 | 140.893 |
| 94 | 111.944 | 117.632 | 122.715 | 128.803 | 142.119 |
| 95 | 113.038 | 118.752 | 123.858 | 129.973 | 143.344 |
| 96 | 114.131 | 119.871 | 125.000 | 131.141 | 144.567 |
| 97 | 115.223 | 120.990 | 126.141 | 132.309 | 145.789 |
| 98 | 116.315 | 122.108 | 127.282 | 133.476 | 147.010 |
| 99 | 117.407 | 123.225 | 128.422 | 134.642 | 148.230 |
| 100 | 118.498 | 124.342 | 129.561 | 135.807 | 149.449 |

# Appendix B

## The Student's *t*-Table

**For a One-Tailed Test:**

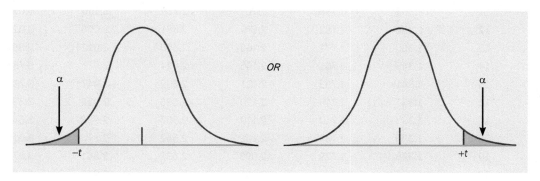

**For a Two-Tailed Test:**

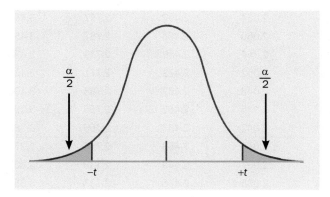

| df | Level of significance for one-tailed test | | | | | |
|---|---|---|---|---|---|---|
| | 0.1 | 0.05 | 0.025 | 0.01 | 0.005 | 0.001 |
| | Level of significance for two-tailed test | | | | | |
| | 0.2 | 0.1 | 0.05 | 0.02 | 0.01 | 0.002 |
| 1 | 3.078 | 6.314 | 12.706 | 31.821 | 63.657 | 318.313 |
| 2 | 1.886 | 2.920 | 4.303 | 6.965 | 9.925 | 22.327 |
| 3 | 1.638 | 2.353 | 3.182 | 4.541 | 5.841 | 10.215 |
| 4 | 1.533 | 2.132 | 2.776 | 3.747 | 4.604 | 7.173 |
| 5 | 1.476 | 2.015 | 2.571 | 3.365 | 4.032 | 5.893 |
| 6 | 1.440 | 1.943 | 2.447 | 3.143 | 3.707 | 5.208 |
| 7 | 1.415 | 1.895 | 2.365 | 2.998 | 3.499 | 4.782 |
| 8 | 1.397 | 1.860 | 2.306 | 2.896 | 3.355 | 4.499 |
| 9 | 1.383 | 1.833 | 2.262 | 2.821 | 3.250 | 4.296 |
| 10 | 1.372 | 1.812 | 2.228 | 2.764 | 3.169 | 4.143 |
| 11 | 1.363 | 1.796 | 2.201 | 2.718 | 3.106 | 4.024 |
| 12 | 1.356 | 1.782 | 2.179 | 2.681 | 3.055 | 3.929 |
| 13 | 1.350 | 1.771 | 2.160 | 2.650 | 3.012 | 3.852 |
| 14 | 1.345 | 1.761 | 2.145 | 2.624 | 2.977 | 3.787 |
| 15 | 1.341 | 1.753 | 2.131 | 2.602 | 2.947 | 3.733 |
| 16 | 1.337 | 1.746 | 2.120 | 2.583 | 2.921 | 3.686 |
| 17 | 1.333 | 1.740 | 2.110 | 2.567 | 2.898 | 3.646 |
| 18 | 1.330 | 1.734 | 2.101 | 2.552 | 2.878 | 3.610 |
| 19 | 1.328 | 1.729 | 2.093 | 2.539 | 2.861 | 3.579 |
| 20 | 1.325 | 1.725 | 2.086 | 2.528 | 2.845 | 3.552 |
| 21 | 1.323 | 1.721 | 2.080 | 2.518 | 2.831 | 3.527 |
| 22 | 1.321 | 1.717 | 2.074 | 2.508 | 2.819 | 3.505 |
| 23 | 1.319 | 1.714 | 2.069 | 2.500 | 2.807 | 3.485 |
| 24 | 1.318 | 1.711 | 2.064 | 2.492 | 2.797 | 3.467 |
| 25 | 1.316 | 1.708 | 2.060 | 2.485 | 2.787 | 3.450 |
| 26 | 1.315 | 1.706 | 2.056 | 2.479 | 2.779 | 3.435 |
| 27 | 1.314 | 1.703 | 2.052 | 2.473 | 2.771 | 3.421 |
| 28 | 1.313 | 1.701 | 2.048 | 2.467 | 2.763 | 3.408 |
| 29 | 1.311 | 1.699 | 2.045 | 2.462 | 2.756 | 3.396 |
| 30 | 1.310 | 1.697 | 2.042 | 2.457 | 2.750 | 3.385 |
| 31 | 1.309 | 1.696 | 2.040 | 2.453 | 2.744 | 3.375 |
| 32 | 1.309 | 1.694 | 2.037 | 2.449 | 2.738 | 3.365 |
| 33 | 1.308 | 1.692 | 2.035 | 2.445 | 2.733 | 3.356 |
| 34 | 1.307 | 1.691 | 2.032 | 2.441 | 2.728 | 3.348 |
| 35 | 1.306 | 1.690 | 2.030 | 2.438 | 2.724 | 3.340 |
| 36 | 1.306 | 1.688 | 2.028 | 2.434 | 2.719 | 3.333 |

| | Level of significance for one-tailed test | | | | | |
|---|---|---|---|---|---|---|
| | **0.1** | **0.05** | **0.025** | **0.01** | **0.005** | **0.001** |
| | Level of significance for two-tailed test | | | | | |
| **df** | **0.2** | **0.1** | **0.05** | **0.02** | **0.01** | **0.002** |
| 37 | 1.305 | 1.687 | 2.026 | 2.431 | 2.715 | 3.326 |
| 38 | 1.304 | 1.686 | 2.024 | 2.429 | 2.712 | 3.319 |
| 39 | 1.304 | 1.685 | 2.023 | 2.426 | 2.708 | 3.313 |
| 40 | 1.303 | 1.684 | 2.021 | 2.423 | 2.704 | 3.307 |
| 41 | 1.303 | 1.683 | 2.020 | 2.421 | 2.701 | 3.301 |
| 42 | 1.302 | 1.682 | 2.018 | 2.418 | 2.698 | 3.296 |
| 43 | 1.302 | 1.681 | 2.017 | 2.416 | 2.695 | 3.291 |
| 44 | 1.301 | 1.680 | 2.015 | 2.414 | 2.692 | 3.286 |
| 45 | 1.301 | 1.679 | 2.014 | 2.412 | 2.690 | 3.281 |
| 46 | 1.300 | 1.679 | 2.013 | 2.410 | 2.687 | 3.277 |
| 47 | 1.300 | 1.678 | 2.012 | 2.408 | 2.685 | 3.273 |
| 48 | 1.299 | 1.677 | 2.011 | 2.407 | 2.682 | 3.269 |
| 49 | 1.299 | 1.677 | 2.010 | 2.405 | 2.680 | 3.265 |
| 50 | 1.299 | 1.676 | 2.009 | 2.403 | 2.678 | 3.261 |
| 51 | 1.298 | 1.675 | 2.008 | 2.402 | 2.676 | 3.258 |
| 52 | 1.298 | 1.675 | 2.007 | 2.400 | 2.674 | 3.255 |
| 53 | 1.298 | 1.674 | 2.006 | 2.399 | 2.672 | 3.251 |
| 54 | 1.297 | 1.674 | 2.005 | 2.397 | 2.670 | 3.248 |
| 55 | 1.297 | 1.673 | 2.004 | 2.396 | 2.668 | 3.245 |
| 56 | 1.297 | 1.673 | 2.003 | 2.395 | 2.667 | 3.242 |
| 57 | 1.297 | 1.672 | 2.002 | 2.394 | 2.665 | 3.239 |
| 58 | 1.296 | 1.672 | 2.002 | 2.392 | 2.663 | 3.237 |
| 59 | 1.296 | 1.671 | 2.001 | 2.391 | 2.662 | 3.234 |
| 60 | 1.296 | 1.671 | 2.000 | 2.390 | 2.660 | 3.232 |
| 61 | 1.296 | 1.670 | 2.000 | 2.389 | 2.659 | 3.229 |
| 62 | 1.295 | 1.670 | 1.999 | 2.388 | 2.657 | 3.227 |
| 63 | 1.295 | 1.669 | 1.998 | 2.387 | 2.656 | 3.225 |
| 64 | 1.295 | 1.669 | 1.998 | 2.386 | 2.655 | 3.223 |
| 65 | 1.295 | 1.669 | 1.997 | 2.385 | 2.654 | 3.220 |
| 66 | 1.295 | 1.668 | 1.997 | 2.384 | 2.652 | 3.218 |
| 67 | 1.294 | 1.668 | 1.996 | 2.383 | 2.651 | 3.216 |
| 68 | 1.294 | 1.668 | 1.995 | 2.382 | 2.650 | 3.214 |
| 69 | 1.294 | 1.667 | 1.995 | 2.382 | 2.649 | 3.213 |
| 70 | 1.294 | 1.667 | 1.994 | 2.381 | 2.648 | 3.211 |
| 71 | 1.294 | 1.667 | 1.994 | 2.380 | 2.647 | 3.209 |
| 72 | 1.293 | 1.666 | 1.993 | 2.379 | 2.646 | 3.207 |

*(continued)*

| df | Level of significance for one-tailed test | | | | | |
|----|------|------|------|------|------|------|
|    | 0.1 | 0.05 | 0.025 | 0.01 | 0.005 | 0.001 |
|    | Level of significance for two-tailed test | | | | | |
|    | 0.2 | 0.1 | 0.05 | 0.02 | 0.01 | 0.002 |
| 73 | 1.293 | 1.666 | 1.993 | 2.379 | 2.645 | 3.206 |
| 74 | 1.293 | 1.666 | 1.993 | 2.378 | 2.644 | 3.204 |
| 75 | 1.293 | 1.665 | 1.992 | 2.377 | 2.643 | 3.202 |
| 76 | 1.293 | 1.665 | 1.992 | 2.376 | 2.642 | 3.201 |
| 77 | 1.293 | 1.665 | 1.991 | 2.376 | 2.641 | 3.199 |
| 78 | 1.292 | 1.665 | 1.991 | 2.375 | 2.640 | 3.198 |
| 79 | 1.292 | 1.664 | 1.990 | 2.374 | 2.640 | 3.197 |
| 80 | 1.292 | 1.664 | 1.990 | 2.374 | 2.639 | 3.195 |
| 81 | 1.292 | 1.664 | 1.990 | 2.373 | 2.638 | 3.194 |
| 82 | 1.292 | 1.664 | 1.989 | 2.373 | 2.637 | 3.193 |
| 83 | 1.292 | 1.663 | 1.989 | 2.372 | 2.636 | 3.191 |
| 84 | 1.292 | 1.663 | 1.989 | 2.372 | 2.636 | 3.190 |
| 85 | 1.292 | 1.663 | 1.988 | 2.371 | 2.635 | 3.189 |
| 86 | 1.291 | 1.663 | 1.988 | 2.370 | 2.634 | 3.188 |
| 87 | 1.291 | 1.663 | 1.988 | 2.370 | 2.634 | 3.187 |
| 88 | 1.291 | 1.662 | 1.987 | 2.369 | 2.633 | 3.185 |
| 89 | 1.291 | 1.662 | 1.987 | 2.369 | 2.632 | 3.184 |
| 90 | 1.291 | 1.662 | 1.987 | 2.368 | 2.632 | 3.183 |
| 91 | 1.291 | 1.662 | 1.986 | 2.368 | 2.631 | 3.182 |
| 92 | 1.291 | 1.662 | 1.986 | 2.368 | 2.630 | 3.181 |
| 93 | 1.291 | 1.661 | 1.986 | 2.367 | 2.630 | 3.180 |
| 94 | 1.291 | 1.661 | 1.986 | 2.367 | 2.629 | 3.179 |
| 95 | 1.291 | 1.661 | 1.985 | 2.366 | 2.629 | 3.178 |
| 96 | 1.290 | 1.661 | 1.985 | 2.366 | 2.628 | 3.177 |
| 97 | 1.290 | 1.661 | 1.985 | 2.365 | 2.627 | 3.176 |
| 98 | 1.290 | 1.661 | 1.984 | 2.365 | 2.627 | 3.175 |
| 99 | 1.290 | 1.660 | 1.984 | 2.365 | 2.626 | 3.175 |
| 100 | 1.290 | 1.660 | 1.984 | 2.364 | 2.626 | 3.174 |
| 120 | 1.289 | 1.658 | 1.980 | 2.358 | 2.617 | 3.373 |
| ∞ | 1.282 | 1.645 | 1.960 | 2.326 | 2.576 | 3.090 |

# Appendix C

## Areas Under the Normal Curve

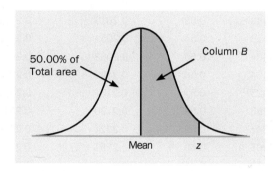

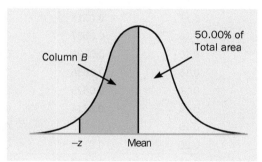

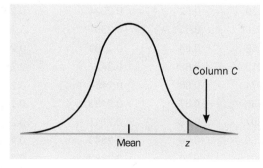

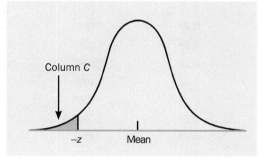

| A | B | C | A | B | C |
|---|---|---|---|---|---|
| z | Area between mean and z | Area beyond z | z | Area between mean and z | Area beyond z |
| 0.00 | 0.0000 | 0.5000 | | | |
| 0.01 | 0.0040 | 0.4960 | 0.41 | 0.1591 | 0.3409 |
| 0.02 | 0.0080 | 0.4920 | 0.42 | 0.1628 | 0.3372 |
| 0.03 | 0.0120 | 0.4880 | 0.43 | 0.1664 | 0.3336 |
| 0.04 | 0.0160 | 0.4840 | 0.44 | 0.1700 | 0.3300 |
| 0.05 | 0.0199 | 0.4801 | 0.45 | 0.1736 | 0.3264 |
| 0.06 | 0.0239 | 0.4761 | 0.46 | 0.1772 | 0.3228 |
| 0.07 | 0.0279 | 0.4721 | 0.47 | 0.1808 | 0.3192 |
| 0.08 | 0.0319 | 0.4681 | 0.48 | 0.1844 | 0.3156 |
| 0.09 | 0.0359 | 0.4641 | 0.49 | 0.1879 | 0.3121 |
| 0.10 | 0.0398 | 0.4602 | 0.50 | 0.1915 | 0.3085 |
| 0.11 | 0.0438 | 0.4562 | 0.51 | 0.1950 | 0.3050 |
| 0.12 | 0.0478 | 0.4522 | 0.52 | 0.1985 | 0.3015 |
| 0.13 | 0.0517 | 0.4483 | 0.53 | 0.2019 | 0.2981 |
| 0.14 | 0.0557 | 0.4443 | 0.54 | 0.2054 | 0.2946 |
| 0.15 | 0.0596 | 0.4404 | 0.55 | 0.2088 | 0.2912 |
| 0.16 | 0.0636 | 0.4364 | 0.56 | 0.2123 | 0.2877 |
| 0.17 | 0.0675 | 0.4325 | 0.57 | 0.2157 | 0.2843 |
| 0.18 | 0.0714 | 0.4286 | 0.58 | 0.2190 | 0.2810 |
| 0.19 | 0.0753 | 0.4247 | 0.59 | 0.2224 | 0.2776 |
| 0.20 | 0.0793 | 0.4207 | 0.60 | 0.2257 | 0.2743 |
| 0.21 | 0.0832 | 0.4168 | 0.61 | 0.2291 | 0.2709 |
| 0.22 | 0.0871 | 0.4129 | 0.62 | 0.2324 | 0.2676 |
| 0.23 | 0.0910 | 0.4090 | 0.63 | 0.2357 | 0.2643 |
| 0.24 | 0.0948 | 0.4052 | 0.64 | 0.2389 | 0.2611 |
| 0.25 | 0.0987 | 0.4013 | 0.65 | 0.2422 | 0.2578 |
| 0.26 | 0.1026 | 0.3974 | 0.66 | 0.2454 | 0.2546 |
| 0.27 | 0.1064 | 0.3936 | 0.67 | 0.2486 | 0.2514 |
| 0.28 | 0.1103 | 0.3897 | 0.68 | 0.2517 | 0.2483 |
| 0.29 | 0.1141 | 0.3859 | 0.69 | 0.2549 | 0.2451 |
| 0.30 | 0.1179 | 0.3821 | 0.70 | 0.2580 | 0.2420 |
| 0.31 | 0.1217 | 0.3783 | 0.71 | 0.2611 | 0.2389 |
| 0.32 | 0.1255 | 0.3745 | 0.72 | 0.2642 | 0.2358 |
| 0.33 | 0.1293 | 0.3707 | 0.73 | 0.2673 | 0.2327 |
| 0.34 | 0.1331 | 0.3669 | 0.74 | 0.2704 | 0.2297 |
| 0.35 | 0.1368 | 0.3632 | 0.75 | 0.2734 | 0.2266 |
| 0.36 | 0.1406 | 0.3594 | 0.76 | 0.2764 | 0.2236 |
| 0.37 | 0.1443 | 0.3557 | 0.77 | 0.2794 | 0.2207 |
| 0.38 | 0.1480 | 0.3520 | 0.78 | 0.2823 | 0.2177 |

| A | B | C | A | B | C |
|---|---|---|---|---|---|
| z | Area between mean and z | Area beyond z | z | Area between mean and z | Area beyond z |
| 0.39 | 0.1517 | 0.3483 | 0.79 | 0.2852 | 0.2148 |
| 0.40 | 0.1554 | 0.3446 | 0.80 | 0.2881 | 0.2119 |
| 0.81 | 0.2910 | 0.2090 | 1.21 | 0.3869 | 0.1131 |
| 0.82 | 0.2939 | 0.2061 | 1.22 | 0.3888 | 0.1112 |
| 0.83 | 0.2967 | 0.2033 | 1.23 | 0.3907 | 0.1093 |
| 0.84 | 0.2995 | 0.2005 | 1.24 | 0.3925 | 0.1075 |
| 0.85 | 0.3023 | 0.1977 | 1.25 | 0.3944 | 0.1056 |
| 0.86 | 0.3051 | 0.1949 | 1.26 | 0.3962 | 0.1038 |
| 0.87 | 0.3078 | 0.1922 | 1.27 | 0.3980 | 0.1020 |
| 0.88 | 0.3106 | 0.1894 | 1.28 | 0.3997 | 0.1003 |
| 0.89 | 0.3133 | 0.1867 | 1.29 | 0.4015 | 0.0985 |
| 0.90 | 0.3159 | 0.1841 | 1.30 | 0.4032 | 0.0968 |
| 0.91 | 0.3186 | 0.1814 | 1.31 | 0.4049 | 0.0951 |
| 0.92 | 0.3212 | 0.1788 | 1.32 | 0.4066 | 0.0934 |
| 0.93 | 0.3238 | 0.1762 | 1.33 | 0.4082 | 0.0918 |
| 0.94 | 0.3264 | 0.1736 | 1.34 | 0.4099 | 0.0901 |
| 0.95 | 0.3289 | 0.1711 | 1.35 | 0.4115 | 0.0885 |
| 0.96 | 0.3315 | 0.1685 | 1.36 | 0.4131 | 0.0869 |
| 0.97 | 0.3340 | 0.1660 | 1.37 | 0.4147 | 0.0853 |
| 0.98 | 0.3365 | 0.1635 | 1.38 | 0.4162 | 0.0838 |
| 0.99 | 0.3389 | 0.1611 | 1.39 | 0.4177 | 0.0823 |
| 1.00 | 0.3413 | 0.1587 | 1.40 | 0.4192 | 0.0808 |
| 1.01 | 0.3438 | 0.1562 | 1.41 | 0.4207 | 0.0793 |
| 1.02 | 0.3461 | 0.1539 | 1.42 | 0.4222 | 0.0778 |
| 1.03 | 0.3485 | 0.1515 | 1.43 | 0.4236 | 0.0764 |
| 1.04 | 0.3508 | 0.1492 | 1.44 | 0.4251 | 0.0749 |
| 1.05 | 0.3531 | 0.1469 | 1.45 | 0.4265 | 0.0735 |
| 1.06 | 0.3554 | 0.1446 | 1.46 | 0.4279 | 0.0721 |
| 1.07 | 0.3577 | 0.1423 | 1.47 | 0.4292 | 0.0708 |
| 1.08 | 0.3599 | 0.1401 | 1.48 | 0.4306 | 0.0694 |
| 1.09 | 0.3621 | 0.1379 | 1.49 | 0.4319 | 0.0681 |
| 1.10 | 0.3643 | 0.1357 | 1.50 | 0.4332 | 0.0668 |
| 1.11 | 0.3665 | 0.1335 | 1.51 | 0.4345 | 0.0655 |
| 1.12 | 0.3686 | 0.1314 | 1.52 | 0.4357 | 0.0643 |
| 1.13 | 0.3708 | 0.1292 | 1.53 | 0.4370 | 0.0630 |
| 1.14 | 0.3729 | 0.1271 | 1.54 | 0.4382 | 0.0618 |
| 1.15 | 0.3749 | 0.1251 | 1.55 | 0.4394 | 0.0606 |
| 1.16 | 0.3770 | 0.1230 | 1.56 | 0.4406 | 0.0594 |

(*continued*)

| A | B | C | A | B | C |
|---|---|---|---|---|---|
| z | Area between mean and z | Area beyond z | z | Area between mean and z | Area beyond z |
| 1.17 | 0.3790 | 0.1210 | 1.57 | 0.4418 | 0.0582 |
| 1.18 | 0.3810 | 0.1190 | 1.58 | 0.4429 | 0.0571 |
| 1.19 | 0.3830 | 0.1170 | 1.59 | 0.4441 | 0.0559 |
| 1.20 | 0.3849 | 0.1151 | 1.60 | 0.4452 | 0.0548 |
| 1.61 | 0.4463 | 0.0537 | 2.01 | 0.4778 | 0.0222 |
| 1.62 | 0.4474 | 0.0526 | 2.02 | 0.4783 | 0.0217 |
| 1.63 | 0.4484 | 0.0516 | 2.03 | 0.4788 | 0.0212 |
| 1.64 | 0.4495 | 0.0505 | 2.04 | 0.4793 | 0.0207 |
| 1.65 | 0.4505 | 0.0495 | 2.05 | 0.4798 | 0.0202 |
| 1.66 | 0.4515 | 0.0485 | 2.06 | 0.4803 | 0.0197 |
| 1.67 | 0.4525 | 0.0475 | 2.07 | 0.4808 | 0.0192 |
| 1.68 | 0.4535 | 0.0465 | 2.08 | 0.4812 | 0.0188 |
| 1.69 | 0.4545 | 0.0455 | 2.09 | 0.4817 | 0.0183 |
| 1.70 | 0.4554 | 0.0446 | 2.10 | 0.4821 | 0.0179 |
| 1.71 | 0.4564 | 0.0436 | 2.11 | 0.4826 | 0.0174 |
| 1.72 | 0.4573 | 0.0427 | 2.12 | 0.4830 | 0.0170 |
| 1.73 | 0.4582 | 0.0418 | 2.13 | 0.4834 | 0.0166 |
| 1.74 | 0.4591 | 0.0409 | 2.14 | 0.4838 | 0.0162 |
| 1.75 | 0.4599 | 0.0401 | 2.15 | 0.4842 | 0.0158 |
| 1.76 | 0.4608 | 0.0392 | 2.16 | 0.4846 | 0.0154 |
| 1.77 | 0.4616 | 0.0384 | 2.17 | 0.4850 | 0.0150 |
| 1.78 | 0.4625 | 0.0375 | 2.18 | 0.4854 | 0.0146 |
| 1.79 | 0.4633 | 0.0367 | 2.19 | 0.4857 | 0.0143 |
| 1.80 | 0.4641 | 0.0359 | 2.20 | 0.4861 | 0.0139 |
| 1.81 | 0.4649 | 0.0351 | 2.21 | 0.4864 | 0.0136 |
| 1.82 | 0.4656 | 0.0344 | 2.22 | 0.4868 | 0.0132 |
| 1.83 | 0.4664 | 0.0336 | 2.23 | 0.4871 | 0.0129 |
| 1.84 | 0.4671 | 0.0329 | 2.24 | 0.4875 | 0.0125 |
| 1.85 | 0.4678 | 0.0322 | 2.25 | 0.4878 | 0.0122 |
| 1.86 | 0.4686 | 0.0314 | 2.26 | 0.4881 | 0.0119 |
| 1.87 | 0.4693 | 0.0307 | 2.27 | 0.4884 | 0.0116 |
| 1.88 | 0.4699 | 0.0301 | 2.28 | 0.4887 | 0.0113 |
| 1.89 | 0.4706 | 0.0294 | 2.29 | 0.4890 | 0.0110 |
| 1.90 | 0.4713 | 0.0287 | 2.30 | 0.4893 | 0.0107 |
| 1.91 | 0.4719 | 0.0281 | 2.31 | 0.4896 | 0.0104 |
| 1.92 | 0.4726 | 0.0274 | 2.32 | 0.4898 | 0.0102 |
| 1.93 | 0.4732 | 0.0268 | 2.33 | 0.4901 | 0.0099 |
| 1.94 | 0.4738 | 0.0262 | 2.34 | 0.4904 | 0.0096 |

| A | B | C | A | B | C |
|---|---|---|---|---|---|
| z | Area between mean and z | Area beyond z | z | Area between mean and z | Area beyond z |
| 1.95 | 0.4744 | 0.0256 | 2.35 | 0.4906 | 0.0094 |
| 1.96 | 0.4750 | 0.0250 | 2.36 | 0.4909 | 0.0091 |
| 1.97 | 0.4756 | 0.0244 | 2.37 | 0.4911 | 0.0089 |
| 1.98 | 0.4761 | 0.0239 | 2.38 | 0.4913 | 0.0087 |
| 1.99 | 0.4767 | 0.0233 | 2.39 | 0.4916 | 0.0084 |
| 2.00 | 0.4772 | 0.0228 | 2.40 | 0.4918 | 0.0082 |
| 2.41 | 0.4920 | 0.0080 | 2.81 | 0.4975 | 0.0025 |
| 2.42 | 0.4922 | 0.0078 | 2.82 | 0.4976 | 0.0024 |
| 2.43 | 0.4925 | 0.0075 | 2.83 | 0.4977 | 0.0023 |
| 2.44 | 0.4927 | 0.0073 | 2.84 | 0.4977 | 0.0023 |
| 2.45 | 0.4929 | 0.0071 | 2.85 | 0.4978 | 0.0022 |
| 2.46 | 0.4931 | 0.0069 | 2.86 | 0.4979 | 0.0021 |
| 2.47 | 0.4932 | 0.0068 | 2.87 | 0.4979 | 0.0021 |
| 2.48 | 0.4934 | 0.0066 | 2.88 | 0.4980 | 0.0020 |
| 2.49 | 0.4936 | 0.0064 | 2.89 | 0.4981 | 0.0019 |
| 2.50 | 0.4938 | 0.0062 | 2.90 | 0.4981 | 0.0019 |
| 2.51 | 0.4940 | 0.0060 | 2.91 | 0.4982 | 0.0018 |
| 2.52 | 0.4941 | 0.0059 | 2.92 | 0.4982 | 0.0018 |
| 2.53 | 0.4943 | 0.0057 | 2.93 | 0.4983 | 0.0017 |
| 2.54 | 0.4945 | 0.0055 | 2.94 | 0.4984 | 0.0016 |
| 2.55 | 0.4946 | 0.0054 | 2.95 | 0.4984 | 0.0016 |
| 2.56 | 0.4948 | 0.0052 | 2.96 | 0.4985 | 0.0015 |
| 2.57 | 0.4949 | 0.0051 | 2.97 | 0.4985 | 0.0015 |
| 2.58 | 0.4951 | 0.0049 | 2.98 | 0.4986 | 0.0014 |
| 2.59 | 0.4952 | 0.0048 | 2.99 | 0.4986 | 0.0014 |
| 2.60 | 0.4953 | 0.0047 | 3.00 | 0.4987 | 0.0013 |
| 2.61 | 0.4955 | 0.0045 | 3.01 | 0.4987 | 0.0013 |
| 2.62 | 0.4956 | 0.0044 | 3.02 | 0.4987 | 0.0013 |
| 2.63 | 0.4957 | 0.0043 | 3.03 | 0.4988 | 0.0012 |
| 2.64 | 0.4959 | 0.0041 | 3.04 | 0.4988 | 0.0012 |
| 2.65 | 0.4960 | 0.0040 | 3.05 | 0.4989 | 0.0011 |
| 2.66 | 0.4961 | 0.0039 | 3.06 | 0.4989 | 0.0011 |
| 2.67 | 0.4962 | 0.0038 | 3.07 | 0.4989 | 0.0011 |
| 2.68 | 0.4963 | 0.0037 | 3.08 | 0.4990 | 0.0010 |
| 2.69 | 0.4964 | 0.0036 | 3.09 | 0.4990 | 0.0010 |
| 2.70 | 0.4965 | 0.0035 | 3.10 | 0.4990 | 0.0010 |
| 2.71 | 0.4966 | 0.0034 | 3.11 | 0.4991 | 0.0009 |
| 2.72 | 0.4967 | 0.0033 | 3.12 | 0.4991 | 0.0009 |

*(continued)*

| A | B | C | A | B | C |
|---|---|---|---|---|---|
| z | Area between mean and z | Area beyond z | z | Area between mean and z | Area beyond z |
| 2.73 | 0.4968 | 0.0032 | 3.13 | 0.4991 | 0.0009 |
| 2.74 | 0.4969 | 0.0031 | 3.14 | 0.4992 | 0.0008 |
| 2.75 | 0.4970 | 0.0030 | 3.15 | 0.4992 | 0.0008 |
| 2.76 | 0.4971 | 0.0029 | 3.16 | 0.4992 | 0.0008 |
| 2.77 | 0.4972 | 0.0028 | 3.17 | 0.4992 | 0.0008 |
| 2.78 | 0.4973 | 0.0027 | 3.18 | 0.4993 | 0.0007 |
| 2.79 | 0.4974 | 0.0026 | 3.19 | 0.4993 | 0.0007 |
| 2.80 | 0.4974 | 0.0026 | 3.20 | 0.4993 | 0.0007 |
| 3.21 | 0.4993 | 0.0007 | 3.61 | 0.4998 | 0.0002 |
| 3.22 | 0.4994 | 0.0006 | 3.62 | 0.4999 | 0.0001 |
| 3.23 | 0.4994 | 0.0006 | 3.63 | 0.4999 | 0.0001 |
| 3.24 | 0.4994 | 0.0006 | 3.64 | 0.4999 | 0.0001 |
| 3.25 | 0.4994 | 0.0006 | 3.65 | 0.4999 | 0.0001 |
| 3.26 | 0.4994 | 0.0006 | 3.66 | 0.4999 | 0.0001 |
| 3.27 | 0.4995 | 0.0005 | 3.67 | 0.4999 | 0.0001 |
| 3.28 | 0.4995 | 0.0005 | 3.68 | 0.4999 | 0.0001 |
| 3.29 | 0.4995 | 0.0005 | 3.69 | 0.4999 | 0.0001 |
| 3.30 | 0.4995 | 0.0005 | 3.70 | 0.4999 | 0.0001 |
| 3.31 | 0.4995 | 0.0005 | 3.71 | 0.4999 | 0.0001 |
| 3.32 | 0.4995 | 0.0005 | 3.72 | 0.4999 | 0.0001 |
| 3.33 | 0.4996 | 0.0004 | 3.73 | 0.4999 | 0.0001 |
| 3.34 | 0.4996 | 0.0004 | 3.74 | 0.4999 | 0.0001 |
| 3.35 | 0.4996 | 0.0004 | 3.75 | 0.4999 | 0.0001 |
| 3.36 | 0.4996 | 0.0004 | 3.76 | 0.4999 | 0.0001 |
| 3.37 | 0.4996 | 0.0004 | 3.77 | 0.4999 | 0.0001 |
| 3.38 | 0.4996 | 0.0004 | 3.78 | 0.4999 | 0.0001 |
| 3.39 | 0.4997 | 0.0003 | 3.79 | 0.4999 | 0.0001 |
| 3.40 | 0.4997 | 0.0003 | 3.80 | 0.4999 | 0.0001 |
| 3.41 | 0.4997 | 0.0003 | 3.81 | 0.4999 | 0.0001 |
| 3.42 | 0.4997 | 0.0003 | 3.82 | 0.4999 | 0.0001 |
| 3.43 | 0.4997 | 0.0003 | 3.83 | 0.4999 | 0.0001 |
| 3.44 | 0.4997 | 0.0003 | 3.84 | 0.4999 | 0.0001 |
| 3.45 | 0.4997 | 0.0003 | 3.85 | 0.4999 | 0.0001 |
| 3.46 | 0.4997 | 0.0003 | 3.86 | 0.4999 | 0.0001 |
| 3.47 | 0.4997 | 0.0003 | 3.87 | 0.4999 | 0.0001 |
| 3.48 | 0.4997 | 0.0003 | 3.88 | 0.4999 | 0.0001 |
| 3.49 | 0.4998 | 0.0002 | 3.89 | 0.4999 | 0.0001 |
| 3.50 | 0.4998 | 0.0002 | 3.90 | 0.5000 | 0.0000 |

| A | B | C | A | B | C |
|---|---|---|---|---|---|
| z | Area between mean and z | Area beyond z | z | Area between mean and z | Area beyond z |
| 3.51 | 0.4998 | 0.0002 | 3.91 | 0.5000 | 0.0000 |
| 3.52 | 0.4998 | 0.0002 | 3.92 | 0.5000 | 0.0000 |
| 3.53 | 0.4998 | 0.0002 | 3.93 | 0.5000 | 0.0000 |
| 3.54 | 0.4998 | 0.0002 | 3.94 | 0.5000 | 0.0000 |
| 3.55 | 0.4998 | 0.0002 | 3.95 | 0.5000 | 0.0000 |
| 3.56 | 0.4998 | 0.0002 | 3.96 | 0.5000 | 0.0000 |
| 3.57 | 0.4998 | 0.0002 | 3.97 | 0.5000 | 0.0000 |
| 3.58 | 0.4998 | 0.0002 | 3.98 | 0.5000 | 0.0000 |
| 3.59 | 0.4998 | 0.0002 | 3.99 | 0.5000 | 0.0000 |
| 3.60 | 0.4998 | 0.0002 | 4.00 | 0.5000 | 0.0000 |

# Answer Key

This answer key provides answers for Set 1: Hand-Calculation Practice Questions in each chapter. See the companion website for answers to Set 2: SPSS Practice Questions.

## Chapter 4

1. 94.6 per cent
2. 5.4 per cent
3.

| Hours | f |
| --- | --- |
| 2 to 5 | 36 |
| 6 to 9 | 73 |
| 10 to 13 | 41 |
| 14 to 17 | 18 |
| Missing | 6 |
| Total | 174 |

4. 168
5.

| Hours | f | Percentage (%) | Valid Per cent | Cumulative Per cent |
| --- | --- | --- | --- | --- |
| 2 to 5 | 36 | 20.7 | 21.4 | 21.4 |
| 6 to 9 | 73 | 42.0 | 43.5 | 64.9 |
| 10 to 13 | 41 | 23.6 | 24.4 | 89.3 |
| 14 to 17 | 18 | 10.3 | 10.7 | 100 |
| Missing | 6 | 3.4 | | |
| Total | 174 | 100 | 100 | |

6. 43.5 per cent
7. 89.3 per cent
8. 35.1 per cent

9a.

| Declared Major | Frequency | Percentage % |
| --- | --- | --- |
| psychology | 8 | 38.10 |
| political science | 3 | 14.29 |
| sociology | 5 | 23.81 |
| anthropology | 4 | 19.05 |
| women's studies | 1 | 4.76 |
| Total | 21 | 100.01* |

*rounding error

9b.

| GPA | Frequency(f) | Per cent (%) | Valid Per cent | Cumulative Per cent |
| --- | --- | --- | --- | --- |
| 2.9 – 3.2 | 4 | 19.0 | 21.1 | 21.1 |
| 3.3 – 3.6 | 5 | 23.8 | 26.3 | 47.4 |
| 3.7 – 4.0 | 8 | 38.1 | 42.1 | 89.5 |
| 4.1 – 4.4 | 2 | 9.5 | 10.5 | 100.0 |
| Missing values | 2 | 9.5 | | |
| Total | 21 | 99.9* | 100.0 | |

* rounding error

## Chapter 5

1. Ford and Toyota are the most frequent categories (f = 4).
2.
   a. 23
   b. 23.5
   c. 22.9

3. mode = 98; mean = 104.9; median = 104

4.

   a. mode = psychology (f = 8)

   b. mean = 3.6

   c. median (i) 3, (ii) 3.5

   d. 3.9

   e. Those students whose GPAs were 3.8 and higher were more likely to apply for graduate school right after graduation (median of 4.5) when compared to the median likelihood for all students in that workshop (median of 3).

## Chapter 6

1. The Johnston family has greater diversity of political affiliations (IQV=0.926) than the Scribner clan (IQV = 0.823).

2. Crater Lake (IQV=0.975) has the greatest heterogeneity, while Lowe Lake has the least (IQV= 0.931).

3. $s^2 = 114.9$; $s = 10.7$

4. Colonel Park has greater variation in class sizes with a standard deviation of 3.1 students compared to 2.6 for Happy Valley.

5. $s^2 = 0.17$; $s = 0.41$

6.

   a. IQV = 0.89

   b. 3

   c. $s = 13.2$; on average respondents' ages were 13.2 years away from the mean age for the sample.

   d.

| Respondent | $x$ | $x - X$ | $(x - X)^2$ |
|---|---|---|---|
| 1 | 36 | −5.45 | 29.70 |
| 2 | 64 | 22.55 | 508.50 |
| 3 | 32 | −9.45 | 89.30 |
| 4 | 25 | −16.45 | 270.60 |
| 5 | 48 | 6.55 | 42.90 |
| 6 | 42 | 0.55 | 0.30 |
| 7 | 31 | −10.45 | 109.20 |
| 8 | 39 | −2.45 | 6.00 |
| 9 | 55 | 13.55 | 183.60 |
| 10 | 39 | −2.45 | 6.00 |
| 11 | 45 | 3.55 | 12.60 |
| Total | 456 | 0 | 1,258.70 |
| $\bar{X} =$ | 41.45 | | S2 = 114.43 |
| | | | S = 10.70 |

7.

   a. West Town $s^2 = 1,593.3$; $S = 39.9$

      East Town $s^2 = 6,155.12$; $S = 78.5$

   b. The mean time for both half marathons was identical. However, measures of dispersion provide further information that tell us about the real character of the two seemingly homogeneous races. It is apparent that the standard deviation of the East Town half marathon is nearly two times that of the West Town half marathon ($s = 78.5$ vs. $s = 39.9$). This tells us that while the runners in the West Town half marathon gave, on average, quite similar performances (the times of runners in West town race were all, on average, relatively close to their group mean), the runners in the East Town half marathon gave relatively diverse performances.

## Chapter 8

1.

   a. 1.37; −1.93

   b. 41.47 per cent; 47.32 per cent

   c. 7.49 per cent of rates fall below 3.0 per cent; 84.13 per cent of rates fall below 4.0

   d. Only 1.32 per cent of rates are above 4.5; 97.98 per cent of rates are above 2.8

   e. 97 per cent

   f. 33.8 per cent

   g. 23.69 per cent

2.

   a. City X: $z = 0.77$; City Y: $z = 2.81$

   b. City X: 22.06 per cent; City Y: 0.25 per cent

   c. 37.31 per cent

   d. 18.85 per cent

   e. 41.68 per cent

   f. 13.14 per cent in City X; 39.74 per cent in City Y

   g. 40.49 per cent

   h. 68.26 per cent

   i. $1,453,000 in market X and $815,000 in market Y

## Chapter 10

1. As *Nationality* shifts from *American* to *Canadian*, *Support for universal health care* changes from low to high. The average difference is 24.7 per cent, and thus *Nationality* makes a modest difference to *Support for universal health care*.

2.

   a. N = 280

b. The univariate distributions are as follows:

| Corporal Punishment | f | Highest Level of Education | f |
|---|---|---|---|
| Never | 173 | Post-secondary | 115 |
| Sometimes | 78 | high school | 107 |
| Often | 29 | <high school | 58 |
| Total | 280 | Total | 280 |

c. Form: Negative (Inverse) → As *Level of education* level increases from less than high school to post-secondary, *Support for corporal punishment* decreases. Strength: Given the average difference of 11.4 per cent, *Level of education* only makes a small difference to *Support for corporal punishment*.

3. Form: Positive (Direct) → *Perceived quality of life* increases as *Level of job satisfaction* increases. Strength: On average, *Job satisfaction* makes only a small difference (14.2 per cent) to *Quality of life*.

4. Form: Positive (direct) → as *Social class* increases, so does *Self-rated health*. Strength: On average, *Social class* makes only small difference to *Self-rated health* (8.4 per cent).

5. Form: Negative (inverse) → As a country's *Health-care spending per capita* increases, its *Infant mortality rate* decreases. Strength: The method of percentage difference shows a modest relationship (26.70 per cent).

6. Conditional distributions are found in the columns.

## Chapter 11

1.

a. (8, 35,000) (17, 50,000) (20, 55,000) (21, 70,000) (12, 40,000) (14, 45,000)

b & c. Scatterplot should look similar to this:

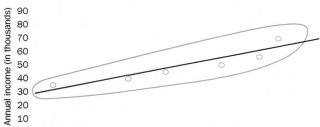

d. The relationship is direct in form: As *Years of schooling* increases, *Annual income* also increases. The

extent of the relationship is moderate: Changing *Years of schooling* makes a moderate difference to *Annual income*. The descriptions of form and extent offer a very precise characterization of the relationship between *Years of schooling* and *Annual income*.

2.

a. (8, 24) (7, 29) (2, 27) (3, 29) (12, 21) (5, 31) (10, 22) (1, 35)

b & c. Scatterplot should look similar to the one below:

d. The relationship is inverse in form: As *Hours of exercise per week* decrease, *BMI* increases. The extent of the relationship is moderate to strong: Changing *Hours of exercise per week* makes a moderate-to-strong difference to *BMI*. The descriptions of form and extent offer a moderately precise characterization of the relationship between *Hours of exercise per week* and *BMI*.

3.

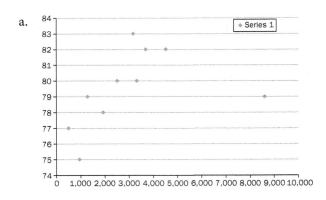

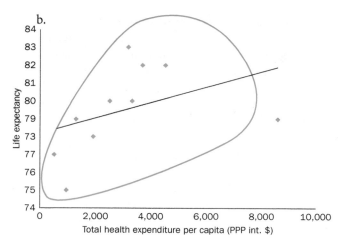

b.

c.  The form of the relationship is positive (direct). As a country's *Total health expenditure per capita* increases, so does its *Life expectancy* for both sexes. The extent of the relationship is moderate: Changing a country's *Total health expenditure per capita* makes a moderate difference to *Life expectancy* in that country. The descriptions of form and extent offer a moderately precise characterization of the relationship between *Total health expenditure per capita* and *Life expectancy*.

d.  When we eliminate the United States from the analysis, our results change markedly. The free-hand best-fit line has steeper slope and fits the data more closely. By eliminating the outlier, the extent and precision change from moderate to strong.

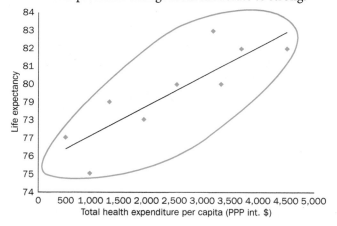

## Chapter 13

1.

  a.  103
  b.  69

c.  0.33

d.  You proportionately reduce your errors by 33 per cent when you know respondents' *Faculty*, as opposed to not knowing it, when predicting their *Math preference*.

e.  Yes, the value changes to 0.38. You proportionately reduce your errors by 38 per cent when you know respondents' *Math preference*, as opposed to not knowing it, when predicting their *Faculty*.

2.

  a.  160
  b.  128
  c.  0.2

d.  You proportionately reduce your errors by 20 per cent when you know respondents' *Nationality*, as opposed to not knowing it, when predicting their *Support for universal health care*.

e.  The average difference in Chapter 10 using the "percentage down, compare across" method was 24.7 per cent. Thus, the results of the two techniques are similar as both indicate a modest relationship between *Nationality* and *Support for universal health care*.

3.

  a.  0.03

  b.  You proportionately reduce your errors by 3 per cent when you know respondents' *Marital status*, as opposed to not knowing it, when predicting their *Life satisfaction*.

4.

  a.  4,151
  b.  9,775
  c.  E1 = 6,963; E2 = 4,151
  d.  −0.40

e.  Form: As *Level of education* increases, *Support for corporal punishment* decreases.
    Strength: You proportionately reduce your errors by 40 per cent when you know respondents' *Level of education*, as opposed to not knowing it, when predicting their *Support for corporal punishment*.

f.  Results of the "percentage down, compare across" technique show an average difference of 14.9 per cent, indicating that *Level of education* makes only a small difference to *Support for corporal punishment*. Why the discrepancy between gamma and "percentage down, compare across" technique? For our purposes, we will simply understand that this discrepancy has to do with the greater precision of gamma (and the relative arbitrariness of the guidelines for interpreting percentage differences).

5.
  a. 27,738
  b. 8,343
  c. E1 = 18,040.5; E2 = 8,343
  d. 0.54
  e. Form: As *Level of job satisfaction* increases, *Quality of life* also increases.
     Strength: You proportionately reduce your errors by 54 per cent when you know respondents' *Level of job satisfaction*, as opposed to not knowing it, when predicting their *Quality of life*.
  f. No, the value for gamma remains the same.

6.
  a. 132
  b. 59
  c. 0.55
  d. Lambda of 0.55 tells us that we can reduce the number of errors we make predicting the dependent variable (*Experience using transmission type*) by 55 per cent when we know the independent variable (*Continent of residence*).

7.
  a. 23,824
  b. 8,713
  c. $E_1$ = 16,268.5; $E_2$ = 8,713
  d. 0.46
  e. The form of the relationship is positive (direct); as *Income inequality* among population within a country rises, so does people's perception about the government's corruption. A gamma value of 0.46 tells us that we can proportionately reduce our errors by 46 per cent when we know the level of *Income inequality* in respondents' countries, as opposed to not knowing it, when predicting their level of *Perceived government corruption*.

8.
  a. 4,460
  b. 1,554
  c. E1 = 3,007; E2 = 1,554
  d. 0.48
  e. The form of the relationship is positive (direct); as an individual's *Income* rises, so does their conservativeness (more right wing). A gamma of 0.48

tells us that we can proportionately reduce our errors by 48 per cent when we know individuals' *Income*, as opposed to not knowing it, when predicting their *Political ideology*.

9.
  a. 147
  b. 137
  c. 0.07
  d. A lambda of 0.07 tells us that we will reduce our errors by 7 per cent when we know the offender's *Gender*, as opposed to not knowing it, when predicting the *Type of traffic offence*.

## Chapter 14

1.
  a. 3.62
  b. 57.98
  c. 94.18; 68.84
  d. $r = 0.91$; $r^2 = 0.83$
  e. Form: The relationship is direct/positive in form, as *Cultural conversation* in the home increases so does student *Reading achievement*.
     Extent: Changing *Cultural conversation* in the home by one hour per week produces a change of 3.62 points in *Reading achievement*.
     Precision: The regression line describing the connection between *Cultural conversation* in the home and *Reading achievement* provides a very accurate ($r = 0.91$) description of the actual connection between these variables.
     Strength: When predicting *Reading achievement*, you proportionately ($r^2 = 0.83$) reduce your errors by 83 per cent ($r^2 = 0.83$) when you know how many hours per week a respondent participates in cultural conversations at home.

2.
  a. 2.86
  b. 14.58
  c. 26.02; 34.6
  d. $r = 0.87$; $r^2 = .76$
  e. Form: The relationship is direct/positive in form, as hours of daily *TV viewing* increases so does *BMI*.
     Extent: Changing hours of *TV viewing* by one hour per day produces a change of 2.86 points in *BMI*.

Precision: The regression line describing the connection between hours of *TV viewing* per day and *BMI* provides a very accurate ($r = 0.87$) description of the actual connection between these variables.

Strength: When predicting *BMI*, you proportionately reduce your errors by 76 per cent ($r^2 = 0.76$) when you know how many hours per day a respondent watches TV.

3.

   a. See website for calculation table.
   b. −0.23
   c. 5.61
   d. 0.78; 3.31
   e. $r = -0.81$; $r^2 = 0.66$
   f. Form: The relationship is negative/inverse in form; *Number of children* decreases as *Years of education* increases.

   Extent: Changing *Years of education* by one year reduces *Number of children* by 0.81.

   Precision: The regression line describing the connection between *Years of education* and *Number of children* provides a very accurate ($r = -0.81$) description of the actual connection between these variables.

   Strength: When predicting *Number of children*, you proportionately reduce your errors by 66 per cent ($r^2 = 0.66$) when you know how many *Years of education* a woman has.

4.

   a. See website for calculation table.
   b. $b = 0.000432$; $a = 78.18$
   c. If $X = 5,000$, $Y = 80.34$; if $X = 1,000$, $Y = 78.61$
   d. $r = 0.41$; $r^2 = 0.41^2 = 0.17$
   e. Form: The relationship is of direct (positive) form. As the country's total health spending per capita increases, so does its average life expectancy for both sexes.

   Extent: Changing total health expenditure per capita by $1000 causes life expectancy to shift by 0.432 years in the same direction.
   f. Precision: The regression line describing the connection between total health expenditure per capita and life expectancy in years for both sexes provides a reasonably accurate ($r = 0.41$) description of the actual connection between these variables.

Strength: When predicting average life expectancy within a country, we proportionately reduce our errors by 17% ($r^2 = 0.17$) when we know the country's total health expenditure per capita.

5.

   a. $b = 0.00164$; $a = 75.57$
   b. If $X = 5,000$, $Y = 83.77$; $Y = $ If $X = 1,000$, $Y = 77.21$
   c. $r = 0.86$; $r^2 = 0.86^2 = 0.74$
   d. Form: The relationship is of direct (positive) form. As country's total health spending per capita increases, so does its average life expectancy for both sexes.

   Extent: Changing total health expenditure per capita by $1000 causes life expectancy to shift by 1.64 years in the same direction.

   Precision: The regression line describing the connection between total health expenditure per capita and life expectancy in years for both sexes provides a very accurate ($r = 0.86$) description of the actual connection between these variables.

   Strength: When predicting average life expectancy within a country, we proportionately reduce the errors by 74% ($r^2 = 0.74$) when we know the country's total health expenditure per capita.
   e. The form of the relationship remained the same in both instances but our conclusions about the extent, precision, and strength of the relationship between life expectancy and health spending have changed. The U.S. is an outlier, and the statistics we used to investigate the relationship between our two variables are easily influenced by extreme values (i.e. outliers). When we exclude the outlier from the analysis, the linear pattern formed by the data becomes much stronger, which explains why the changes we notice in extent, precision, and strength of the relationship. After excluding the U.S. from the analysis, the independent variable produces more change on the dependent variable ($b = .00164$ vs. $b = .000432$), the regression line becomes more precise at predicting the values on the dependent variable from values on the independent variables ($r = .86$ vs. $r = .41$), and we manage to significantly reduce the amount of errors when making such predictions ($r^2 = .074$ vs. $r^2 = .17$).

# Chapter 16

1.

a.

| Support Gun Registry | Male | Female | Total |
|---|---|---|---|
| Yes | 5 (41.7%) | 10 (55.6%) | 15 |
| No | 7 (58.3%) | 8 (44.4%) | 15 |
| Total | 12 | 18 | 30 |

b.

| Support Gun Registry | Rural | | Total |
|---|---|---|---|
| | Male | Female | |
| Yes | 1 (20%) | 2 (25%) | 3 |
| No | 4 (80%) | 6 (75%) | 10 |
| Total | 5 | 8 | 13 |

| Support Gun Registry | Urban | | Total |
|---|---|---|---|
| | Male | Female | |
| Yes | 4 (57%) | 8 (80%) | 12 |
| No | 3 (43%) | 2 (20%) | 5 |
| Total | 7 | 10 | 17 |

c. There is a 13.9 per cent difference, indicating a small relationship between *Gender* and *Support for gun registry*; females are slightly more likely to support the gun registry.

d. There is 5 per cent *Gender* difference in the *rural* partial table, indicating a small relationship between *Gender* and *Support for gun registry* among *rural* respondents; rural females are only slightly more likely to support the registry. There is a 23 per cent difference in the *urban* partial table, indicating a modest relationship between *Gender* and *Support for gun registry*; urban females are more likely to support the gun registry.

e. Yes. The effect of *Gender* is greater in the *urban* partial (23 per cent) than in the original table (13.9 per cent) or the *rural* partial table (5 per cent).

f. Logic dictates that it is more likely that where one lives (urban vs. rural residence) came prior in time to one's attitudes toward the gun registry. It is much less likely that someone would choose their place of residence based on their attitudes toward the gun registry. Therefore it makes sense to understand the *Rural or urban resident* variable as antecedent (prior in time) to attitudes toward the gun registry, and thus according to the scientific criterion of making causal attribution, to understand place of residence as possibly exerting an effect on attitudes toward the gun registry.

g. Interaction. This analysis shows that the connection between *Gender* and *Support for the gun registry* is complicated by the effects of residence. For urbanites the connection between the between *Gender* and *Support for the gun registry* is modest (with females more likely to support the gun registry). By contrast, for rural residents there is no connection between *Gender* and *Support for the gun registry* (i.e. *Gender* makes no difference).

2.

a.

| Mental Well-being | Marital Status | | Total |
|---|---|---|---|
| | Married | Single | |
| Good | 11 (73.3%) | 8 (53.3%) | 19 |
| Poor | 4 (26.7%) | 7 (46.7%) | 11 |
| Total | 15 | 15 | 30 |

b.

**MALE**

| Mental Well-Being | Marital Status | | Total |
|---|---|---|---|
| | Married | Single | |
| Good | 6 (75%) | 3 (37.5%) | 9 |
| Poor | 2 (25%) | 5 (62.5%) | 7 |
| Total | 8 | 8 | 16 |

**FEMALE**

| Mental Well-Being | Marital Status | | Total |
|---|---|---|---|
| | Married | Single | |
| Good | 5 (71.4%) | 5 (71.4%) | 10 |
| Poor | 2 (28.6%) | 2 (28.6%) | 4 |
| Total | 7 | 7 | 14 |

c. Percentage difference = 20 (indicating a small/modest relationship)

d. Male: percentage difference = 37.5 (moderate relationship)

Female: percentage difference = 0 (no relationship)

e. Yes, the variable does influence the original relationship. In the case of males, the original relationship between *Marital status* and *Mental well-being* is stronger than in the zero-order table, but in case of females the original relationship disappears entirely.

f. Interaction: The relationship is conditioned/specified by the control variable (i.e. dependent on the level/category of the third variable).

g. Is the original relationship genuine? Yes and no. The results of the elaboration analysis show that being married has genuine benefits for one's mental health, but only in the case of men. Our data do not support the existence of such health benefits for women. That is, the health benefits of being married are conditioned by gender: men experience benefits, women do not.

## Chapter 17

1.

a. *BMI* = 19.5 + 1.37 (*TV viewing*) + 1.04 (*Sodas*) − 0.72 (*Exercise*)

b. 23.53

c. Can't be determined as the coefficients are in original units; coefficients need to be standardized to make this determination.

d. *BMI* increases by 1.37 points for every additional hour of *TV viewing* per day, controlling for *Number of sodas* and amount of exercise. *BMI* increases 1.04 units for every additional soda per day, controlling for *TV viewing* and exercise. *BMI* decreases by 0.72 units for every additional hour of exercise per week, controlling for *TV viewing* and *Number of sodas*.

e. The predicted *BMI* (19.5) for someone who watches zero hours of TV per week, drinks zero sodas per day, and exercises zero hours per week.

2.

a. Alcoholic drinks per week = 6 − 0.4 (*GPA*) − 0.35 (*Hours studying*) + 0.09 (*Twitter followers*)

b. $Y_1 = 2.11; Y_2 = 3.54; Y_3 = 2.90$

c. *Number of drinks consumed per week* decreases by 0.4 drinks for every tenth of a grade-point increase in *GPA*, controlling for *Hours studying* per week and number of *Twitter followers*. *Number of drinks consumed per week* decreases by 0.35 drinks for every additional hour of studying per week,

controlling for *GPA* and number of *Twitter followers*. *Number of drinks consumed per week* increases by 0.09 drinks for every additional Twitter follower, controlling for *GPA* and *Hours studying*.

3.

a. Y = 112 + 0.0018 (*Population*) − 4.2 (*Temperature*) + 0.35 (*Number of roundabouts*)

b. $Y_A = 1{,}590.8; Y_B = 2{,}935.3; Y_C = 1{,}034.1$

c. The number of traffic collisions in a city will increase by 1.8 for every 1,000 increase in *Population*. The number of traffic accidents per month will decrease by 4.2 for every degree above zero, but will increase by 4.2 for every degree below zero. The number of collisions in a city will increase by 3.5 for every additional 10 roundabouts. Again, each coefficient gives the effect of that variable while controlling for (holding constant) the effects of all the other variables in the equation.

d. The intercept of 112 means that for cities with a *Population* of zero, with *Average monthly temperature* of zero, and with zero roundabouts, the number of accidents will be 112. This does not make sense, because a city of population zero will not have 112 accidents per month, but in step 3a we did specify that this was a prediction model for cities with *Population* of at least 100,000.

4.

a. Y = 73 + 0.83 (*Hours of exercise per week*) + 1.2 (*Number of servings of vegetables per day*) − 0.35 (*Years of smoking*)

b. $Y_A = 74.39; Y_B = 84.81; Y_C = 62.82$

c. Life expectancy will increase by 0.83 years for every additional *Hour of exercise per week*. Life expectancy will increase by 1.2 years with each additional *Serving of vegetables per day*. Life expectancy will decrease by 0.35 years for every additional *Year of smoking*.

d. Someone who never exercises, eats no vegetables, and doesn't smoke will have a life expectancy of 73 years.

## Chapter 18

1. 0.0983

2.

a. 168.82

b. It decreases to 150.52

c. It increases to 265.87

3. 0.016 or 1.6 per cent

**4.**

    a.  0.015 or 1.5 per cent

    b.  It decreases to 1.3 per cent.

    c.  It increases to 2.2 per cent.

**5.**

    a.  0.10 hours; 0.14 hours

    b.  0.06 hours; 0.09 hours

**6.**

    a.  3.09 per cent

    b.  2.33 per cent

**7.**

    a.  0.41 points

    b.  0.28 points; 0.23 points; 0.20 points

    c.  No, the standard error does not change proportionately to the changing sample size. If it did, then in our example the standard error of 0.38 points for a sample size of 250 would have decreased by half (to 0.19) when we doubled the sample size to 500 individuals. Instead, the standard error decreased only to 0.28 points. Similarly, when we tripled the sample size, we would have expected the standard error to decrease to 0.13 (0.38 ÷ 3), but instead it decreased only to 0.23 points.

**8.**

    a.  We need to compute standard error for sample proportions. The medication produced the desired results in 74.98 per cent of the sample.

    b.  0.91 per cent

## Chapter 19

**1.** $CI_{95} = 78.8 \pm 0.18$

The sample result showed that the average high-school grade average among university undergraduates was 78.8 per cent, with a margin of error ± 0.18 per cent and a level of confidence of 95 per cent.

or

The results indicate we can be 95 per cent confident that the actual average high-school grade average lies between 78.62 per cent and 78.98 per cent.

**2.** $846.18

The sample result showed that average household credit card debt was $13,577, with a margin of error ± 423.09 and a level of confidence of 99 per cent.

or

The results indicate we can be 99 per cent confident that the actual average household credit card debt lies between $13,153.91 and $14,000.09.

**3.**

    a.  $CI_{95} = 65 \pm 3.27$; lower limit = 61.73; upper limit = 68.27

The sample result showed that 65 per cent of university students report eating a fast-food meal at least once per week, with a margin of error of ± 3.27 per cent and a level of confidence of 95 per cent.

or

The results indicate we can be 95 per cent confident that the actual percentage of university students who eat fast food at least once a week is between 61.73 per cent and 68.27 per cent.

    b.  44 per cent ± 4.3 per cent; Interval width: 8.6

The sample result showed that 44 per cent of university students report being happy with the variety of food offerings on their campus with a margin of error of ± 4.3 per cent and a level of confidence of 99 per cent.

or

The results indicate we can be 99 per cent confident that the actual percentage of university students who are happy with the variety of food offerings on their campus is between 39.7 and 48.3 per cent.

    c.  $CI_{95} = 35$ per cent ± 11.72 per cent; Lower limit = 23.28 per cent; Upper limit = 46.72 per cent.

The sample result showed that 35 per cent of LBGTQ university students report having experienced some form of harassment on their campus with a margin of error of ± 11.72 per cent and a level of confidence of 95 per cent.

or

The results indicate we can be 95 per cent confident that the actual percentage of *LBGTQ* university students who have experienced some form of harassment on their campus is between 23.28 per cent and 46.72 per cent.

    d.  $CI_{99} = 67$ per cent ± 14.68 per cent Interval width = 29.36

The sample result showed that 67 per cent of international university students report being satisfied with their university experience in Canada, with a margin of error of ± 14.68 per cent and a level of confidence of 99 per cent.

or

The results indicate we can be 99 per cent confident that the actual percentage of international university students who are satisfied with their university experience in Canada is between 52.32 per cent and 81.68 per cent.

e. $CI_{95} = 685 \pm 22.03$; lower limit = 662.97; upper limit = 707.03

The sample result showed that the average monthly rent paid by international university students in Canada is $685, with a margin of error of ± $22.03 and a level of confidence of 95 per cent.

or

The results indicate we can be 95 per cent confident that the actual average monthly rent paid by international university students in Canada is between $662.97 and $707.03.

f. $CI_{99} = 8 \pm 0.86$

Interval width = 1.72

The sample result showed that the average number of months international students had been attending university in Canada was 8, with a margin of error of ± 0.86 months and a level of confidence of 99 per cent.

or

The results indicate we can be 99 per cent confident that the actual average number of months international students have been attending university in Canada is between 7.14 months and 8.86 months.

4.

a. $CI_{95} = 21$ per cent ± 4.39 per cent; Lower limit = 16.61 per cent; Upper limit = 25.39 per cent

The sample result showed that 21 per cent of the population support the Green party, with a margin of error ± 4.39 per cent and a level of confidence of 95 per cent.

or

The results indicate we can be 95 per cent confident that the actual voters' support for the Green party lies between 16.61 per cent and 25.39 per cent.

b. $CI_{95} = 21$ per cent ± 3.10 per cent

Lower limit: 17.90 per cent

Upper limit: 24.10 per cent

*The sample result showed that 21 per cent of the population support the Green party, with a margin of error ± 3.10 per cent and a level of confidence of 95 per cent.*

or

The results indicate we can be 95 per cent confident that the actual voters' support for the Green party lies between 17.90 per cent and 24.10 per cent.

c. $CI_{95} = 21$ per cent ± 2.55 per cent; Lower limit = 18.45 per cent; Upper limit = 23.55 per cent

The sample result showed that 21 per cent of the population support the Green party, with a margin of error ± 2.55 per cent and a level of confidence of 95 per cent.

or

The results indicate we can be 95 per cent confident that the actual voters' support for the Green party lies between 18.45 per cent and 23.55 per cent.

d. $CI_{95} = 21$ per cent ± 2.23 per cent; Lower limit = 18.77 per cent; Upper limit = 23.23 per cent

The sample result showed that 21 per cent of the population support the Green party, with a margin of error ± 2.23 per cent and a level of confidence of 95 per cent.

or

The results indicate we can be 95 per cent confident that the actual voters' support for the Green party lies between 18.77 per cent and 23.23 per cent.

e. No, the size of the confidence interval does not decrease proportionately to the growing sample size. If it did, then quadrupling the sample size would decrease the size of the confidence interval from 8.78 per cent for 500 individuals to 2.20 per cent for 2,000 individuals. Instead, the size of the confidence interval decreases only to 4.46 per cent for a sample of 2,000.

f. No, due to the law of diminishing returns. A sample estimate with a margin of error of 8.78 per cent (n = 500) is too imprecise, but by doubling the sample we manage to lower it by 2.58 per cent to a margin of error of 6.20 per cent. By adding an additional 500 (for a total of 1,500) people we only shave off another 1.1 per cent but increase our financial cost by one-third.

5.

a.

$CI_{95} = 23.1 \pm 1.04$

$CI_{99} = 23.1 \pm 1.38$

b.

$CI_{95}$: Lower limit: 22.06; Upper limit: 24.14

$CI_{99}$: Lower limit: 21.72; Upper limit: 24.48

c.

The sample result showed that the average MCAT score of all test takers was 23.1 points, with a margin of error ± 1.04 points and a level of confidence of 95 per cent.

or

The results indicate we can be 95 per cent confident that the average MCAT score for all test takers lies between 22.06 and 24.14 points. The sample result showed that the average MCAT score of all test takers was 23.1 points, with a margin of error ± 1.38 points and a level of confidence of 99 per cent.

or

The results indicate we can be 99 per cent confident that the average MCAT score for all test takers lies between 21.72 and 24.48 points.

d. The 95 per cent confidence interval becomes slightly wider: 23.1 ±1.81.

6.

a. $CI_{95} = 3.8 \pm 0.18$
Lower limit: 3.62 hours
Upper limit: 3.98 hours
The sample result showed that children between ages six and nine watch, on average, 3.8 hours of TV per day, with a margin of error ± 0.18 points and a level of confidence of 95 per cent.

or

The results indicate we can be 95 per cent confident that the average number of hours of TV watching lies between 3.62 and 3.98 hours for children of ages between six and nine.

b. $CI_{99} = 3.8 \pm 0.36$
Lower limit: 3.44 hours
Upper limit: 4.16 hours
The sample result showed that children between ages of six and nine watch, on average, 3.8 hours of TV per day, with a margin of error ± 0.36 points and a level of confidence of 99 per cent.

or

The results indicate we can be 99 per cent confident that the average number of hours of TV watching lies between 3.44 and 4.16 hours for children of ages between six and nine.

7.

a. $CI_{99.9}$ = 68 per cent ± 10.19 per cent
b. Lower limit: 57.81 per cent

Upper limit: 78.19 per cent
*The sample result showed that the proportion of children between ages six and nine who own at least one violent video game is 68 per cent, with a margin of error of ± 10.19 per cent and a level of confidence of 99.9 per cent.*

or

The results indicate we can be 99.9 per cent confident that the proportion of children between ages six and nine who own at least one violent video game lies between 57.81 per cent and 78.19 per cent.

c. $CI_{99.9}$ = 68 per cent ± 30.84 per cent
Lower limit: 37.16 per cent
Upper limit: 98.84 per cent
The sample result showed that the proportion of children between ages six and nine who own at least one violent video game is 68 per cent, with a margin of error ± 30.84 per cent and a level of confidence of 99.9 per cent.

or

The results indicate we can be 99.9 per cent confident that the proportion of children between ages six and nine who own at least one violent video game lies between 37.16 per cent and 98.84 per cent.

When we decrease the sample size from 228 to only 35, the 99.9 per cent confidence interval widens (i.e. precision decreases) markedly, by 39.86 per cent (10.91 × 2 – 30.84 × 2).

8.

a. $CI_{99}$ = 118 ± 0.98
Lower limit: 117.02 points
Upper limit: 118.98 points

b. *The sample result showed that the IQ of all undergraduate university students was 118 points, with a margin of error ± 0.98 points and a level of confidence of 99 per cent.*

or

The results indicate we can be 99 per cent confident that the average undergraduate university students' IQ lies between 117.02 and 118.98 points.

c. The margin of error for a 99 per cent confidence interval increases by 3.18 points when we decrease the sample size from 250 to 41.

## Chapter 20

1.

a. There is no relationship between daily caloric intake and BMI.

b. There is no relationship between female literacy rates and birth rates in the developing world.

c. Americans are no more likely to own a gun than are Canadians.

d. The monetary return on years of education obtained is no different for visible minority Canadians than for Caucasian Canadians.

e. The life expectancy for Aboriginal Canadians is no different than for other Canadians.

f. Wealth inequality is not rising in Canada.

g. University students consume no more alcohol per capita than the general population.

h. Public transit usage rates do not increase with increasing gas prices.

i. Divorce rates are no different among dual-earner marriages than among single-earner marriages.

j. Home sales do not decrease with increasing interest rates.

k. There is no different between males and females in mean number of Facebook friends.

l. There is no association between early childhood nutrition and school readiness at age five.

2.

a. There is a relationship between parental educational attainment and children's educational attainment (non-directional).
or
Children whose parents have higher educational have, on average, higher educational attainment themselves (directional).

b. There is a relationship between socio-economic position and one's health status (non-directional).
or
People with higher socio-economic status have, on average, better health than those with lower socio-economic status (directional).

c. There is a relationship between gender and personal income (non-directional).
or
Men have, on average, higher personal income than women (directional).

d. There is a relationship between one's income/wealth and the average number of disability-free life expectancy (non-directional).
or
People with higher income/wealth have higher disability-free life expectancy than poorer people (directional).

e. Women are more likely to perpetrate violence against their own children than men.

f. There is a relationship between one's nationality (Canadian/American) and one's attitude towards beavers (non-directional).
or
Canadians have a more positive attitude toward beavers than do Americans (directional).

g. There is a relationship between hours spent partying per week and one's GPA (non-directional).
or
The more hours per week one spends by partying, the lower their GPA (directional).

h. There is a relationship between one's marital status and one's mental well-being (non-directional).
or
Married people experience more mental well-being than people who are single.

i. Divorced older women are more likely than divorced older men to experience poverty (directional).

j. Couples with small children experience higher strain on their relationship than couples with older children (directional).

k. There is no relationship between gender and mathematical abilities.

l. The number of hours you spend studying this textbook is (somehow) related to your grade in your research methods course (non-directional).
or
The more hours you spend studying this textbook, the better your grade in your research methods course is going to be (directional).

3.

a.

| Own Gun | Nationality | | Total |
|---|---|---|---|
| | Canadian | American | |
| Yes | 75 (104.1) | 135 (105.9) | 210 |
| No | 153 (123.9) | 97 (126.1) | 250 |
| Total | 228 | 232 | 460 |

Expected frequencies in parentheses

b. 29.59

c. df = 1

d. Yes, chi-square$_{obt}$ > chi-square$_{crit}$ : 29.68 > (α = 0.05, df, 1) = 3.84;  it also exceeds 6.64 (α = 0.01, df, 1)

e. The results indicate that the null hypothesis stating that Americans are no more likely to own a gun than Canadians is rejected at the 0.01 level of significance. This result supports the hypothesis that Americans are more likely to own a gun than are Canadians. These statistically significant results suggest that we are 99 per cent confident that a greater proportion of Americans than Canadians own guns.

4.

a. Degrees of freedom = (3 − 1)(3 − 1) = 4
Critical chi-square value (α = 0.05, df, 4) = 9.5; (α = 0.01, df, 4) = 13.3
The obtained chi-square of 20.91 exceeds both critical values.

b. The results indicate that the null hypothesis stating that there is no association between place of residence and political affiliation is rejected at the 0.01 level of significance.  This result supports the hypothesis that place of residence is associated with political party affiliation. These statistically significant results suggest that we are 99 per cent confident that place of residence does matter for political party affiliation.

5.

a.

| Mental Well-Being | Marital Status | | Total |
|---|---|---|---|
| | Married | Single | |
| Good | 6 (4.5) | 3 (4.5) | 9 |
| Poor | 2 (3.5) | 5 (3.5) | 7 |
| Total | 8 | 8 | 16 |

Expected frequencies in parentheses

b. 2.28

c. df = 1; Chi-square$_{critical}$ = 3.841

d. Conclusion: The results indicate that the null hypothesis stating that there is no relationship between marital status of men and their mental well-being cannot be rejected at the 0.05 significance level. The research hypothesis that married men experience, on average, more mental well-being is not supported by our results.

**Note:** Although the percentage difference technique (Chapter 16) suggests that there indeed is a relationship between men's marital status and their mental well-being, the results of our significance test simply imply that at a 95 per cent significance level and with the current sample size (of mere 16 individuals) we cannot rule out the possibility that the observed connection between the two variables is not merely due to chance alone. The next step, therefore, would be to increase the sample size and to repeat the test of significance.

6.

a.

| Mental Well-Being | Marital Status | | Total |
|---|---|---|---|
| | Married | Single | |
| Good | 24 (18) | 12 (18) | 36 |
| Poor | 8 (14) | 20 (14) | 28 |
| Total | 32 | 32 | 64 |

Expected frequencies in parentheses

b. 9.14

c. df = 1; Chi-square$_{critical}$ = 3.841

d. Conclusion: The results indicate that the null hypothesis stating that there is no relationship between marital status of men and their mental well-being is rejected at the 0.05 level of significance. This result supports the hypothesis that married men experience, on average, more mental well-being than single men. We are 95 per cent confident that our results were not due to chance and that they reflect real differences observable in the population to which we generalize.

7.

a.

| Infant Mortality Rate | Health Care Spending per Capita | | | Total |
|---|---|---|---|---|
| | Low | Medium | High | |
| Low | 2 (4.9%) | 10 (29.4%) | 16 (64%) | 28 |
| Medium | 13 (31.7%) | 18 (52.9%) | 8 (32%) | 39 |
| High | 26 (63.4%) | 6 (17.7%) | 1 (4%) | 33 |
| Total | 41 | 34 | 25 | 100 |

From observing the conditional distributions of the dependent variable (the columns in this table), it becomes apparent that the relationship is an inverse/negative one. As health care spending per capita increases, a country's infant mortality rate decreases. Research hypothesis: Countries with higher health care spending per capita will have lower infant mortality rate than countries with lower health care spending.

b.

| Infant Mortality Rate | Health Care Spending per Capita | | | Total |
|---|---|---|---|---|
| | Low | Medium | High | |
| Low | 2 (11.48) | 10 (9.52) | 16 (7) | 28 |
| Medium | 13 (15.99) | 18 (13.26) | 8 (9.75) | 39 |
| High | 26 (13.53) | 6 (11.22) | 1 (8.25) | 33 |
| Total | 41 | 34 | 25 | 100 |

Expected frequencies in parentheses

c. 42.27

d. df = 4; Chi-square$_{critical}$ = 13.277

e. Conclusion: The results indicate that the null hypothesis stating that there is no relationship between a country's health spending per capita and its infant mortality rate is rejected at the 0.01 level of significance. This result supports the hypothesis that countries with higher health spending per capita tend to have lower infant mortality rates than countries with lower health care spending. We are 99 per cent confident that our results are not due to chance alone and that they reflect real differences observable in the population of countries to which we generalize.

8.

a.

| Self-Rated Health | Social Class | | | Total |
|---|---|---|---|---|
| | Low | Medium | High | |
| Poor | 28 (22%) | 18 (15.5%) | 7 (9.9%) | 53 |
| Fair | 54 (42.5%) | 32 (27.6%) | 15 (21.1%) | 101 |
| Good | 32 (25.2%) | 45 (38.8%) | 28 (39.4%) | 105 |
| Excellent | 13 (10.2%) | 21 (18.1%) | 21 (29.6%) | 55 |
| Total | 127 | 116 | 71 | 314 |

By comparing the conditional distributions of the dependent variables we find that the relationship is direct/positive. That is, as one's social class increases, so does their self-rated health. Research hypothesis: Social class is positively related to self-rated health.

b.

| Self-Rated Health | Social Class | | | Total |
|---|---|---|---|---|
| | Low | Medium | High | |
| Poor | 28 (21.44) | 18 (19.58) | 7 (11.98) | 53 |
| Fair | 54 (40.85) | 32 (37.31) | 15 (22.84) | 101 |
| Good | 32 (42.47) | 45 (38.79) | 28 (23.74) | 105 |
| Excellent | 13 (22.25) | 21 (20.32) | 21 (12.44) | 55 |
| Total | 127 | 116 | 71 | 314 |

Expected frequencies in parentheses

c. 25.98

d. df = 6; Chi-square$_{critical}$ = 22.457

e. Conclusion: The results indicate that the null hypothesis stating that there is no relationship between socio-economic class and self-rated health is rejected at the 0.001 level of significance. This result supports the hypothesis that people with higher social status will have, on average, higher self-rated health than people with lower socioeconomic status. We are 99.9 per cent confident that our results were not due to chance alone and that they reflect real differences observable in the population to which we generalize.

9.

a. Degrees of freedom = (4 − 1)(4 − 1) = 9
Critical chi-square value (α = 0.05, df, 9) = 16.9; (α = 0.01, df, 9) = 21.7
The obtained chi-square value of 14.05 does not exceed either of the critical values of alpha at 0.05 or 0.01.

b. The results indicate that the null hypothesis stating that there is no association between level of education attained and maternal education must be retained. This result contradicts the hypothesis that there is an association between level of education attained and maternal education.

## Chapter 21

1.
   a.  $H_0$: $\mu 1 = \mu 2$ *or* $\mu 1 - \mu 2 = 0$
   Major Smith students do not differ from the general high school population in Central School Division in terms of their average number of Facebook friends.
   $H_1$: $\mu 1 \neq \mu 2$
   Major Smith students do differ from the general high school population in Central School Division in terms of their average number of Facebook friends.
   b.  $t_{obt} = 2.43$
   c.  $t_{crit} = 2.035$ (df = 33, $\alpha = 0.05$, two-tailed)
   d.  The results indicate a significant difference, and therefore we reject the null hypothesis at $\alpha = 0$.05. We are 95 per cent confident that the average number of Facebook friends is significantly lower among Major Smith students than the general population of high school students in Central School Division.

2.
   a.  $H_0$: $P_u = 0.35$
   The proportion of Central School Division high school students who log onto Facebook daily does not differ from the proportion of Canadian high school students who log on to Facebook daily.
   $H_1$: $P_u \neq 0.35$
   The proportion of Central School Division high school students who log onto Facebook daily differs from the proportion of Canadian high school students who log on to Facebook daily.
   b.  $z_{obt} = -4.79$
   c.  $z_{crit} = 2.58$ ($\alpha = 0.01$, two-tailed)
   d.  The obtained z-value exceeds the critical z-value; therefore, we reject the null hypothesis at $\alpha = 0.01$. We are 99 per cent confident that a significantly smaller proportion of Central School Division high school students log onto Facebook daily than do Canadian high school students.

3.
   a.  $H_0$: $\mu 1 = \mu 2$ *or* $\mu 1 - \mu 2 = 0$
   There is no difference in the average number of sick days per year between shift workers and workers on straight day shifts.
   $H_1$: $\mu 1 > \mu 2$

Shift workers use a significantly higher average number of sick days per year than straight day-shift workers.
   b.  $s_{x_1 - x_2} = 0.210$
   c.  $t_{obt} = 3.81$
   d.  $t_{crit} = 2.63$ (df = 92, $\alpha = 0.01$, one-tailed)
   e.  As our obtained t-value exceeds the critical t-value, we reject the null hypothesis at $\alpha = 0.01$. The results indicate that shift workers average significantly more sick days per year than straight day-shift workers. In other words, we are 99 per cent confident that shift workers average more sick days annually than straight day-shift workers do.

4.
   a.  $H_0$: $P_{u_1} = P_{u_2}$ *or* $P_{u_1} - P_{u_2} = 0$
   There is no difference between sociology majors and psychology majors in the proportion of students who indicate buying fair-trade products is important to them.
   $H_1$: $P_{u_1} \neq P_{u_2}$
   There is a difference between sociology majors and psychology majors in the proportion of students who indicate buying fair-trade products is important to them.
   b.  $P_u = 0.6584$
   c.  $\sigma_{p-p} = 0.0364$
   d.  $Z_{obt} = 1.10$
   e.  $Z_{crit} = 1.96$ ($\alpha = 0.05$, two-tailed)
   f.  The obtained z-value does not exceed the critical z-value; therefore, we retain the null hypothesis. We are 95 per cent confident that there is no significant difference between sociology majors and psychology majors in the proportion of students who report buying fair-trade products is important to them.

5.
   a.  $H_0$: $\mu 1 = \mu 2$ *or* $\mu 1 - \mu 2 = 0$
   There is no difference between rural and urban households in the mean number of motor vehicles owned.
   $H_1$: $\mu 1 > \mu 2$
   Rural households will, on average, own more motor vehicles than urban households.
   $s_{x_1 - x_2} = 0.076$
   $t_{obt} = = 6.58$
   $t_{crit} = 1.988$ (df = 85; $\alpha = 0.05$, one-tailed)

The obtained $t$-score (6.58) exceeds the critical $t$-score (1.99); therefore we reject the null hypothesis of no difference at alpha = 0.05. We are 95 per cent confident that rural households possess significantly more vehicles, on average, than urban households.

b. $H_0$: $\mu1 = \mu2$ *or* $\mu1 - \mu2 = 0$
There is no difference in the percentage of males and females who hold university degrees.
$H_1$: $\mu1 \neq \mu2$
There is a significant difference in the percentage of males and females who hold university degrees.
$P_u = 0.5885$
$\sigma_{p-p} = 0.042$
$Z_{obt} = 1.67$
$Z_{crit}$ ($\alpha = 0.05$, two-tailed) = 1.96
The obtained $t$-score (1.67) fails to exceed the critical $t$-score (1.96); therefore we retain the null hypothesis of no difference at alpha = 0.05. We are 95 per cent confident that there is no significant difference between the percentage of males and females who hold university degrees.

c. $H_0$: $\mu1 = \mu2$   *or* $\mu1 - \mu2 = 0$
There is no difference in the percentage of males and females who hold university degrees.
$H_1$: $\mu1 > \mu2$
A significantly higher percentage of females than males hold university degrees.
$Z_{obt} = 1.67$
$Z_{crit}$ ($\alpha = 0.05$, one-tailed) = 1.65
The obtained $t$-score (1.67) exceeds the critical $t$-score (1.65); therefore we reject the null hypothesis of no difference at alpha = 0.05. We are 95 per cent confident that the percentage of females who hold university degrees is significantly higher than the percentage of males who hold university degrees.

d. $H_0$: $\mu1 = \mu2$ or $\mu1 - \mu2 = 0$
There is no difference between Alberta and the rest of Canada in the proportion of voters who approve of Stephen Harper's performance as prime minister.
$H_1$: $\mu1 > \mu2$
A significantly higher proportion of Albertan voters than Canadian voters approve of Stephen Harper's performance as prime minister.

$$Z_{obt} = \frac{0.59 - 0.52}{\sqrt{0.52(0.48)/255}} = \frac{0.07}{\sqrt{0.2496/255}}$$

$$= \frac{0.07}{\sqrt{0.00098}} = \frac{0.07}{0.031} = 2.26$$

$$Z_{crit}(a = 0.01, \text{one-tailed}) = 2.33$$

The obtained $z$-score (2.26) fails to exceed the critical $z$-score (2.33) at the 0.01 level of significance; therefore we retain the null hypothesis. We are 99 per cent confident that there is no significant difference between the percentage of Albertans and Canadians who approve of Stephen Harper's performance.

6.
a. $H_0$: $\mu1 = \mu2$ OR $\mu1 - \mu2 = 0$
There is no difference in average MCAT score between those who took the private preparatory course in 2013 and all MCAT test takers that year.
$H_1$: $\mu1 > \mu2$
Students of the private preparatory course do, on average, significantly better on MCAT than the average of all MCAT takers.

b. $t_{obt} = 3.16$; $t_{crit} = 1.684$ (df = 40 , $\alpha = 0.05$, one-tailed)

c. The obtained $t$-score (3.16) exceeds the critical $t$-score (1.684); therefore we reject the null hypothesis of no difference at alpha = 0.05. We are 95 per cent confident that those who take the private preparatory course will, on average, do better on MCAT than all MCAT test takers on average.

7.
a. $H_0$: $\mu1 = \mu2$ OR $\mu1 - \mu2 = 0$
There is no relationship between men watching "unrealistically" beautiful women from music videos and magazines and their perception of "real" women as less attractive.
$H_1$: $\mu1 > \mu2$
Men who watch "unrealistically" beautiful women from music videos and magazines will, on average, perceive "real" women as less attractive than men who do not watch/read such material.

b. $s_{x_1 - x_2} = 0.344$; $t_{obt} = 6.69$; $t_{crit} = \pm1.987$ (df = 88, $\alpha = 0.05$, two-tailed)

c. The obtained $t$-score (6.69) exceeds the critical $t$-score ($\pm1.987$); therefore we reject the null hypothesis of no difference at alpha = 0.05. We are 95 per cent confident that men who watch "unrealistically" beautiful women from music videos and magazines do, on average, perceive "real" women as significantly less attractive than men who do not watch/read such material.

8.

a. $H_0$: $P_{u_1} = P_{u_2}$ *or* $P_{u_1} - P_{u_2} = 0$
There is no difference between patients who received a placebo and those who received the tested medication.
$H_1$: $P_{u_1} \neq P_{u_2}$
There is a difference between those patients who received a placebo and those who received the tested medication.

b. $P_u = 0.375$; $\sigma_{p-p} = 0.0174$; $Z_{obt} = -1.72$; $Z_{crit}$ ($\alpha = 0.05$, one-tailed) $= -1.65$

c. The obtained z-score (−1.72) exceeds the critical z-score (−1.65); therefore we reject the null hypothesis of no difference at alpha = 0.05. We are 95 per cent confident that those patients who used the tested medication were more likely to notice significant improvement in their insomnia than those patients who only received placebo.

d. The result of the z-test did indicate that the difference between the control and experimental group was statistically significant; however, we can easily observe that the difference between the groups is extremely small (only an additional 3 per cent of the patients who used the tested medication showed improvement over those who were given a placebo). This is a typical case of a situation when a difference or result is statistically significant due to use of a large sample size, the one-tailed test, or a low level of significance, but the substantive significance of that result is very small. Is a medication that promises a chance of improvement over a placebo as small as 3 out of 100 clinically useful (will it have real practical impact on improving patients' condition)? This question is not answered by statistical significance testing.

9.

a. $H_0$: $P_u = 0.23$
The proportion of children with an anxious–ambivalent attachment to their mother is no different for children whose mothers suffer from a bipolar affective disorder compared to all other children in the population.

$H_1$: $P_u \neq 0.23$
The proportion of children with an anxious–ambivalent attachment to their mother is higher for children of mothers suffering from a bipolar affective disorder compared to all other children in the population.

b. $Z_{obt} = -5.11$; $Z_{crit} = $ ($\alpha = 0.01$, one-tailed) $= -2.06$

c. The obtained z-score (−5.11) exceeds the critical z-score (−2.06); therefore we reject the null hypothesis of no difference at alpha = 0.01. We are 99 per cent confident that a higher proportion of children of mothers who suffer from bipolar affective disorder have anxious–ambivalent attachment to their mother compared to all other children in the population.

10.

a. $H_0$: $P_{u_1} = P_{u_2}$ *or* $P_{u_1} - P_{u_2} = 0$
There is no relationship between scientists' opinion about scientific consensus around GMO and the source of their research funding.
$H_1$: $P_{u_1} \neq P_{u_2}$
Scientists whose funding comes from private biotech corporations are different in their opinion about scientific consensus around GMO than those members of the scientific community whose funding does not come from such source.

b. $P_u = 0.54$; $\sigma_{p-p} = 0.11$; $Z_{obt} = 5.64$; $Z_{crit}$ ($\alpha = 0.05$, two-tailed) $= \pm 1.96$

c. The obtained z-score (5.64) exceeds the critical z-score (±1.96); therefore we reject the null hypothesis of no difference at alpha = 0.05. We are 95 per cent confident that there is a difference in opinion about scientific consensus around GMO between scientists whose funding comes from private biotech corporations and those scientists whose research is funded through other sources.

11.

a. $H_0$: $\mu 1 = \mu 2$ *or* $\mu 1 - \mu 2 = 0$
There is no difference in waiting time for an MRI scan between people from the upper socio-economic class and the country's average.
$H_1$: $\mu 1 < \mu 2$
The average waiting time for an MRI scan is shorter for people from the upper socio-economic class compared to the national average.

b. $t_{obt} = -10.80$; $t_{crit} = \pm 1.995$ (df = 69, $\alpha = 0.05$, two-tailed)

c. The obtained $t$-score (-10.80) exceeds the critical $t$-score ($\pm1.995$); therefore we reject the null hypothesis of no difference at alpha = 0.05. We are 95 per cent confident that people from the upper socio-economic class have a shorter average waiting time for an MRI scan than the national average.

12.

a. $H_0$: $\mu1 = \mu2$ OR $\mu1 - \mu_2 = 0$
   There is no difference between graduates of the honours programs in psychology and sociology and their average cumulative GPA at graduation.
   $H_1$: $\mu1 \neq \mu_2$

There is a difference between cumulative GPA at graduation of graduates of the psychology honours program and of the sociology honours program.

b. $s_{x_1 - x_2} = 0.0213$; $t_{obt} = 2.817$; $t_{crit} = \pm1.996$ (df = 68, $\alpha = 0.05$, two-tailed)

c. The obtained $t$-score (2.817) exceeds the critical $t$-score ($\pm1.996$); therefore we reject the null hypothesis of no difference at alpha = 0.05. We are 95 per cent confident that the honours undergraduate psychology program at this university is more competitive, based on the statistically significant difference in cumulative GPA between the psychology and sociology graduates.

# Glossary

**absolute zero** The condition in which a measurement of zero literally means there is no amount of the variable present.

**abstract** Experience that occurs in the mind; imaginary.

**actuarial prediction** Forecasts based on repetitive empirical patterns.

**alpha level** The probability of Type I error in a study (level of statistical significance).

**alternative hypothesis** See *research/alternative/substantive hypothesis*.

**association** A relationship where a change in one variable is systematically connected to change in another variable.

**authentic** A declaration that the apparent relationship is real (in the sense that it is unaffected by other variables).

**bar chart** A graph used to visually represent crosstabulations. The bars represent values of the dependent variable and the separations between them represent the independent variable.

**bimodal** A distribution containing two modes.

**bivariate analysis** Statistical analysis procedures for summarizing the relationship between two variables.

**case** An object measured on one or more variables.

**categorical variable** A variable at the nominal or ordinal level of measurement whose score represents a discrete value.

**causal connection** A relationship where change in an independent variable produces genuine change in the dependent variable.

**cell** The intersection of a row and a column in a data matrix.

**census** The collection of information from all members of a population.

**central tendency** Measures that indicate the location(s) where typical scores of a variable are found.

**ceteris paribus** A Latin phrase meaning "all things being equal"; a disclaimer stating the conditional nature of a reported relationship.

**charts and graphs** Visual displays of data.

**clinical prediction** Forecasts based on theoretical understanding.

**codebook** A tool for translating respondents' answers into numerical scores and vice versa.

**coefficient of determination ($r^2$)** A measure of the strength of the relationship between variables in a scatterplot.

**coefficient of multiple determination ($R^2$)** A PRE measure of how well several independent variables account for variation in a dependent variable.

**column** A vertical line in a data matrix.

**concept** An abstract category for organizing sensory (empirical) experience.

**conceptual definition** Establishing the meaning of an abstract term by expressing it in other terms.

**conceptualization** The abstract process employed to create concepts.

**concrete** Experience-based physical sensations (touch, taste, smell, sight, hearing).

**conditional distribution** A frequency distribution of observations on one variable for a specific value of another variable.

**confidence interval** A range of scores within which the population parameter likely occurs.

**confidence level** The likelihood that a specific confidence interval captures the population parameter.

**continuous variable** A variable at the interval or ratio levels of measurement that potentially includes an infinite number of scores.

**control variable** A variable defining the context for relationships between independent and dependent variables.

**critical value** The value in a sampling distribution that determines whether or not the null hypothesis is rejected.

**crosstab (crosstabulation)** A table displaying the frequencies of intersecting values of an independent and dependent variable.

**curvilinear relationship** A relationship in which the form changes across the connection.

**data** An aggregation of quantitative measurements (observations) across several variables and/or cases.

**data matrix** Organization of data into rows including cases, columns including variables, and cells identifying scores.

**deduction** A logical reasoning process that proceeds from specific instances to a general case.

**degrees of freedom** The number of independent pieces of information used in significance testing.

**denominator** The base of a fraction, which reports the total number of cases.

**dependent variable** A variable that is changed as a consequence of change in some other (independent) variable(s).

**descriptive statistics** Procedures that summarize information about variables collected from a sample.

**direct relationship** A form in which the variables change in the same direction.

**dispersion** Measure that indicates how spread out the cases are across a variable.

$E_1$ The number of errors in predicting dependent variable scores using the marginal prediction rule.

$E_2$ The number of errors in predicting dependent variable scores using the relational prediction rule.

**elaboration** The process of examining the effects of control variables on an original (zero-order) relationship.

**empirical** A term that describes observations based on concrete experience.

**empirical deduction** The logical process for transforming abstract propositions into empirically testable hypotheses.

**epistemic correspondence** The adequacy of operationalization, the fit between concept and indicator variable(s).

**extent** A measure that identifies the impact of the independent variable on the dependent variable.

**fact** An empirical statement about which there is general agreement.

**false negative** A result that reports the absence of a condition when it actually exists.

**false positive** A result that reports the presence of a condition that does not exist.

**false zero** A statistical result indicating no relationship between two variables when a relationship actually exists.

**falsifiability** A criterion for rejecting propositions lacking empirical support.

**first-order relationship** A reported relationship between two variables that takes account of one control variable.

**form** The specification of the kind of connection in a relationship.

**frequency distribution** A statistical description on a single variable that reports the occurrence of each score in a sample.

**frequency polygon** A special form of line graph that illustrates the midpoint of each category of a continuous level variable which are connected with a continuous line.

**gamma** A proportionate reduction in error statistic useful for measuring the association between ordinal variables.

**heterogeneity** Indication of the level of dissimilarity among a set of scores.

**histogram** Although it looks like a bar chart, it is not. Histograms provide visual representations of continuous level variables.

**homogeneity** Indication of the level of similarity among a set of scores.

**hypothesis** A statement of the expected empirical relationship between two or more variables.

**hypothesis testing** A set of procedures for determining the statistical significance of a set of findings.

**hypothetico-deductive method** A methodological process that deduces hypotheses from a scientific theory and empirically tests them using quantitative methods.

**ideal type** A theoretical model representing a perfect version of something; used for comparison and assessment of actual versions.

**ideology** A theory that is not falsifiable.

**independent variable** A variable whose change initiates change in some other variable(s).

**induction** A logical reasoning process that proceeds from specific instances to a general case.

**inferential statistics** Procedures that determine the generalizability of descriptive results from a probability sample.

**instrumentation** The process of creating a tool to measure a variable.

**interaction** A declaration that an apparent relationship is enhanced or supressed dependent upon the specific values of other variables.

**intercept** The point on the $Y$-axis where the regression line crosses.

**interval** The level of measurement that distinguishes fixed unit differences between objects.

**interval estimate** A range within which a sample statistic represents a population parameter.

**intervening** A declaration that an apparent relationship is due to a variable that mediates the connection between the independent and dependent variables.

**inverse relationship** A form in which the variables change in opposite directions.

**joint (conditional) frequency** The number of observations that occur at the intersection of a specific value of the independent variable and a specific value of the dependent variable.

**knowledge** Empirical rules expressing the connections between facts.

**kurtosis** A characteristic referring to how peaked a distribution is.

**lambda** A proportionate reduction in error statistic useful for measuring the association between nominal variables.

**leptokurtic** A distribution whose high kurtosis indicates it is sharp.

**levels of measurement** Procedures for determining the complexity of information extracted from an object.

**level of statistical significance** See *alpha level*.

**linear least squares regression line** The straight regression line on a scatterplot that comes closest to all the data points.

**linear relationship** A relationship in which the same form occurs throughout the connection.

**line graph** See *frequency polygon*.

**margin of error** The range around a point estimate that forms an interval estimate.

**marginal prediction rule** A technique for predicting scores on the dependent variable based only on information about the distribution of scores on the dependent variable.

**mean** A measure of central tendency indicating equal shares of a measured variable.

**measurement** The application of an instrument to an object in a ruleful way.

**median** A measure of central tendency indicating where the middle of the distributed scores occurs.

**mesokurtic** A distribution whose kurtosis indicates it is moderately peaked.

**mode** A measure of central tendency indicating the most commonly occurring value(s) of a variable.

**multimodal** A distribution containing three or more modes.

**multivariate analysis** Statistical analysis procedures for summarizing the relationship between three or more variables.

**negative skew** A skewed distribution constraining more negative scores.

**nominal** The level of measurement that only distinguishes differences between objects.

**non-probability sample** A sample where the likelihood of cases being selected is unknown.

**non-spurious** A connection between two variables that is unaffected by other variables.

**null hypothesis** An assumption about a population parameter that a researcher hopes to reject through hypothesis testing.

**numerator** The top of a fraction, which reports number of cases of interest.

**observation** The empirical measurement of an object in terms of some variable.

**one-sample test** A univariate statistical significance test determining whether the value of a single variable in the sample can be generalized to the population.

**one-tailed test** A significance test in which directional differences are predicted.

**operational definition** The use of operationalization, instrumentation and measurement to establish an empirical understanding of an abstract concept.

**operationalization** The process of translating an abstract concept into one or more variables that indicate its meaning.

**ordinal** The level of measurement that rank-orders differences between objects.

**parameter** The value of a variable in the population.

**partial correlation** A coefficient that describes the connection between two variables when the effects of all control variables are removed.

**pattern** A collection of percepts.

**Pearson's r** A measure of the precision of a regression line's fit to a scatterplot.

**percept** A single component of concrete experience.

**platykurtic** A distribution whose low kurtosis indicates it is flat.

**pie chart** A graph used to visually represent frequency distributions. It is formulated in a circle with wedges equal to the proportion each value represents as a per cent.

**point estimate** A precise generalization from a sample statistic to a population parameter.

**population** A complete set including all cases on a particular type.

**positive skew** A skewed distribution constraining more positive scores.

**possibility space** The range of potential outcomes that can occur when two variables interact.

**precision** A measure of how well the regression line fits the actual data points in a scatterplot.

**primary research** The original collection of data for specific research purposes.

**probability sample** A sample in which each case has a known likelihood of selection.

**proposition** An abstract statement expressing the relationship between concepts.

**qualitative methods** Procedures for gathering empirical evidence that rely on the intensive examination of selected cases to induce insight into the meaning of people's experience.

**quantitative methods** Methods for gathering empirical evidence that examine selected variables on a representative sample of cases to test general propositions.

**range** A measure of dispersion indicating the difference between the highest and lowest reported scores.

**ratio** The level of measurement that employs an absolute zero point and distinguishes fixed unit differences between objects.

**regression line** A line that specifies the form of a relationship in a scatterplot.

**relational prediction rule** A technique for predicting scores on the dependent variable based on information about the distribution of scores on the independent variable.

**relationship** A connection between two objects in which a change on one object is associated with a systematic change in the other.

**research/alternative/substantive hypothesis** A statement that contradicts the null hypothesis and is the researcher's best prediction.

**row** A horizontal line in a data matrix.

**sample** A selection of cases intended to represent all cases in a population.

**sample distribution** A display of the scores for a variable from a specific sample.

**sampling distribution** A theoretical frequency distribution of a statistic collected from an infinite number of samples.

**sampling frame** A list identifying all members of a population of interest.

**scientific theory** A theory that includes empirically testable propositions.

**second-order relationship** A reported relationship between two variables that takes account of two control variables.

**secondary research** The re-analysis of primary research evidence for new research purposes.

**sequence** A criterion that establishes that a change in one variable occurred prior to change in another variable.

**skewness** A distribution whose halves are not identical shapes.

**slope** The degree of steepness of the regression line.

**spurious** An apparent connection between two variables that is not genuine.

**standard deviation** The square root of the mean of the squared differences from the mean.

**standard error** A statistical measure of the amount of variation in a sampling distribution.

**standard score (z-score)** A respondent's score on a variable expressed in standard deviation units (rather than the original measurement unit).

**statistics** A set of mathematical techniques for analyzing, summarizing, and interpreting quantitative information from a sample.

**strength** A measure that specifies how much difference one object makes to another in a relationship.

**substantive hypothesis** See *research/alternative/substantive hypothesis.*

**symmetry** A distribution whose halves are identical shapes (mirror images).

**t-distribution** A family of probability distributions used for hypothesis testing on small samples.

**t-test** A statistical significance test employed on smaller sample sizes.

**theory** An explanatory narrative composed on a set of logically interrelated propositions.

**tied pair** In the calculation of gamma, a pair of cases that has either the same score on the independent variable, the same score on the dependent variable, or the same score on both the independent and dependent variables.

**two-sample test** A bivariate test determining if a difference between two sample statistics is statistically significant.

**two-tailed test** A significance test in which the predicted differences are non-directional.

**Type I error** The rejection of a null hypothesis that is actually true.

**Type II error** The failure to reject a null hypothesis that is actually false.

**unbiased estimator** Statistics from probability samples that, on average, accurately reflect the population parameter.

**unimodal** A distribution containing a single mode.

**univariate analysis** Statistical analysis procedures for summarizing data on single variables.

**valid per cent** A percentage calculation that excludes all missing cases.

**value** A specific score on a variable.

**variable** A property of an object that can change.

**variance** A measure of dispersion appropriate for interval and ratio variables.

**zero-order relationship** A reported relationship between two variables that does not take account of any control variables.

**z-test** A statistical significance test which assumes an approximately normal sample distribution.

# References

Berger, Peter, and Kellner, Hansfried. 1964. "Marriage and the Social Construction of Reality: An Exercise in the Microsociology of Knowledge." *Diogenes*. 46:1–23.

Berger, Peter, and Luckmann, Thomas. 1967. *The Social Construction of Reality: A Treatise in the Sociology of Knowledge*. New York: Anchor.

Braroe, Niels. 1975. *Indian and White: Self Image and Interaction in a Canadian Plains Community*. New York: Stanford University Press.

Ellis, Carolyn. 1986. *Fisher Folk: Two Communities on Chesapeake Bay*. Lexington: University of Kentucky Press.

George, Darren, and Mallery, Paul. 2012. *IBM SPSS Statistics 19 Step by Step: A Simple Guide and Reference*. Boston: Pearson.

Harris, Sam. 2005. *The End of Faith: Religion, Terror, and the Future of Reason*. New York: W.W. Norton.

Leech, N.L., Barrett, K.C., and Morgan, G.A. 2011. *IBM SPSS for Intermediate Statistics: Use and Interpretation*. 4th edition. New York: Routledge.

Lofland, John, and Stark, Rodney. 1965. "Becoming a World-Saver: A Theory of Conversion to a Deviant Perspective" *American Sociological Review* 30: 862–875.

Morgan, G.A., Leech, N.L., Gloeckner, G.W., and Barrett, K.C. 2010. *IBM SPSS for Introductory Statistics: Use and Interpretation*. 4th Edition. New York: Routledge.

Radloff, S.L. 1977. The CES-D Scale: A Self-Report Depression Scale for Research in the General Population. *Applied Psychological Research* 1(3): 385–401.

Rokeach, Milton. 1964. *The Three Christs of Ypsilanti*. New York: Vintage.

Scheff, Thomas. 1990. *Microsociology: Discourse, Emotion, and Social Structure*. Chicago: The University of Chicago Press.

Schmitt, D. P., and Allik, J. 2005. "Simultaneous Administration of the Rosenberg Self-Esteem Scale in 53 Nations: Exploring the Universal and Culture-Specific Features of Global Self-Esteem." *Journal of Personality and Social Psychology*, 89(4): 623–642.

Statistics Canada 2006. "University enrolment." *The Daily*. www.statscan.gc.ca/daily/061107/d061107a.htm. Retrieved 25 July 2009.

Szafran, Robert. 2012. *Answering Questions with Statistics*. Thousand Oaks, California: Sage.

Websdale, Neil. 2010. *Familicidal Hearts: The Emotional Styles of 211 Killers*. Toronto: Oxford University Press.

# Index